# Flowering and Flower Development in Plants

# Flowering and Flower Development in Plants

Guest Editors

**Jinzhi Zhang**
**Avi Sadka**

MDPI

Basel • Beijing • Wuhan • Barcelona • Belgrade • Novi Sad • Cluj • Manchester

*Guest Editors*
Jinzhi Zhang
National Key Laboratory for Germplasm Innovation & Utilization of Horticultural Crops
Huazhong Agricultural University
Wuhan
China

Avi Sadka
Department of Fruit Trees Sciences
Institute of Plant Sciences
Bet Dagan
Isreal

*Editorial Office*
MDPI AG
Grosspeteranlage 5
4052 Basel, Switzerland

This is a reprint of the Special Issue, published open access by the journal *Agronomy* (ISSN 2073-4395), freely accessible at: https://www.mdpi.com/journal/agronomy/special_issues/Flowering_Flower.

For citation purposes, cite each article independently as indicated on the article page online and as indicated below:

| Lastname, A.A.; Lastname, B.B. Article Title. *Journal Name* **Year**, *Volume Number*, Page Range. |
|---|

**ISBN 978-3-7258-3801-1 (Hbk)**
**ISBN 978-3-7258-3802-8 (PDF)**
**https://doi.org/10.3390/books978-3-7258-3802-8**

# Contents

# About the Editors

**Jinzhi Zhang**

Jinzhi Zhang is a professor at Huazhong Agricultural University. He graduated from the Department of Fruit Trees at Huazhong Agricultural University with a doctoral degree, and visited and exchanged ideas at the University of California, San Diego, from 2013 to 2014. As of now, he has published over 60 articles as the first author or corresponding author in SCI source journals; He has also hosted sub projects of the National Natural Science Foundation, International Science Foundation, National Major Research and Development Program, and the New Teacher Fund of the Ministry of Education, served as a topic editor for journals such as *Horticulture*, *Agronomy*, and *Forests*, and is currently serving as an editor for journals such as *BMC Plant Biology*, *Horticulture*, and *Journal of Plant Growth Regulation*. Recently, he has been engaged in molecular biology research on fruit tree development.

**Avi Sadka**

Avi Sadka is a research scientist at the ARO, The Volcani Center, at the Institute of Plant Sciences. They were promoted to Rank B in 2002 and Rank A in 2017. They served as the head of the Department of Fruit Trees Sciences at ARO between 2007 and 2011; between 2002 and 2003, as a visiting scientist at the University of California, Davis, Plant Sciences (E. Blumwald); and between 2016 and 2017, as a visiting scientist at the University of California, Davis, Plant Sciences (E. Blumwald). Their editorial responsibilities and experience include their 2012 experience as Chief Editor, *Acta Hort*, Proceedings of the International CIPA Conference; 2013 experience as a member on the *Plant Science* Editorial Board; and 2014 experience as an editor of the *Horticultural Journal* (formerly, *J. Japanese Society Hort. Sci.*). "Physiology: the lively learning of the logic of life" (the Physiological Society of Japan). Avi Sadka is a plant physiologist, and throughout their career, they have aimed at characterizing and understanding questions related to various aspects of plant growth and development, and plants' responses to external stimuli.

# Preface

The transition from vegetative to reproductive growth is crucial in plant life cycles, regulated by five key pathways: photoperiod, vernalization, age, autonomy, and gibberellin, alongside environmental stressors like drought, salinity, and nutrient deficiency. Flower quality/quantity determines crop yield and fruit quality, with developmental mechanisms described by the ABCDE model. Although numerous flowering-related genes have been identified in annual and perennial plants, and molecular understanding has advanced significantly, deeper exploration of flowering mechanisms remains essential for future crop breeding improvements. This Special Issue of *Agronomy* is focused on the latest fundamental discoveries in the field of flowering and flower development in plants.

**Jinzhi Zhang and Avi Sadka**
*Guest Editors*

*agronomy*

MDPI

Editorial

# Flowering and Flower Development in Plants

**Min Chen and Jin-Zhi Zhang** *

National Key Laboratory for Germplasm Innovation & Utilization of Horticultural Crops, College of Horticulture and Forestry Science, Huazhong Agricultural University, Wuhan 430070, China; minchen970402@webmail.hzau.edu.cn

* Correspondence: jinzhizhang@mail.hzau.edu.cn; Tel.: +86-27-6201-80231; Fax: +86-27-8728-2010

**Citation:** Chen, M.; Zhang, J.-Z. Flowering and Flower Development in Plants. *Agronomy* **2024**, *14*, 256. https://doi.org/10.3390/agronomy14020256

Received: 17 January 2024
Revised: 23 January 2024
Accepted: 24 January 2024
Published: 25 January 2024



In the life cycle of a plant, flowering marks the transition from vegetative growth to reproductive development. The optimal flowering time of different plants is an important agricultural trait that ensures the yield and quality of their harvested products. In the case of ornamental plants, the period of time between anthesis and floral senescence facilitates pollination and ensures the seed production necessary for the survival of the species. In addition, the development of flower organs also plays a crucial role in the reproductive development of plants. Many plant species have evolved multiple ways of regulating flowering to improve their adaptability to endogenous factors and complex environmental conditions. The importance of flowering and flower development, as well as the more comprehensive understanding of this issue that we have developed in recent years, make a systematic summary of the progress of research related to this topic quite necessary.

Extensive physiological and molecular genetic analyses have revealed that plant flowering is largely regulated by six major floral regulatory pathways in *Arabidopsis*, namely, the vernalization, photoperiod, autonomous, gibberellin, thermosensory (ambient temperature perception), and age pathways [1–5]. These pathways are ultimately intertwined with several central transcription factors, including *FLOWERING LOCUS T* (*FT*), *TERMINAL FLOWER1* (*TFL1*), *LEAFY* (*LFY*), and *SUPPRESSOR OF OVEREXPRESSION OF CONSTANS1* (*SOC1*) [6]. Many plant species have demonstrated that *FT* promotes their flowering and *TFL1* inhibits it [7–10]. An updated article on the role of transcription factors in the regulation of flowering highlights the involvement of *PsFT*, *PsTFL1*, and *PsFD* in the continuous flowering of tree peony (*Paeonia* × *lemoinei* 'High Noon') [11]. The authors of this article also discuss how this understanding of the molecular regulatory mechanisms underlying continuous flowering will be conducive to further investigation, and will be of great importance for the breeding of tree peonies and other perennial woody plants. In many species, the role of flowering genes in regulating plant flowering is highly conserved. A recent publication on the function of MADS-box family genes detailed two MADS-box genes—*LsAG2* and *LsDEF1*, isolated from *Lagerstroemia speciosa*—that were ectopically expressed in *Arabidopsis*, and how those transgenic plants exhibited early flowering and floral organ aberrations, during which the expression of genes associated with flowering (e.g., *AP1*, *LFY*, and *FT*) were up-regulated [12].

Floral organs are the basis of plant classification and evolution [13]. Studies of the MADS-box family have shown that members of this family are extensively involved in the floral transition and floral organ morphogenesis [14,15]. However, there are still many members of the MADS-box gene family that have not yet been discovered, and the molecular regulatory mechanisms of flower organ development require further exploration. Qin et al. 2023 recently established the evolution and divergence patterns of MIKC-type MADS-box gene family members in six *Rosaceae* plants [13]. This is important for predicting functional regions of the MIKC-type MADS-box genes and will be extremely useful in the discovery of other genes that regulate key traits.

Flower development was first summarized using the ABC model [16], which classifies the genes involved in the differentiation of floral organs into three categories—A, B, and

C—which individually or cooperatively regulate the development of stamens, pistils, petals, and the calyx. Following the cloning and functional identification of a large number of transcription factors related to flower development in both annual plants and perennials, flower development was then summarized using the ABCDE model [17]. The newly identified D-class genes are involved in the regulation of ovule development and formation, while E-class genes are considered to be independent from ABC genes, able to regulate the expression of ABC genes, and ultimately participate in the regulation of flower organs during the floral transition and flower development. It is well known that studies on the molecular mechanisms of flower development are well established in angiosperms, while research into gymnosperms is relatively scarce [18,19]. In this Special Issue, a new publication compares the transcriptomic profiles of male and female *Pinus halepensis* cones at different developmental stages in order to uncover the genes involved in their reproductive induction and development [20]. The authors demonstrate that the expression of the C-class genes *MADS1* and *DAL14* correlate with female cone development, whereas male cone development is associated with the expression of the B-class genes *PhDAL11* and *PhDAL13* and the C-class gene *PhMADS2*. These results can provide important reference values for further investigation into the developmental stages of reproduction in gymnosperms.

Floral longevity is an important functional trait that reflects the ornamental value of flowering plants. Numerous researchers have conducted studies on the flowering process, including the morphological structure of flowers, and their mineral elements, nutrients, and related genes, in order to elucidate the biological and morphological characteristics that regulate the flowering period of various plants [21,22]. Another cutting-edge article in this Issue depicts the relationship between the water balance and different floral structural traits in six tree peonies, and investigates the relevant mechanism that causes the discrepancies seen in their flower longevity [23]. It is particularly important to understand the morphological development and physiological characteristics of flower organs during different stages of flower development, which makes the development of effective scientific techniques for investigating these traits so significant. A recent publication performed experiments to explore the structure of petals, their nutrient content, hormones, and mineral elements, and the activity of antioxidant enzymes at different flowering stages of *Styrax japonicus*, in order to indicate which specific factors are important for flowering, and thus lay the foundation for extending the flowering period [24]. Similar to studies on the extension of flower longevity, research on the flower color of ornamental plants is also important for future selective breeding. A recently updated study was the first to elucidate the molecular mechanisms of flavonoid biosynthesis and flower coloring in *Loropetalum chinense* and *Loropetalum chinense* var. *rubrum* via flavonoid metabolomics, transcriptomics, and full-length sequencing [25]. The authors demonstrated that significant differences in flavonoid content were observed among four cultivars with similar genetic backgrounds, and that the expression of the flavonoid synthesis gene correlates to a high proportion of flavonoids in *Loropetalum chinense* var. *rubrum*.

It is well known that plant flowering is also affected by various environmental factors [26]. A new review summarizes our current knowledge on the role of environmental factors, such as drought stress, low-temperature stress, and high-temperature stress, in the regulation of annuals and perennials' flowering time [27]. Recently, plant growth regulators have been widely used in agricultural production due to their significant ability to promote flowering and control the branching of fruit trees [28]. Of these regulators, the inhibitory effect of gibberellin (GA) on flowering has been demonstrated in fruit trees such as pear, plum, peach, and apple [29]. Conversely, the application of paclobutrazol (PBZ) can promote the flowering of plants [30]. A publication in this Issue records the results of spraying lemons with different concentrations of GA and PBZ in order to study the effects of these two plant growth regulators on the nutritional physiology and flower formation of lemons [29]. The authors discovered that the optimal PBZ concentration for promoting lemon flowering was 600 mg/L, and that the quality of the lemons was also improved accordingly. The physiological effect of GA on lemon flowering was the opposite to that of PBZ. The selection of appropriate concentrations of plant growth regulators is of great significance in controlling the flowering and fruit setting of fruit trees; it can also provide a theoretical basis for the high quality, high yield, and stable production of fruit trees.

Flowering and flower development are critical to the reproductive success of plants and still require further investigation for future plant breeding. The series of articles presented in this Special Issue clearly illustrate that there are both conservative and highly-specific ways to regulate the flowering and flower development of different plants. We look forward to future scientific advancements in the reproductive development of plants, which may be influenced by our understanding of these Special Issue articles, as well as to the development of eco-friendly and innovative strategies for the regulation of flowering and flower development.

**Funding:** This research received no external funding.

**Conflicts of Interest:** The authors declare no conflicts of interest.

## References

1. Bastow, R.; Mylne, J.S.; Lister, C.; Lippman, Z.; Martienssen, R.A.; Dean, C. Vernalization requires epigenetic silencing of *FLC* by histone methylation. *Nature* **2004**, *427*, 164–167. [CrossRef]
2. Wu, Z.; Fang, X.; Zhu, D.; Dean, C. Autonomous pathway: *FLOWERING LOCUS C* repression through an antisense-mediated chromatin-silencing mechanism. *Plant Physiol.* **2020**, *182*, 27–37. [CrossRef]
3. Blázquez, M.A.; Ahn, J.H.; Weigel, D. A thermosensory pathway controlling flowering time in *Arabidopsis thaliana. Nat. Genet.* **2003**, *33*, 168–171. [CrossRef]
4. Osnato, M.; Castillejo, C.; Matías-Hernández, L.; Pelaz, S. *TEMPRANILLO* genes link photoperiod and gibberellin pathways to control flowering in *Arabidopsis. Nat. Commun.* **2012**, *3*, 808. [CrossRef] [PubMed]
5. Wang, J.-W. Regulation of flowering time by the miR156-mediated age pathway. *J. Exp. Bot.* **2014**, *65*, 4723–4730. [CrossRef]
6. Wickland, D.P.; Hanzawa, Y. The *FLOWERING LOCUS T/TERMINAL FLOWER 1* gene family: Functional evolution and molecular mechanisms. *Mol. Plant* **2015**, *8*, 983–997. [CrossRef] [PubMed]
7. Coelho, C.P.; Minow, M.A.A.; Chalfun-JÃºnior, A.; Colasanti, J. Putative sugarcane *FT/TFL1* genes delay flowering time and alter reproductive architecture in *Arabidopsis. Front. Plant Sci.* **2014**, *5*, 221. [CrossRef] [PubMed]
8. Bai, M.; Liu, J.; Fan, C.; Chen, Y.; Chen, H.; Lu, J.; Sun, J.; Ning, G.; Wang, C. *KSN* heterozygosity is associated with continuous flowering of *Rosa rugosa* Purple branch. *Hort. Res.* **2021**, *8*, 26. [CrossRef]
9. Iwata, H.; Gaston, A.; Remay, A.; Thouroude, T.; Jeauffre, J.; Kawamura, K.; Oyant, L.H.S.; Araki, T.; Denoyes, B.; Foucher, F. The *TFL1* homologue *KSN* is a regulator of continuous flowering in rose and strawberry. *Plant J.* **2011**, *69*, 116–125. [CrossRef]
10. Taoka, K.-I.; Ohki, I.; Tsuji, H.; Furuita, K.; Hayashi, K.; Yanase, T.; Yamaguchi, M.; Nakashima, C.; Purwestri, Y.A.; Tamaki, S.; et al. 14-3-3 proteins act as intracellular receptors for rice Hd3a florigen. *Nature* **2011**, *476*, 332–335. [CrossRef]
11. Zhang, L.; Cheng, F.; Huang, H.; Geng, Z.; He, C. *PsFT*, *PsTFL1*, and *PsFD* are involved in regulating the continuous flowering of tree peony (*Paeonia* × *lemoinei* 'High Noon'). *Agronomy* **2023**, *13*, 2071. [CrossRef]
12. Yang, L.; Li, Z.; Zheng, T.; Wang, J.; Cheng, T.; Zhang, Q. Overexpression of two MADS-Box genes from *Lagerstroemia speciosa* causes early flowering and affects floral organ development in *Arabidopsis. Agronomy* **2023**, *13*, 976. [CrossRef]
13. Qin, Y.; Zhu, G.; Li, F.; Wang, L.; Chen, C.; Zhao, H. MIKC-Type MADS-Box gene family discovery and evolutionary investigation in *Rosaceae* plants. *Agronomy* **2023**, *13*, 1695. [CrossRef]
14. Schilling, S.; Kennedy, A.; Pan, S.; Jermiin, L.S.; Melzer, R. Genome-wide analysis of MIKC-type MADS-box genes in wheat: Pervasive duplications, functional conservation and putative neofunctionalization. *New Phytol.* **2020**, *225*, 511–529. [CrossRef]
15. He, C.M.; Si, C.; da Silva, J.A.T.; Li, M.Z.; Duan, J. Genome-wide identification and classification of MIKC-type MADS-box genes in Streptophyte lineages and expression analyses to reveal their role in seed germination of orchid. *BMC Plant Biol.* **2019**, *19*, 223. [CrossRef] [PubMed]
16. Irish, V. The ABC model of floral development. *Curr. Biol.* **2017**, *27*, R887–R890. [CrossRef] [PubMed]
17. Liu, J.Y.; Fu, X.D.; Dong, Y.W.; Lu, J.; Ren, M.; Zhou, N.N.; Wang, C.Q. MIKC$^{C}$-type MADS-box genes in *Rosa chinensis*: The remarkable expansion of ABCDE model genes and their roles in floral organogenesis. *Hort. Res.* **2018**, *5*, 25. [CrossRef] [PubMed]
18. Silva, C.S.; Puranik, S.; Round, A.; Brennich, M.; Jourdain, A.; Parcy, F.; Hugouvieux, V.; Zubieta, C. Evolution of the plant reproduction master regulators LFY and the MADS transcription factors: The role of protein structure in the evolutionary development of the flower. *Front. Plant Sci.* **2016**, *6*, 1193. [CrossRef]
19. Theissen, G.; Melzer, R.; Rümpler, F. MADS-domain transcription factors and the floral quartet model of flower development: Linking plant development and evolution. *Development* **2016**, *143*, 3259–3271. [CrossRef]
20. Reisfeld, G.; Faigenboim, A.; Fox, H.; Zemach, H.; Eshed Williams, L.; David-Schwartz, R. Differentially expressed transcription factors during male and female cone development in *Pinus halepensis. Agronomy* **2022**, *12*, 1588. [CrossRef]
21. Luo, H.H.; Wang, Q.; Zhang, J.K.; Wang, L.S.; Li, Y.B.; Yang, G.Z. Minimum fertilization at the appearance of the first flower benefits cotton nutrient utilization of nitrogen, phosphorus and potassium. *Sci. Rep.* **2020**, *10*, 6815. [CrossRef]

22. Li, X.M.; Fan, T.; Zou, P.; Zhang, W.H.; Wu, X.J.; Zhang, Y.X.; Liao, J.P. Can the anatomy of abnormal flowers elucidate relationships of the androecial members in the ginger (*Zingiberaceae*)? *Evodevo* **2020**, *11*, 12. [CrossRef]
23. Guo, Y.; Qiu, Y.; Hu, H.; Wang, Y.; Mi, Z.; Zhang, S.; He, S.; Jia, W. Petal morphology is correlated with floral longevity in *Paeonia suffruticosa*. *Agronomy* **2023**, *13*, 1372. [CrossRef]
24. Chen, C.; Chen, H.; Ni, M.; Yu, F. A study on petal morphological and physiological characteristics of *Styrax japonicus* during the flowering period. *Agronomy* **2021**, *11*, 1498. [CrossRef]
25. Zhang, X.; Zhang, L.; Zhang, D.; Liu, Y.; Lin, L.; Xiong, X.; Zhang, D.; Sun, M.; Cai, M.; Yu, X.; et al. Transcriptomic and metabolomic profiling provides insights into flavonoid biosynthesis and flower coloring in *Loropetalum chinense* and *Loropetalum chinense* var. *rubrum*. *Agronomy* **2023**, *13*, 1296. [CrossRef]
26. Zhu, J.K. Abiotic stress signaling and responses in plants. *Cell* **2016**, *167*, 313–324. [CrossRef] [PubMed]
27. Chen, M.; Zhang, T.-L.; Hu, C.-G.; Zhang, J.-Z. The role of drought and temperature stress in the regulation of flowering time in annuals and perennials. *Agronomy* **2023**, *13*, 3034. [CrossRef]
28. Goldberg-Moeller, R.; Shalom, L.; Shlizerman, L.; Samuels, S.; Zur, N.; Ophir, R.; Blumwald, E.; Sadka, A. Effects of gibberellin treatment during flowering induction period on global gene expression and the transcription of flowering-control genes in *Citrus* buds. *Plant Sci.* **2013**, *198*, 46–57. [CrossRef] [PubMed]
29. Du, Y.; Li, J.; Dong, J.; Hu, C.; Li, D.; Tan, Q.; Zhang, J.; Li, J.; Zhou, X.; Zhu, C.; et al. Effects of different regulating measures on the floral and nutritional physiology of lemon. *Agronomy* **2022**, *12*, 2381. [CrossRef]
30. Upreti, K.K.; Reddy, Y.T.N.; Prasad, S.R.S.; Bindu, G.V.; Jayaram, H.L.; Rajan, S. Hormonal changes in response to paclobutrazol induced early flowering in mango cv. Totapuri. *Sci. Hortic.* **2013**, *150*, 414–418. [CrossRef]

*agronomy*

MDPI

*Review*

# The Role of Drought and Temperature Stress in the Regulation of Flowering Time in Annuals and Perennials

**Min Chen, Tian-Liang Zhang, Chun-Gen Hu and Jin-Zhi Zhang** *

National Key Laboratory for Germplasm Innovation & Utilization of Horticultural Crops, College of Horticulture and Forestry Science, Huazhong Agricultural University, Wuhan 430070, China; minchen970402@webmail.hzau.edu.cn (M.C.); tianliangzhang@webmail.hzau.edu.cn (T.-L.Z.); chungen@mail.hzau.edu.cn (C.-G.H.)

* Correspondence: jinzhizhang@mail.hzau.edu.cn; Tel.: +86-27-6201-80231; Fax: +86-27-8728-2010

**Abstract:** Plants experience a variety of adverse environments during their vegetative growth and reproductive development, and to ensure that they complete their life cycle successfully, they have evolved specific defense mechanisms to cope with unfavorable environments. Flowering is a vital developmental stage and an important determinant of productivity in the lifetime of plants, which can be vulnerable to multiple abiotic stresses. Exposure to stress during this period can have dramatic effects on flower physiological and morphological development, which may ultimately lead to a substantial loss of yield in seed-producing plants. However, there has been increasing research evidence that diverse abiotic stresses, ranging from drought, low temperature, and heat stress can promote or delay plant flowering. This review focuses on how plants alter developmental direction to balance between survival and productivity under drought and extreme temperature conditions. Starting from the perspective of the functional analysis of key flowering-regulated genes, it is of great help for researchers to quickly gain a deeper understanding of the regulatory effects of abiotic stress on the flowering process, to elucidate the molecular mechanisms, and to improve the regulatory network of abiotic-stress-induced flowering. Additionally, the important agronomic significance of the interaction between abiotic stress and the flowering regulation of perennial plants under climate change conditions is also discussed after summarizing studies on the mechanisms of stress-induced flowering in annual plants. This review aims to clarify the effects of abiotic stresses (mainly drought and temperature) on plant flowering, which are significant for future productivity increase under unfavorable environmental conditions.

**Keywords:** flowering; abiotic stress; drought; temperature; perennial plants

**Citation:** Chen, M.; Zhang, T.-L.; Hu, C.-G.; Zhang, J.-Z. The Role of Drought and Temperature Stress in the Regulation of Flowering Time in Annuals and Perennials. *Agronomy* **2023**, *13*, 3034. https://doi.org/10.3390/agronomy13123034

Academic Editors: Youssef Rouphael and Fengjie Sun

Received: 1 November 2023
Revised: 24 November 2023
Accepted: 8 December 2023
Published: 11 December 2023



## 1. Introduction

Plants are sessile and cannot move to escape from adverse environmental conditions. Hence, the developmental process of many plants is highly changeable in response to the environmental stresses they encounter. Abiotic stresses, including drought, salinity, devastating temperature (extreme high or low), and nutrient (mainly nitrogen, phosphorus, and potassium) starvation [1], can have a dramatic impact on plant growth and productivity, such as cellular water scarcity, cell membrane damage, enzyme inactivation, and other defects, ultimately leading to severe yield reductions and huge economic losses [2–5]. Therefore, abiotic stress has been an important issue in plant vegetative [6,7] and reproductive [8,9] development, for which the study of abiotic stress effects during reproductive development is of great significance for the maintenance of food production as well as for the world economy.

Flowering is an important agricultural trait in the successful transition of plants from vegetative to reproductive growth, as the optimal flowering time is critical for maximizing reproductive success and ensuring seed production, which is a key step in the evolutionary success of plants. Due to the continuously fluctuated environmental conditions, plants have evolved specific defense mechanisms to ensure maximum reproductive success [10]. For example, when individuals cannot survive under severe stress conditions, they produce seeds by adjusting the timing of flowering. In recent years, the effects of abiotic stresses on flowering induction have been documented in numerous plant species [11–19]. The ability of abiotic stresses to regulate flowering, with drought and temperature being important stress factors, suggests that plants can combine abiotic stresses effects with flowering signaling pathways. Thus, stress-induced flowering has been recognized as a new means of flowering response due to its important biological benefits throughout the plant life cycle [20].

Extensive physiological and molecular genetic analyses have revealed six major floral regulatory pathways in *Arabidopsis thaliana*, namely, photoperiod, autonomous, thermosensory (ambient temperature perception), vernalization, gibberellin, and age pathways [21–26]. These pathways are ultimately integrated through several flowering genes that regulate flowering in plants. Under long-day conditions, the transcription factor *CONSTANS (CO)* (1, Table 1) acts upstream to activate the expression of *FLOWERING LOCUS T (FT)* (2, Table 1), a core floral integrator gene transcribed in the leaves. Plants sense seasonal changes in day length via leaves, and subsequently, when the optimum environmental conditions are achieved, FT protein is translated into the bast cells as a long-distance signal (florigen), which is eventually delivered to the shoot apical meristem to activate phloem tissue genes, such as *LEAFY (LFY)* (3, Table 1) and *APETALA1 (AP1)* (4, Table 1), to induce flowering in *Arabidopsis* [27,28]. The study of the *Arabidopsis* flower formation pathway is a crucial step in revealing the regulatory network of flower-forming in plants. With the continuation of related research, accumulating evidence indicates that the key integrating genes of the flowering network can also be induced by abiotic stress in regulating flowering time. Abiotic-stress-induced flowering has become a research hotspot, which has attracted extensive attention from worldwide researchers. For example, drought as well as UV-C stress induced the expression of *FT* in *Arabidopsis* [11,29], resulting in early flowering, while the expression of *FT* in *pharbitis* [30,31] was induced by low temperature and nutrient deficiency. Early flowering in response to drought stress in *Arabidopsis* requires the combined function of the flowering gene *GIGANTEA (GI)* (5, Table 1) and the floral integration factor *SUPPRESSOR OF OVEREXPRESSION OF CONSTANS 1 (SOC1)* (6, Table 1) in addition to the upregulation of *FT* [11]. *Arabidopsis* responds to salt stress by inhibiting *FT* and *CO* expression with delayed flowering. Meanwhile, *GI* is also involved in salt-stress tolerance in *Arabidopsis* [32]. These findings suggest that the key integrating genes of the flower-forming network have dual roles in regulating the flowering time and stress tolerance response, and they may be potential genes involved in stress-induced flowering. Taken together, flowering is subject to a combination of an endogenous gene regulatory network as well as external environmental stimuli, which also supports the importance and necessity of further research on abiotic-stress-induced flowering.

Despite its importance, there have been fewer studies concerning the effects of abiotic stresses on plant reproductive development compared to those in the vegetative growth process. Moreover, abiotic-stress-induced flowering has been reported mainly in annual plants, especially during seed germination, seedling growth, and yield. In contrast to annual plants, there are fewer cases of abiotic-stress-induced flowering in perennials, especially woody plants. This may be partly due to the fact that studies involving reproductive development, such as flowering, require longer plant growth cycles and years of study, which can be a great challenge for relevant materials acquisition as well as for researchers. Seasonal flowering is a typical feature that distinguishes perennials from annuals [33]. Citrus is representative of perennial fruit trees that generally bloom once in the spring or several times a year, depending on the variety and genotype, and flowering induction is affected by drought and low-temperature stresses [34,35]. Drought-induced flowering provides an applicable method for shortening the vegetative growth of woody plants, which can be utilized in genetic research and breeding [36]. These cases of stress-induced vegetative growth and flowering indicate that the vegetative growth and reproductive developmental programs of perennial plants can also be affected by abiotic stress, which might conceal some biological significance [13,36,37]. Therefore, we collected articles on the topic of drought and temperature affecting plant flowering over the last decade through PubMed. The study discusses thc similarities and differences between annual and perennial plants. Some insights are provided for further research on the mechanisms by which drought and temperature affect flowering in perennial plants. Accordingly, this review summarizes current knowledge showing that diverse abiotic stresses (mainly drought and temperature) modify the flowering time of plants and examines physiological alterations in their response to stress. Meanwhile, the role of critical flowering regulation genes in response to stress was identified with consideration of the cross-talk molecular mechanisms underlying flowering time regulation and stress response. Finally, valuable agronomic implications of the cross-talk between abiotic stress and flowering regulation under climate change conditions is discussed in perennial plants.

**Table 1.** Specialized terms and their abbreviations appearing in this review.

| Number | Abbreviations | Full-Name | Number | Abbreviations | Full-Name |
|---|---|---|---|---|---|
| 1 | *CO* | *CONSTANS* | 24 | *HDA6* | *HISTONE DEACETYLASE 6* |
| 2 | *FT* | *FLOWERING LOCUS T* | 25 | *FES1* | *FRI ESSENTIAL 1* |
| 3 | *LFY* | *LEAFY* | 26 | *FRL1* | *FRI-LIKE 1* |
| 4 | *AP1* | *APETALA1* | 27 | *FLX* | *FLC EXPRESSOR* |
| 5 | *GI* | *GIGANTEA* | 28 | *SUF4* | *SUPPRESSOR OF FRI 4* |
| 6 | *SOC1* | *SUPPRESSOR OF OVEREXPRESSION OF CONSTANS 1* | 29 | *VRN1* | *VERNALIZATION1* |
| 7 | ROS | reactive oxygen species | 30 | *FUL* | *FRUITFULL* |
| 8 | ABA | abscisic acid | 31 | *VAL* | *CAULIFLOWER* |
| 9 | SD | short-day conditions | 32 | *PIF4* | *PHYTOCHROME-INTERACTING TRANSCRIPTION 4* |
| 10 | LD | long-day conditions | 33 | *MAF2* | *MADSAFFECTING FLOWERING 2* |
| 11 | *TSF* | *TWIN SISTER OF FT* | 34 | *FCA* | *FLOWERING CONTROL LOCUS A* |
| 12 | *SVP* | *SHORT VEGETATIVE PHASE* | 35 | *FVE* | *FLOWERING LOCUS VE* |
| 13 | *FLC* | *FLOWERING LOCUS C* | 36 | *FLM* | *FLOWERING LOCUS M* |
| 14 | *Hd3a* | *HEADING DATE 3a* | 37 | *HOS1* | *HIGH EXPRESSION OF OSMOTICALLY RESPONSIVE GENE 1* |
| 15 | *RFT1* | *RICE FLOWERING LOCUS T1* | 38 | HSR | heat stress response |
| 16 | *Ehd1* | *EARLY HEADING DATE 1* | 39 | HSPs | heat shock proteins |
| 17 | *RCN1* | *RICE CENTRORADIALIS 1* | 40 | HSFs | heat stress transcription factors |
| 18 | *FD* | *FLOWERING LOCUS D* | 41 | *BOB1* | *BOBBER1* |
| 19 | FAC | florigen activation complex | 42 | *FTL3* | *FLOWERING LOCUS T-like 3* |
| 20 | *FRI* | *FLOWERING CONTROL LOCUS A* | 43 | *PRR* | *PSEUDO RESPONSE REGULATOR* |
| 21 | *OST1* | *OPEN STOMATA 1* | 44 | *LUX* | *LUX ARRHYTHMO* |
| 22 | *VOZ1* | *VASCULAR PLANT ONE-ZINC FINGER 1* | 45 | Eps-D1 | Earliness *per se* locus |
| 23 | *RFS* | *Regulator of Flowering and Stress* | 46 | *EG1* | *EXTRA GLUME 1* |

## 2. Drought-Induced Flowering

Global warming and the continued increase in the world's population have led to a shortage of freshwater resources and a further decline in groundwater levels, posing a major challenge to agriculture worldwide [38]. Drought, which is defined as being in a state of water shortage for several consecutive weeks [39], is the most common abiotic stress around the world and severely affects flowering time, flower morphological developmental processes, and the seed productivity of several plant species. It is particularly noteworthy that drought stress can also cause flower abortion and eventually plant sterility by altering the expression levels of various genes critical to flowering regulation pathways, which regulate both flowering time and response to drought stress [40]. In the following, we discuss how plants perceive and respond to drought stress, further summarize the differential physiological phenotypes of various plant species under drought stress conditions, and, finally, we focus on the potential molecular mechanisms underlying drought-stress-induced flowering.

### 2.1. Perception and Coping Strategies of Drought Stress

Plants perceive drought stress signals mainly through the leaves and the root system. Stomatal movement can be observed in leaves, and drought stress can lead to the accumulation of reactive oxygen species (ROS) (7, Table 1) and of abscisic acid (ABA) (8, Table 1) in leaves, which, in turn, regulates the movement of guard cells and ultimately determines the opening and closing state of stomata [41]. However, it is difficult to determine how plants respond to drought stress in the root system [42,43]. A deficiency of water can constrain the growth and development process of plants, and can even have a significant impact on plant survival [1,44]. As a result, plants have evolved a variety of strategies to cope with damage caused by drought stress. The process by which plants sense water deficit signals and further initiate coping strategies in response to drought stress is known as drought resistance. The adaptability of plants to drought stress mainly consists of three different coping strategies, namely, drought escape, drought avoidance, and drought tolerance [45]. Drought escape, a common strategy exploited in response to drought stress, refers to plants that accelerate flowering and shorten their entire life cycle before severe drought stress hinders their survival [46,47]. However, in order to achieve early flowering, a drought escape strategy will terminate vegetative growth in advance, which can severely influence the growth and development of vegetative organs, and eventually lead to a dramatic decrease in seed yield. Drought avoidance (also known as drought dehydration) is another strategy for plants to cope with external drought conditions by increasing the internal water content (by reducing water loss or maximizing water uptake) [48]. The drought tolerance strategy is the ability of plants to tolerate low internal water content and to adapt to the drought stress while initiating reproduction [49].

Under drought stress conditions, plants can respond by early or late flowering, depending on the onset, duration, and severity of drought [20,50]. A bibliometrics analysis showed that plant response to drought has become an important research topic [51]. When plants are adequately supplied with water, the stomata remain open to a large extent to enable the plants to fully photosynthesize, while under mild drought stress, plants will appropriately regulate stomatal closure to minimize water loss by reducing transpiration, but this will result in a decrease in the rate of photosynthesis. When subjected to severe drought stress, the stomata are generally in a minimally open state to ensure that some photosynthesis can take place, thus guaranteeing the normal survival needs of plants [52]. This is one of the main approaches for plants to avoid damage caused by drought stress in the short term [53]. Influenced by geography, many terrestrial plants are frequently affected by drought stress and have developed various drought-tolerant mechanisms to adapt to or to resist the drought environment through a long-term evolutionary process [54,55]. The adaptation of plants to the drought environment is mainly reflected both morphologically and biochemically [56–58]. Morphologically, adaptation is manifest in the presence of a very thick cuticle on the leaf surface, with the fenestrated cells being tightly arranged, while some leaves have a tomentum on the surface, which can effectively control water loss, and can also absorb dew at night to replenish the plant's own water [59]. Generally, there is a very well-developed root structure with greater water and nutrient absorption capacity and a poorly developed aboveground branching structure with weaker transpiration and better water retention capacity [60]. The more drought-resistant the plant, the greater the root–crown ratio. These morphological adaptations are closely related to the cell division, elongation, and differentiation of the root apex. Plant vascular tissue systems, such as the xylem and phloem, are involved in the transport of substances, while their developmental status also affects plant drought resistance [61]. In *Arabidopsis*, drought-escape-induced early flowering is associated with the phloem tissue transport of the florigen FT protein from the leaves to the shoot apical meristems [62]. Biochemically, it is manifested in the high expression of some drought resistance genes that positively increase the content of amino acids and sugars in plants (such as proline and trehalose), the enhanced activity of antioxidant-related enzymes, and inhibition of the activity of enzymes involved in degradation pathways to ensure that normal metabolic homeostasis is maintained under drought stress condition [63]. ROS, including superoxide radicals ($O_2^-$), hydrogen peroxide ($H_2O_2$), and hydroxyl radicals ($OH^-$), regulate plant growth and development at lower concentrations [64,65]. Excessive accumulation of ROS under drought stress leads to membrane lipid peroxidation [66,67]. Previous studies have shown that excessive ROS in plants will be scavenged by antioxidant mechanisms, including the enzymatic antioxidants, SOD (superoxide dismutase), CAT (catalase), POD (peroxidase), and the non-enzymatic antioxidants, ascorbic acid, proline, flavonoids, and polyphenols, which ultimately improve the plant's drought resistance [68–71]. Ascorbic acid has also been reported to play a role in controlling the flowering time in plants [72]. Understanding the perception of drought stress and the coping strategies (early or late flowering) used by plants in response to drought provides a physiological basis for subsequent studies on the molecular mechanisms of drought-stress-induced flowering.

### 2.2. *Flowering Time of Various Plant Species in Response to Drought Stress*

Most plants have evolved and adapted to the frequent fluctuations in the natural environment, especially the severe damage caused by droughts due to water deficit. Drought stress can lead to alterations in the flowering time of various plants, effects on flower development (including reduced flower number, restricted filament elongation, and delayed anther development), immature seed development, and reduced yields [40,73,74]. Therefore, the flowering time is an important agricultural characteristic for the development of adaptation to drought stress in a wide range of plants (Table 2). The discrepancy between early and late flowering resulting from the effects of drought stress depends on the plant species [20,75]. For example, forage and biofuel crops generally have delayed flowering as a desirable target for them due to the importance of the plant vegetative biomass. Cereal crops, by contrast, usually exhibit early flowering as an ideal trait to shorten the harvest time, and, thus, increase the number of plantings during the growing season, while growing as fast as possible to minimize damage from drought stress caused by environmental fluctuations [73,76]. Drought stress caused by water deficit delays the flowering time of *Arabidopsis* under short-day (SD) (9, Table 1) conditions but accelerates flowering time under long-day (LD) (10, Table 1) conditions [11,46]. Drought stress causes premature bolting of Chinese cabbage in the growing season, which leads to insufficient vegetative growth and influences yield and quality [77]. By studying the effect of drought stress on flowering time in *Brassica rapa* offspring, it was found that materials with seeds collected after drought flowered earlier than those collected before drought, suggesting that *Brassica rapa* responds to drought stress and evolves towards earlier flowering [78]. Studies of drought-induced flowering in the chickpea suggest that drought stress generally accelerates the flowering time of temperate grain legumes [79,80]. The role of drought in influencing the flowering time varies among plant species and with environmental conditions, so that drought regulation of flowering is the result of multiple factors.

Recently, there have been studies on drought-induced flowering in several other plant species [12,31,81–83]. In contrast to early flowering induced by drought stress, the flowering time of rice is delayed under water deficit conditions to avoid reproductive growth in unfavorable environmental conditions, but water shortage still results in the retardation of plant growth and spikelet development, which leads to reduced crop yield and ultimately economic losses [84–86]. Water deficiency can also delay the first flowering of *Medicago polymorpha*. The effect of drought stress on early and late flowering in plants is closely related to the intensity and duration of the water deficit, in addition to varying by plant species. The artificial control of the duration and intensity of drought treatment in agricultural production plays an important role in accelerating plant development, especially regarding flowering. It was found that the response of rice to drought stress was dependent on the intensity of drought, and mild water deficit in the early development stage triggered a drought escape response with accelerated flowering and reduced tillering [87,88]. In a study of wheat response to drought stress, it was also found that the flowering time showed a nonlinear relationship with the plant water content status, with mild water deficit shortening the flowering time, while severe drought stress delayed flowering [89].

Drought-induced flowering is a phenomenon more common in annual plants. The drought stress regulation of plant flowering has been less well studied in woody plants and remains poorly understood, mainly due to the long duration of vegetative growth in perennials and the excessive long period of research. Drought stress is one of the major environmental factors inducing flowering in adult citrus in subtropical regions. Different citrus species are induced to flower by different environmental conditions, with lemon, four-season orange, and kumquat being mainly affected by drought stress, while sweet orange, trifoliate orange, mandarin orange, grapefruit, and tangerine are mainly affected by seasonal low temperature [39,90]. Notably, drought-induced flowering in citrus was also accompanied by the upregulation of the *CiFT* expression level [37,90]. Earlier studies have also shown that *Citrus latifolia* flowering is also induced by drought stress, and that all citrus plant species share this same flowering mechanism [91,92]. Additionally, the perennial woody plant *Sapium sebiferum* takes 3–5 years to flower normally, but one-year-old seedlings under drought stress flower early, which provides a feasible way to shorten the vegetative growth years of woody plants in genetic research and breeding efforts [36]. These indicate that the effect of drought stress on the flowering time is not specific to a particular plant species but is conserved in annuals as well as perennials. There are few studies on the regulation of flowering time by drought stress in perennials due to the long study time and the lack of phenotypes, but the available evidence supports the feasibility of researching this topic in perennials. Moreover, studies on drought-induced flowering in perennial plants can be carried out on the basis of sufficient theoretical evidence in annual plants.

### 2.3. Molecular Regulatory Mechanisms of Flowering Involved in Drought Stress

Drought-stress-induced flowering, as well as the traditional flower formation pathway, accomplish the same flowering purpose, but the traditional pathway is the primary option for flowering under normal environmental conditions, whereas drought-induced flowering is an emergency response under stressful conditions [20,93]. Compared with the in-depth studies of the flower formation regulatory pathways in plants, the molecular regulatory mechanisms of flowering involved in drought stress are still obscure. Drought stress triggers the differential expression of a variety of genes, including flowering time regulation genes and transcription factors associated with the stress response. Among them, the key genes of flowering regulation in response to drought stress tolerance are *FT*, *CO*, *LFY*, *GI*, *SOC1*, and *TWIN SISTER OF FT (TSF)* (11, Table 1) [11,40,46,94] (Figure 1).

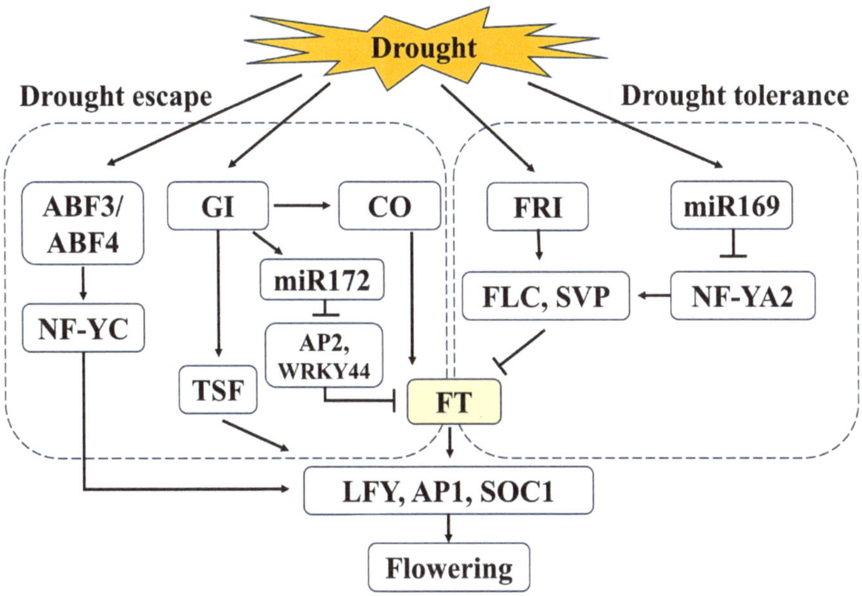


**Figure 1.** Simplified regulatory pathways linking drought stress and flowering in *Arabidopsis thaliana*. Drought escape: *ABF3/ABF4* further activates the expression of *LFY*, *AP1*, and *SOC1* by targeting *NF-YC* [95]. *GI* accelerates flowering under drought conditions by positively regulating the expression of *CO* and *miR172* [94], which, in turn, activate the expression of *FT*, or directly activate the transcription of *TSF*, which ultimately upregulates the expression levels of *LFY*, *AP1*, and *SOC1* [11,32,46]. Drought tolerance: *miR169* targets *NF-YA2* to reduce its transcriptional abundance [96], which attenuates the repressive effect on downstream genes *FLC* and *SVP* [97], while *FRI* positively regulates the expression of *FLC* and *SVP*, resulting in the repression of *FT* transcription and delayed flowering under drought conditions [98]. Solid lines indicate identified associations, arrows indicate positive regulation, and horizontal bars indicate negative regulation.

The molecular mechanisms by which drought stress regulates flowering time in *Arabidopsis* have been partially elucidated. Emerging evidence suggests that *GI*, a photoperiodic pathway gene that promotes flowering, is a pivotal regulator of the abiotic stress response and can influence plant tolerance to abiotic stresses, especially drought [11,99]. Under long-day environmental conditions, water deprivation achieves drought-induced early flowering in *Arabidopsis* through ABA-dependent control of *GI* signaling that activates expression of the florigen genes *FT* and *TSF*. Under short-day conditions, the drought and plant stress hormone ABA is considered to inhibit the transcription of *FT* and *TSF* via activating repressors of floral formation, which, in turn, leads to late flowering in Arabidopsis [11,20,32,46,100]. It has been confirmed that the *GI-miR172* pathway is involved in drought-induced early flowering by downregulating *WRKY44* (directly repressed by miR172) [94] (Figure 1). Several other flowering inhibition genes are also induced by drought stress. For example, water deficit induces the flowering repressor gene *SHORT VEGETATIVE PHASE (SVP)* (12, Table 1), which represses the transcription of genes related to ABA catabolism, and increases ABA accumulation, which improves drought tolerance in *Arabidopsis*, but flowering is delayed [97]. Similarly, *FLOWERING LOCUS C (FLC)* (13, Table 1), a flowering suppressor gene, also plays a role in the drought stress pathway, and the loss of *FLC* function leads to early flowering and decreased drought tolerance in *Arabidopsis* [101]. In rice (*Oryza sativa*), activation of the florigen genes *HEADING DATE 3a (Hd3a)* (14, Table 1), flowering integration factor *OsMADS50* (an orthologue gene of *SOC1* in *Arabidopsis*), and *RICE FLOWERING LOCUS T1* (*RFT1*) (15, Table 1) (*AtFT*-like gene) coordinates the modulation of the drought escape response [87]. Meanwhile, the CCT domain protein Ghd7 plays an important role in delaying the rice heading date and regulating drought stress tolerance under long-day conditions [102,103]. The transcription levels of *Hd3a*, *RFT1*,and *EARLY HEADING DATE 1* (*Ehd1*, upstream of the florigen genes) (16, Table 1) are drastically decreased under drought environmental conditions, which eventually leads to delayed floral transition [85]. *RICE CENTRORADIALIS 1* (*RCN1*, an orthologue of *TFL* in *Arabidopsis*) (17, Table 1) is reported in rice as a flowering time regulation gene in the pathway of drought-regulated floral transition that interacts with the 14-3-3 protein and OsFD1 to repress Hd3a protein function but not its transcriptional level, causing delayed flowering in rice under drought stress [84,104] (Table 2). These results suggest that when plants are subjected to drought stress, a large number of genes are induced to be expressed, including genes critical to the flowering pathway, and that differences in the expression of these genes between species ultimately lead to different flowering outcomes.

**Table 2.** Some examples of stress-induced flowering associated with flowering pathway genes. The yellow areas show examples of flowering induced by drought stress, the blue areas represent low temperature stress, and the red areas represent heat stress.

| Abiotic Stress Factors | Species | Flowering Response | Related Flowering Pathway Genes | References |
|---|---|---|---|---|
| Drought (LD) | *Arabidopsis* | early flowering | *FT, GI, SOC1, TSF* | [11] |
| Drought (SD) | *Arabidopsis* | delayed flowering | *FT, TSF* | [46] |
| Drought | Rice | early flowering | *Hd3a (AtFT), OsMADS50 (AtSOC1), RFT1, Ehd1, OsTIR1, OsABF2, OsmiR393* | [85,87,105] |
| | | delayed flowering | *RCN1 (AtFT)* | [84] |
| | Maize | early flowering | *ZmNF-YA3* | [106] |
| | Barley | early flowering | *miR172, AP2-like* | [107] |
| | Citrus | induction | *CiNF-YA1* | [92] |
| | | | *CiFD* | [13] |
| | *Brachypodium* | delayed flowering | *BdRFS* | [17] |
| | *Solanum lycopersicum* | early flowering | *SlOST1, SlVOZ1* | [18] |
| | *Arabidopsis* | delayed flowering | *OXS3, AP1* | [49] |
| | | | *OXS2, SOC1* | [82] |
| | | early flowering | *ABF3/4, NF-YC, SOC1* | [95] |
| | | | *miR169d, AtNF-YA2* | [96] |
| | *Sapium sebiferum* | induction | *GA1, AP2, CYR2* | [36] |
| Low temperature | *Arabidopsis* | delayed flowering | *FCA, FVE, SVP, FLM* | [23,108] |
| | | | *MAF2* | [109,110] |
| | | | *HOS1, CO, FLC* | [111–113] |
| | | | *HOS15, GI* | [114] |
| | | early flowering | *miR169d, AtNF-YA2* | [96] |
| | *Phaibitis* | induction | *PnFT1, PnFT2* | [30,31] |
| | Chrysanthemum | induction | *MAF2 (AtFLC)* | [110] |
| | Poplar | induction | *FT1* | [115] |
| | Citrus | induction | *CiNF-YA1, CiFT* | [92] |
| | | | *CiFD* | [13] |
| | | | *CiFT, CsFT* | [39,116] |
| | *Medicago sativa* | delayed flowering | *MsFRI-L* | [117] |
| | *Barley/Wheat* | induction | *VRN1, VIN2, VRN3* | [118–120] |
| | *Cymbidium goeringii* | induction | *CgSVP* | [121] |
| Heat stress | *Arabidopsis* | early flowering | *FT* | [122] |
| | | | *PIF4, PIF5* | [123,124] |
| | Soybean | induction | *GmFT2a, GmFT5a* | [125] |
| | Barley | delayed flowering | *FLC* gene family | [126] |
| | Rice | early flowering | *EG1, OsGI* | [127] |
| | Maize | early flowering | *ZmNF-YA3, ZmFTL12* | [106] |
| | Chrysanthemum | delayed flowering | *FTL3 (AtFT)* | [128] |
| | *Brassica rapa* | delayed flowering | *H2A.Z, FT* | [129] |

A nuclear factor-Y (NF-Y) transcription factor, *ZmNF-YA3*, has the dual function of promoting maize flowering while increasing plant drought tolerance, but there is a lack of evidence on the specific mechanism of *ZmNF-YA3* in drought-affected maize flower formation [106]. Also, *Arabidopsis ABF3* and *ABF4* act with *NF-YCs* to mediate drought-accelerated flowering by regulating *SOC1* [95] (Figure 1). In citrus, *CiNF-YA1* was also found to promote drought-induced flowering by forming a complex with *CiNF-YB2* and *CiNF-YC2* to activate *CiFT* expression, and overexpression of *CiNF-YA1* in citrus increased plants drought-sensitivity [92]. It is evident that *NF-YAs* are likely to be functionally conserved in regulating flowering in annual and perennial plants, with functional diversity resulting from physiological differences in response to stress. These studies support a critical role for NF-YAs in promoting not only the flowering time but also drought response (tolerance/sensitivity). However, future studies are needed to clarify whether NF-YAs are directly involved in regulating drought-affected flowering. The bZIP transcription factor *FLOWERING LOCUS D (FD)* (18, Table 1), together with FT and the 14-3-3 proteins, is the florigen activation complex (FAC) (19, Table 1) that regulates plant flowering [130]. *CiFD* was found to form two distinct proteins through alternative splicing, *CiFDα* and *CiFDβ*, that both initiate flowering in citrus. Among them, *CiFDα* was induced by low temperature while *CiFDβ* was induced by drought stress. The regulatory mechanism of *CiFDβ* promoting drought-induced flowering is independent of FAC and interacts directly with *AP1* [13] (Table 2). *FRIGIDA (FRI)* (20, Table 1) is an essential regulator of flowering in various plant species, including *Populus balsamifera* [131], *Medicago sativa* [117], *Brassica napus* [132], and *Vitis vinifera* [133]. Importantly, *FRI* modulates drought tolerance through the *FLC–OST1* regulatory module [101] (Figure 1). *CiFRI*, a homologue of *FRI* in citrus, was drought-induced, and overexpression of *CiFRI* enhanced drought tolerance in *Arabidopsis* and citrus, whereas silenced plants showed drought sensitivity, and the ectopic expression in *Arabidopsis* exhibited late flowering. The citrus dehydrogenase gene *CiDHN* may maintain the stability of the CiFRI protein during drought-induced degradation [14]. Therefore, the drought-induced flowering regulation genes are conserved between annual and perennial plants. Together, these studies strongly support the pivotal roles of flowering-time-regulated genes in drought stress response and tolerance. The main challenge in woody plants is that the regulatory role of genes in drought-induced flowering can be demonstrated, but there is no phenotypic evidence of flowering, which is related to its own longer developmental process.

In addition, drought-induced transcription factors are closely related to the existing flowering regulatory pathways, and these TFs affect the flowering process of plants by regulating the transcription level of flowering-regulated genes [12,17,134]. The tomato *OPEN STOMATA 1 (SlOST1)* (21, Table 1) loss-of-function mutant causes reduced drought tolerance in plants, and the *slost1* mutant exhibits a late-flowering phenotype under both normal and drought environmental conditions. *SlOST1* combines with the flowering integrated gene *VASCULAR PLANT ONE-ZINC FINGER 1 (SlVOZ1)* (22, Table 1) to form a regulatory module, and then interacts with the promoter of *SINGLE FLOWER TRUSS* to regulate tomato flowering under drought stress [18]. A conserved and specific gene family in plants, the *Regulator of Flowering and Stress (RFS)* family (23, Table 1), produces dramatic alterations in transcriptional levels in response to drought environmental stimuli. Overexpression of *BdRFS* in *Brachypodium distachyon* not only substantially delayed flowering but also promoted drought tolerance. The *rfs* mutants in *Arabidopsis* and *Brachypodium distachyon* displayed an early flowering phenotype and were susceptible to water deprivation [17].

Studies have reported that epigenetic mechanisms, including histone acetylation as well as methylation, are involved in the plant stress response and flowering time regulation. Histone deacetylase *HISTONE DEACETYLASE 6 (HDA6)*-deficient mutant plants (24, Table 1) exhibited a phenotype of reduced drought stress tolerance and delayed flowering with the repression of *FLC* expression [135–137]. The histone H4 gene *BrHIS4.A04*, which interacts with *BrVIN3.1*, is overexpressed in Chinese cabbage and reduces plant susceptibility to drought stress and accelerates flowering under normal growth conditions, whereas under water deficit environmental conditions, the histone H4 gene represses the expression of photoperiodic flowering genes to prevent premature bolting [77]. The regulation of flowering time under drought stress is also related to microRNAs (miRNAs). miRNAs are considered to be important suppressors of gene expression at the transcriptional and post-transcriptional levels. The involvement of miRNAs in regulating drought-stress-induced plant flowering responses has been found in many species, such as the annual plants *Arabidopsis* [138], rice [139], wheat [140], and maize [141], as well as in perennial plant species [142]. miR172 acts in the process of drought tolerance and flowering time regulation. miR172b-3p and miR172b-5p, derived from a common precursor, promoted flowering and enhanced drought tolerance in barley (Table 2). miR172b-3p expression was upregulated under drought stress treatment, which suppressed the four *AP2*-like transcription factors in barley to accelerate flowering. The expression of miR172b-5p was inhibited under drought conditions; thus, trehalose-6-phosphate synthase (TPS), a key enzyme for trehalose biosynthesis targeted by miR172b-5p, was significantly accumulated to enhance drought tolerance in barley [107]. miR156, which is in the same age pathway regulating plant flowering as miR172, is also induced by drought stress and delays flowering of *Arabidopsis* and tobacco [138,143]. miR169 family members play an important role in stress-induced flowering by inhibiting *NF-YA2*, which, in turn, decreases *FLC* expression, allowing the promotion of flowering [96] (Figure 1). In OsmiR393-overexpressing rice plants, miR393 responds to drought stress by targeting and, thus, repressing the expression of the auxin receptor genes *OsTIR1* and *OsAFB2* for early flowering [105]. In addition, differential expression of key proteins and post-translational modifications, such as SUMOylation, act in regulating plants' flowering processes under drought stress conditions [144]. Taken together, when plants are subjected to drought, a variety of molecular regulatory mechanisms can be activated, suppressed, and integrated to maintain survival through the adjustment of flowering time.

## 3. Temperature-Induced Flowering

With global climate fluctuation, extreme temperature conditions become more intense and more frequent, and temperature becomes another abiotic stress factor that has an enormous impact on flowering time [145]. Temperature stress, including heat and cold stress, are a serious threat to the physiological and developmental processes of plants, particularly in terms of floral transition and crop productivity [20,93,146]. For many plants, the process of prolonged exposure to non-freezing temperatures to allow for a successful floral transition the following spring is called vernalization. Heat and cold acclimation have been considered as a possible strategy for plants to mitigate the damage caused by temperature stress [12]. Clearly, understanding the response of plants to diverse environmental temperatures is essential to discriminate between the effects of vernalization and temperature stress on plants' floral transition. Below, we discuss the vernalization pathway in model plants, further summarize the differential physiological phenotypes of various plant species under cold and heat stress conditions, and, finally, we focus on the potential molecular mechanisms underlying temperature-induced flowering.

### 3.1. Vernalization-Mediated Floral Transition

The successful transition of plants from vegetative growth to reproductive development relies on the completion of flowering at an appropriate time to cope with the hazards caused by adverse environmental conditions. Accordingly, several plant species have evolved mechanisms to integrate diverse environmental cues, thereby coordinating flowering time with favorable seasonal conditions, and the vernalization pathway constitutes a typical example of such a process [147–149]. Vernalization is an essential adaptation of plants to natural environmental temperatures, ensuring the acquisition of the competence to flower in spring by prolonged exposure to low winter temperatures so that reproductive developmental processes can be carried out under appropriate temperature conditions [150]. This programmed physiological response does not directly initiate flowering, but rather provides meristematic tissue with the ability to perceive environmental flowering signals. In temperate regions, many annual winter crops, such as wheat (*Triticum aestivum*) and barley (*Hordeum vulgare*), flower in the spring of the following year, which is conducive to successful reproduction, and adequate exposure to low temperatures for several months in winter (0–10 °C for about a month or more) is one of the determinants of their flowering [151–153]. Vernalization is also widely found in biennial and perennial plant species, such as vegetables and fruit trees, which require a prolonged period of cold to break dormancy and initiate flowering [154]. Plant perception of low temperature during the vernalization response consists of two distinct but continuous processes, namely, cold perception to induce tolerance and output of a vernalization dosage to accelerate the developmental transition [154,155].

The major molecular mechanisms and diverse genetic networks controlling vernalization have been intensively studied in several plant species, especially in *Arabidopsis*, wheat, and barley. In the model plant *Arabidopsis*, the vernalization requirement is, to a large extent, conferred by the interaction of two determinant proteins: the MADS-box protein FLC [156,157] and the scaffold protein FRI [158]. In the autumn prior to winter cold exposure, *FRI* transcriptionally activates *FLC* by forming complexes with *FRI ESSENTIAL 1 (FES1)*, *FRI-LIKE 1 (FRL1)*, *FLC EXPRESSOR (FLX)*, and *SUPPRESSOR OF FRI 4 (SUF4)* (25–28, Table 1) [98]. High levels of *FLC* expression subsequently act both in leaves and meristems to delay flowering through the transcriptional repression of genes encoding the flowering promoters, such as *FT*, *FD*, *SOC1*, and *LEAFY* [159,160] (Figure 2). Therefore, winter-annual *Arabidopsis* undergoes vegetative growth in the fall, and the vernalization response leads to the inhibition of the floral repressor *FLC* expression, which ultimately initiates flowering in the following spring [161]. In contrast, *Arabidopsis* summer annuals with lower *FLC* expression can flower rapidly within a single growing period, thus completing the entire life cycle [98].

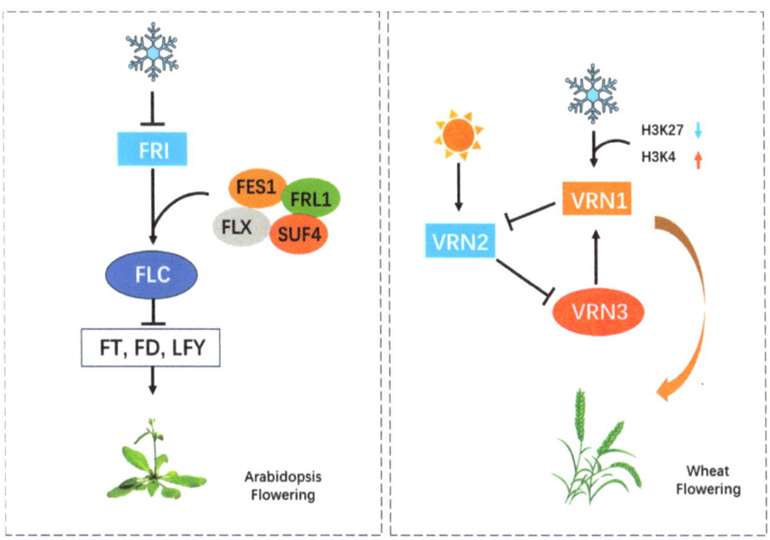


**Figure 2.** Vernalization pathways in *Arabidopsis* and wheat. Solid lines indicate identified connections, arrows indicate positive regulation, horizontal bars indicate negative regulation, red arrows indicate increases, and blue arrows indicate decreases. Left: *FRI* interacts with *FES1*, *FRL1*, *FLX*, and *SUF4* to activate the transcription of the downstream gene *FLC*, which regulates flowering in *Arabidopsis* by repressing the flowering integrative genes *FT*, *FD*, and *LFY*, and low temperature suppresses *FRI* expression [98,157,159,162–164]. Right: prior to vernalization, *VRN1* expression levels are low, and high levels of *VRN2* inhibit *VRN3* expression, thereby preventing flowering. During prolonged cold winter exposure, *VRN1* expression is activated by chromatin modification via decreased H3K27 methylation and increased H3K4 methylation, which downregulates *VRN2* expression and promotes the accumulation of *VRN3* in the leaf, and *VRN3* moves to the apical meristem to maintain high levels of *VRN1* and accelerate flowering [118,120,147,165–167].

Comparatively, genes associated with vernalization requirements vary from *Arabidopsis* to the winter cereals wheat and barley. Thus far, a model of vernalization has been established in wheat and barley in which *VERNALIZATION1 (VRN1)* (29, Table 1), *VRN2*, and *VRN3* are the core regulatory genes in the molecular framework of vernalization-responsive flowering time control (Figure 2). *VRN1* encodes a MADS-box transcription factor and acts as a plant flowering activator in grasses by vernalization-induced regulation of the plant transition from vegetative growth to reproductive development and is putatively presumed to be orthologous to the *Arabidopsis FRUITFULL (FUL), AP1,* and *CAULIFLOWER (VAL)* (30–31, Table 1) genes. *VRN2,* a floral repressor encoding a CCT domain and zinc-finger containing protein, delays flowering until the plant's vernalization. *VRN3* is the orthologous gene of *Arabidopsis FT* encoding a polyethanolamine-binding protein whose expression is induced by the photoperiod and vernalization, and acts as a mobile florigen to accelerate the flowering of plants [168–171]. Prior to vernalization, which is the stage of vegetative growth in winter cereals, *VRN1* expression levels are low, whereas high levels of *VRN2* repress *VRN3* expression, thereby preventing flowering [172]. During prolonged winter cold exposure, *VRN1* expression is activated via chromatin modifications of reduced H3K27 methylation and increased H3K4 methylation, thereby downregulating the expression of *VRN2* and ultimately promoting *VRN3* accumulation in the leaves after vernalization [119,120,165]. Subsequently, *VRN3* in the leaves then moves to the shoot apical meristem in the form of a mobile florigen to maintain high levels of *VRN1* and accelerates flowering [165]. In recent years, the molecular mechanisms of the vernalization response in *Brachypodium distachyon,* which belongs to the same subfamily as wheat and barley, have been gradually revealed. The regulatory mechanisms of *VRN1* and *VRN3* are similar to those of wheat and barley, but *VRN2* can be induced under long-term low-temperature exposure, indicating that the molecular mechanisms of the vernalization response are not conserved despite being in the same subfamily [148]. The vernalization requirement is a crucial trait of the transition from the vegetative to the reproductive stage in many crop species, and, therefore, the studies of the molecular mechanisms underlying vernalization-regulated flowering in plants are of great significance for the understanding of low-temperature-induced flowering.

### 3.2. Cold-Stress-Responsive Flowering

The vernalization response in plants can only be achieved by exposure to winter cold for a sufficiently long period of time, whereas cold stress acclimation involves a rapid response to low, non-freezing temperatures. The vernalization pathway accomplishes developmental transitions mainly by accumulating prolonged cold exposure, whereas cold stress induces low-temperature tolerance in plants mainly by sensing low temperatures for a short period of time. Thus, unlike the extensive studies of the vernalization pathways in annual plants, the regulatory pathway of cold stress acclimation may be distinct. Cold acclimation is the cold or freeze stress tolerance acquired by plants exposed to low positive temperatures of 0–5 °C for short periods of time [173,174]. The adaptation of diverse temperature ranges is specific in various plant species, with 0–15 °C during rice growth considered to be cold or chilling stress, while near or below 0 °C is considered to be freezing stress. Additionally, *Arabidopsis* is adapted to growth in a different temperature range with rice, and the temperatures for cold (also referred to as chilling) and freezing stresses are 4–10 °C and below 0 °C, respectively [175–177].

During cold and freezing stress, plant growth and developmental processes are inhibited, but plants have evolved the competence to resist, tolerate, escape, or adapt to the stresses through biochemical, morphological, and transcriptional alterations to protect themselves from stressful environmental conditions. The biological membrane is the main damaged part of plants under low-temperature stress. Low temperature will reduce the fluidity of the membrane, thus enhancing the permeability of the membrane to electrolytes and other small molecules, resulting in an imbalance in ion exchange [66,178]. Also, low-temperature stress triggers lipid peroxidation. Malondialdehyde (MDA) is the final decomposition product of membrane lipid peroxidation, which reflects the degree of damage to the plant and serves as one of the indicators of cold tolerance, and the accumulation of MDA will cause the disruption of the integrity of biological membranes, thereby altering the membrane permeability and affecting the normal physiological and biochemical reactions of plants [66,179,180]. Cold or chilling stress refers to the low, but non-freezing, temperature that occurs frequently in nature, which has a tremendous restrictive effect on the geographic location and crop yield of plants (especially tropical plants), leading to reduced membrane fluidity and cellular dysfunction, which can cause plant wilting, etiolation, and even necrosis, ultimately constraining growth and development [181–183]. Freezing stress refers to the subzero temperatures in nature that have a dramatic impact on plant photosynthesis, respiration, and metabolic processes, and mainly involves the formation of ice, which leads to cell dehydration and membrane damage, accompanied by alterations in calmodulin, intracellular $Ca^{2+}$ levels, ROS signaling, and phytohormones [173]. It should also be noted that the influence of low-temperature stress depends not only on the actual temperature, but also on whether the cold or freezing stress appears gradually or suddenly, and on the duration of the low-temperature stress. Additionally, the regulation of flowering time by the ambient temperature has been reported in multiple plants range from annuals to perennials, where short-term low-temperature stress can cause delayed flowering in a wide range of plants [23] (Table 2).

Plants regulate the activity of many classical floral pathway regulators, such as *AP1* [184], *FT* [46,164], *PHYTOCHROME-INTERACTING TRANSCRIPTION 4 (PIF4)* [123], *PIF5* [124], *MADSAFFECTING FLOWERING 2 (MAF2)* [109] (32–33, Table 1), and *SVP* [121], by sensing changes in the ambient temperature, thereby optimizing flowering time and improving cold acclimation or freezing tolerance (Figure 3). The exposure of vernalization-sensitive *Arabidopsis* to prolonged cold promotes floral transition via the vernalization pathway. By contrast, cold stress caused by short-term cooler temperatures delays flowering through activation of *FLC*. In *Arabidopsis*, *FLOWERING CONTROL LOCUS A (FCA)*, *FLOWERING LOCUS VE (FVE)*, *SVP*, and *FLOWERING LOCUS M (FLM)* (34–36, Table 1) are the cross-talk regulators between flowering time and the cold stress response. *SVP* prefers to bind to *FLM-β* to regulate low-temperature-responsive flowering [108]. *FCA* and *FVE* are floral-autonomous pathway genes and act to repress the expression of *FLC*, while *SVP* forms a flowering repressor complex with *FLC* to become a floral repressor, which ultimately co-regulates *FT* expression (Figure 3). The mutants *fca*, *fve*, and *svp* are insensitive to ambient low-temperature-induced flowering [23,185–187]. More recent studies have shown that the photoperiodic flowering pathway plays an important role in temperature perception, mediated by the circadian clock genes *GI* and *CO*, which, in turn, regulate the floral integrators *SOC1* and *FT*. The transcript levels of *GI* are elevated together with upregulation of cold-responsive genes under low-temperature stress (Figure 3). Compared with the control, *gi* mutants with a genetic background of the *Col-0* ecotype are more resistant to freezing stress [100], whereas *gi* mutants with a genetic background of the *Ler* ecotype have increased susceptibility to freezing stress and are defective in freezing tolerance [188]. Also, *GI* responds to temperature-regulated flowering time in *Medicago truncatula* [189]. Secondly, *CO* serves as a molecular link that combines cold signaling with the photoperiodic flowering pathway. The degradation of CO induced by low-temperature stress is mediated by the E3 ubiquitin ligase *HIGH EXPRESSION OF OSMOTICALLY RESPONSIVE GENE 1 (HOS1)* (37, Table 1), which, in turn, inhibits the *CO*-mediated activation of *FT* and ultimately regulates the delayed flowering of *Arabidopsis* [111] (Figure 3). Additionally, *HOS1* and *FVE* negatively regulate the cold stress response, with *fve* and *hos1* mutants having enhanced cold stress tolerance and *hos1* mutants exhibiting an early flowering phenotype. The chromatin modification factor *FVE* forms a histone repressor complex with *HDA6*, which inhibits *FLC* transcription by directly binding to *FLC* chromatin, ultimately leading to altered flowering time. Under short-term cold stress, *HOS1* induces *FLC* expression at the chromatin level by antagonizing the actions of *FVE* and *HDA6*, interfering with the association of *HDA6* with *FLC* chromatin in an *FVE*-dependent manner, which, in turn, leads to late flowering [112,113]. Moreover, other transcription factors are also important for plants to regulate flowering in response to cold stress. *HOS15* can interact with *GI* and mediate GI degradation to repress flowering in response to low ambient temperature in *Arabidopsis* [114] (Figure 3). The *atho9* mutant exhibits a delayed flowering phenotype and is extremely sensitive to cold stress.

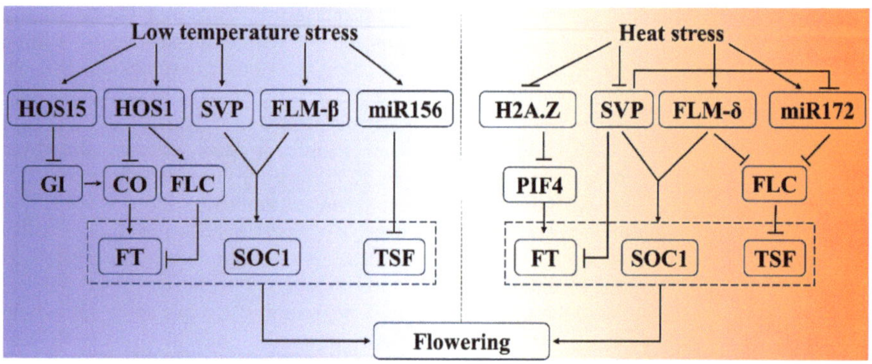


**Figure 3.** A simplified regulatory pathway for the link between ambient temperature and flowering in *Arabidopsis thaliana*. Low-temperature stress: *HOS15* and *HOS1* were able to degrade GI and CO proteins, respectively, indirectly inhibiting the expression of *FT* [111,114], in addition to *HOS1* positively regulating the expression of *FLC* [113], whereas *FLM-β* regulated the expression of *FT*, *SOC1*, and *TSF* by enhancing the stabilization of *SVP* [108,186]. miR156 directly targeted *TSF*, reducing its transcript abundance, and delayed flowering in *Arabidopsis* under low temperature [138]. Heat stress: under higher temperature conditions, miR172 negatively regulated the expression of *FLC* and deregulated the inhibitory effect of *FLC* on *TSF*. H2A.Z was repressed at high temperature, which attenuated its inhibitory effect on PIF4, resulting in an increase in the transcript level of *FT* [129]. *FLM-δ* accelerated the degradation of SVP at high temperatures, which deregulated the inhibitory effect on genes such as *FT*, *SOC1*, and *TSF* [108]. Solid lines indicate identified links, arrows indicate positive regulation, and horizontal bars indicate negative regulation.

In contrast, the short-day plant *Pharbitis nil*, when exposed to a temperature range of 16–18 °C shows restricted vegetative growth and is induced to flower; thus, the process is considered to be cold-stress-induced flowering [190]. *PnFT1* and *PnFT2*, two homologs of *Arabidopsis FT* in *Pharbitis* (Table 2), are major regulatory genes involved in low-temperature-induced flowering [30,31]. Further studies are required to elucidate the gene networks that regulate flowering under cold stress environmental conditions. The effect of cold stress on the floral transition has also been reported in perennials, such as chrysanthemum (*Chrysanthemum morifolium*) [110] and poplar (*Populus* spp.) trees [115] (Table 2). Similarly, among subtropical and tropical tree species, including macadamia, mango, avocado, lychee, lime, trifoliate orange, and satsuma mandarin, flowering is also induced by low-temperature stress [116,191–193]. The increase in flower density under cold stress is associated with upregulation of *CiFT* expression in citrus buds and leaves. The citrus transcription factor *CiNF-YA1* has been reported to control the response process to low-temperature stress by forming a complex with *CiNF-YB2* and *CiNF-YC2* as an upstream regulator of *CiFT* [92]. The CiFDα protein produced by *FD* alternative splicing is induced by low temperature, together with FT and the 14-3-3 proteins, forming a complex that regulates plant flowering [13]. Furthermore, two *FT* homologues in poplar trees have been shown to play important roles in coordinating the transition between vegetative growth and reproductive development, with *FT1* determining the initiation of reproductive development in response to low winter temperatures [115] (Table 2). However, the molecular mechanisms involved in the regulation of flowering by cold stress in perennials, especially in woody plants, remain rudimentary compared to those in annual plants. Further studies are necessary to elucidate the specific gene networks that regulate flowering in perennial plants under cold-stress environmental conditions by referring to the relevant mechanisms involved in annual plants.

### 3.3. Heat-Stress-Responsive Flowering

With predicted global warming and climate change approaching, crop growth and development can be limited by high ambient temperature [194–196]. Heat stresses are considered to be above the normal optimal temperature and exceed the critical threshold of plant tolerance, which is ultimately sufficient to cause irreversible damage to plant growth and development [197]. The temperature threshold is defined as the optimal temperature range for regulating growth throughout the plant life cycle, which varies according to the plant species, developmental stage, and organ type. Warm-season crops, such as maize, soybean, tomato, and rice, generally exhibit higher temperature thresholds than barley and wheat from temperate zones. Many crops display slightly lower temperature thresholds at the early seed development and fruit-setting stages compared to other developmental phases [198,199]. Reproductive organs are more sensitive to moderate or extreme temperatures during plant growth and development compared to vegetative organs [200].

A recent review also indicates that how plants respond to high-temperature stress has become a hot topic of research [51]. Plants are sensitive to temperature, and heat stress affects a range of physiological, biochemical, morphological, and developmental processes, which, in turn, mitigate excessive damage from high temperature through such alterations [122,201]. Heat stress interferes with cellular homeostasis, mainly manifested as increased fluidity of lipid bilayers at high temperatures, increased membrane permeability, protein denaturation, organelle function destruction, and even cell death [202,203]. It has been reported that the nucleus and cytoplasm oxidation of *Arabidopsis thaliana* is increased under heat stress, as evidenced by elevated levels of $H_2O_2$, which, in turn, induces various reactive oxygen scavenging enzymes at the transcriptional level [178,204,205]. Moreover, heat stress can also induce accelerated plant growth, which, in turn, generates plant thermal adaptation by moving susceptible parts away from the stress, a process known as thermomorphogenesis [202,203,206–208]. For example, *Arabidopsis* display a tendency for hypocotyl and petiole elongation under moderately high ambient temperature conditions [201], whereas the stomata of barley, tomato, and *Arabidopsis* are induced to remain open under extreme heat conditions [194,209]. Similarly, monocotyledonous crops, such as rice, wheat, barley, and maize, or dicotyledonous crops, such as soybean and tomato, showed elongation of leaves and hypocotyls at moderately high temperatures, with tomato also exhibiting hypoplastic leaves [194,210–212]. Importantly, the series of changes that occur when plants are gradually or briefly exposed to high temperatures are conducive to the enhancement of heat stress tolerance, whereas the sudden occurrence and prolonged duration of heat stress can be detrimental to plant growth and development.

The reproductive development stage, especially the floral transition process, is greatly affected by high-temperature fluctuations compared to the adaptive growth of vegetative development under heat stress conditions [213–215]. Plants avoid high-temperature exposure during fertilization by altering flowering time under heat stress [216–218] (Table 2). Wheat, sorghum, and rice, for example, flower in the morning or in the cooler evening to complete fertilization before extreme temperatures lead to sterility [76,219,220]. Plant species that cannot change their flowering time in response to high-temperature fluctuations gradually disappear from their former natural habitats and eventually migrate to higher altitudes and latitudes [201]. Moreover, heat stress also has a tremendous effect on floral organ development in addition to flowering time. The adaptability of plants to high-temperature stress in the natural environment is different, which can be reflected in the great differences in flowering responses to heat stress. Heat stress has been reported to accelerate flowering in *Arabidopsis* and to delay flowering in *Brassica rapa* [129], suggesting that the effect of high ambient temperatures on flowering time is not a universal outcome.

High temperatures induce early flowering in the monocotyledonous bulb plant *Narcissus tazetta* [221,222], in contrast to elevated summer temperatures that delay flowering in *Chrysanthemum morifolium* [128,223]. The accelerated flowering of crops in response to high ambient temperature fluctuations does not translate into higher yields, due to the fact that accelerated flowering in response to heat stress is also accompanied by faster senescence, which leads to stunted grain growth, ultimately causing lower crop yields [224]. Therefore, understanding the specific mechanisms of plant flowering in response to heat stress is of great significance for improving adaptive growth at high temperatures and for more effectively balancing the relationship between plant tolerance to heat stress and normal growth at critical developmental stages.

The plant heat stress response (HSR) acquires high-temperature tolerance through the induction of heat shock proteins (HSPs) and heat stress transcription factors (HSFs) [225] (38–40, Table 1). Transcriptional regulation is extremely important in the plant response to heat stress, especially HSFs, which are responsible for the rapid transcriptional activation of downstream genes. The HSF-encoding genes are defined by 21 members in *Arabidopsis* and 25 members in rice, which are differentiated by the structural features of the oligomerized domains into A, B, and C categories [226]. Based on an analysis of HSFs in *Arabidopsis* and tomato, it was shown that the thermal-response-related genes of the HSF family are also closely related to plant reproductive development [227]. The development of gametophytes under heat stress in *Arabidopsis* is regulated by the homeostasis between *AtHSFB2a* and its natural antisense RNA [228]. In tomato (*Solanum lycopersicum*), moderate heat treatment (37.5 °C) leads to the accumulation of *SlHSFA2* and enhances the tolerance of seedlings under extreme heat stress conditions (47.5 °C). Meanwhile, *SlHSFA2* expression was inhibited when pollen was subjected to heat stress in the early developmental stage of pollen formation, thus reducing pollen viability and the germination rate, suggesting that *SlHSFA2* is an important regulator of heat tolerance during pollen development [229]. HSPs were initially recognized as proteins strongly induced by heat stress and which protected plants from stress by re-establishing normal protein conformation and cellular homeostasis under the control of HSFs [230]. The *Arabidopsis* BOBBER1 (AtBOB1) (41, Table 1) protein belongs to a class of small HSPs (sHSPs), and *atbob1* mutants have been reported to exhibit developmental defects in flowers and the inflorescence meristem and lower tolerance to high temperature, indicating that HSPs play a central role in both heat tolerance and the floral development of plants [231].

The molecular regulatory mechanisms of floral transition in response to heat stress differ in plants. The accelerated flowering in *Arabidopsis* under moderately high temperatures is due to increased *AtFT* expression [122,232]. *ft* mutants did not exhibit accelerated flowering in response to heat stress [233]. Flowering response to heat stress was also achieved in soybean by promoting *GmFT2a* and *GmFT5a* expression and repressing the upstream negative regulators *E1* and *E2* [125]. Similarly, high temperature, which delays flowering in chrysanthemum, is also associated with downregulation of *FLOWERING LOCUS T-like 3* (*FTL3*, an *FT* homologue) (42, Table 1) expression [128]. However, there is no difference in the expression levels of *FT1* in barley and wheat subjected to high-temperature treatment, indicating that the flowering of cereals responds to heat stress in an *FT*-independent manner [214] (Table 2). Knowledge about the specific mechanism of floral transition in *Arabidopsis* in a fluctuating high-temperature environment provides a framework for further research on flowering response to heat stress in different plants. At the molecular level, the histone variant *H2A.Z* decreases the incorporation of nucleosomes during flowering at higher temperatures, which, in turn, permits the binding of the *bHLH* transcription factor *PIF4* to the *FT* promoter, and ultimately accelerates flowering in a temperature-dependent manner. Neither *pif4* nor *pif5* mutants can accelerate flowering under heat stress, and *pif4 pif5* double mutants flower later compared with the single mutation, suggesting that *PIF4* and *PIF5* are related to high-temperature-responsive flowering [123,234]. Additionally, *PIF4* has been reported to be involved in controlling temperature-sensing memory and reproductive transition in *Arabidopsis* [123], and the *PIF4* homologue in soybean exhibits unique high-temperature adaptability [235]. *SVP* tends to bind *FLM-δ* to regulate the expression of *FT*, *SOC1*, and *TSF*, ultimately mediating flowering in response to high temperature [108] (Figure 3).

Due to the influence of extreme high temperatures, the alteration of flowering time can have a huge impact on crop productivity. In barley, high-temperature-dependent *ELF3* increased transcription levels of the core circadian clock genes *GI*, *PSEUDO RESPONSE REGULATOR (PRR)*, and *LUX ARRHYTHMO (LUX)* [236,237] (43–44, Table 1). A late-flowering mutant allele, *Ppd-1*, was reported to retard flower development and reduce the flower and seed number in spring barley at high temperatures, in contrast to the *Ppd-1* and *elf3* mutants, both of which showed earlier flower development and unchanged seed number [237]. Moreover, an allelic variation of *ELF3* in wheat may underlie an Earliness *per se* locus (Eps-D1) (45, Table 1), which regulates flowering time depending on temperature [238,239]. High-temperature-delayed short-day flowering was also accompanied by increased expression of *HvODDSOC2*, the MADS box flowering suppressor associated with the *FLC* gene family [126]. Heat stress during flowering of rice, a short-day plant, can lead to sterility. *EXTRA GLUME 1 (EG1)* (46, Table 1), a gene encoding a lipase primarily located in mitochondria, acts upstream of flower identity genes (*OsG1*, *OsMADS1*, and *OsMADS6*) in a high-temperature-dependent manner, thereby promoting robust flower development [127]. In contrast, the *eg1* mutants exhibit high plasticity in spikelet development at high temperature and have a detrimental effect on maintaining flower development [127]. In maize, *ZmNF-YA3* binds to the *FT-like12* promoter to accelerate maize flowering, and the *Zmnf-ya3* mutant exhibits different high-temperature tolerance compared with the control, indicating that NF-Y transcription factors play an important role in maize flowering response to heat stress [106] (Table 2). In summary, studies on the mechanisms of flowering in response to heat stress vary widely between model plants and crops, and the regulatory networks that have been studied are still relatively basic. Therefore, future research on the floral transition caused by heat stress can enrich the overall regulatory network by cross-referencing the components of molecular mechanism studies in different plant species, laying an extraordinarily important foundation for research on perennial plants.

The effects of heat stress on the reproductive development of perennial plants, especially woody plants, are poorly understood. Woody plants growing in the temperate zone follow an annual growth cycle of sprouting in spring, flowering in summer, followed by the emergence of new buds in the fall [240]. However, the current global warming has a great impact on the timing of spring bud germination, so the control of temperature on the timing of bud burst in woody plants has become a new trend in research. In sexually mature poplar, *FT1* has been reported to initiate reproductive development in response to low temperatures, whereas *FT2* promotes vegetative growth as well as bud set inhibition under high-temperature conditions [115]. Additionally, *GtFT2* was found to be involved in dormancy regulation induced by temperatures in the perennial herb gentian, peaking at the time of dormancy release [241]. Due to the lack of available model plant species, the current understanding underlying the molecular regulation mechanisms of the bud dormancy cycle in woody plants is still incomplete. Similarly, due to the long study period and the fact that the effects of heat stress on reproductive growth in woody plants are not immediately apparent, the analysis of the mechanisms of this aspect in woody plants is limited. Taken together, future studies on the molecular mechanisms of flowering response to heat stress in woody plants can be further expanded by utilizing *Arabidopsis* as the framework. A detailed knowledge of the molecular mechanisms by which heat stress affects flowering is critical for breeding plant species with higher yields and greater tolerance to heat and provides crucial information for mitigating the effects of climate alteration.

## 4. Conclusions and Future Perspectives

The optimal flowering time is crucial for maximizing reproductive success, a process that is precisely regulated by a wide variety of environmental and internal factors. Accordingly, this review outlines the current knowledge showing that diverse abiotic stresses, such as drought, low-temperature and high-temperature stress, alter the flowering time of plants. Meanwhile, the roles of the critical flowering regulation genes and stress-response-related genes are identified with consideration of the cross-talk molecular mechanisms underlying flowering time regulation and the stress response. Compared with annual plants, especially the model plant *Arabidopsis*, perennials, especially woody plants, involve a longer period of time, and a heavier workload, to study the molecular mechanisms related to reproductive development, resulting in the fact that little is known about abiotic stresses controlling flowering time in perennial plants. Hence, it is important to understand the molecular mechanisms of flowering time regulation in perennials under environmental stress. Although our understanding underlying the effects of abiotic stress on flowering timing has increased considerably in perennials, most of the regulatory mechanisms associated with abiotic stress have been identified in *Arabidopsis* and currently remain elusive in perennials. More importantly, the molecular pathways by which abiotic stress affects flowering time in different plants are similar, but also exhibit many unique aspects. In future studies, it will be both challenging and necessary to utilize the molecular network obtained in *Arabidopsis* to advance the understanding of the molecular pathways involved in perennial plants flowering time in response to abiotic stresses, as well as to conduct comparative studies on different plant species.

Flowering time is affected by abiotic stresses through a complex network, but most of the existing studies focus on the role of individual genes in the stress-regulated flowering process, which makes it difficult to localize the key factors and is insufficient to show that the whole process is controlled by one pathway only. Currently, multi-omics combined analysis can provide in-depth analysis of complex regulatory networks, including gene expression, protein function, and metabolites [242–244]. Comparative transcriptomics has revealed differentially expressed genes for drought-stress-induced flower formation in *Curcuma kwangsiensis* [74], as well as a regulatory cascade for flower development under drought stress in *Arabidopsis thaliana* [40]. Meanwhile, the detection of protein changes under abiotic stresses by comparative proteomics analysis is essential for revealing stress-induced signal perception as well as response, and this information cannot be revealed by transcriptomic analysis [245]. Certainly, metabolomics is also an important molecular tool for identifying plant responses to different abiotic stresses [246,247]. However, the application of omics approaches to abiotic-stress-regulated flowering time in plants is relatively rare, so the combination of different omics tools is a new trend for future research to identify complex regulatory networks involved in the abiotic stress regulation of plant flowering. Since environmental alteration generally entails the simultaneous existence of multiple abiotic stress factors, it is also necessary to focus on the effects of the combination of various abiotic stresses on plant flowering in the future, so as to provide better tools for breeders to cultivate reproductive traits in response to environmental alterations.

**Author Contributions:** M.C. and T.-L.Z. performed different aspects of the research; J.-Z.Z. and C.-G.H. designed the research; M.C. wrote the manuscript. All authors have read and agreed to the published version of the manuscript.

**Funding:** This research was supported financially by the National Natural Science Foundation of China (grant nos. 319702356, 32072521, and 31872045).

**Conflicts of Interest:** The authors have no conflict of interest to declare.

## References

1. Zhu, J.-K. Abiotic stress signaling and Responses in plants. *Cell* **2016**, *167*, 313–324. [CrossRef] [PubMed]
2. Hu, H.H.; Xiong, L.Z. Genetic engineering and breeding of drought-resistant crops. *Annu. Rev. Plant Biol.* **2014**, *65*, 715–741. [CrossRef] [PubMed]
3. Webber, H.; Ewert, F.; Olesen, J.E.; Müller, C.; Fronzek, S.; Ruane, A.C.; Bourgault, M.; Martre, P.; Ababaei, B.; Bindi, M.; et al. Diverging importance of drought stress for maize and winter wheat in Europe. *Nat. Commun.* **2018**, *9*, 4249. [CrossRef] [PubMed]
4. Kopecká, R.; Kameniarová, M.; Cerny, M.; Brzobohaty, B.; Novák, J. Abiotic stress in crop production. *Int. J. Mol. Sci.* **2023**, *24*, 6603. [CrossRef]

5. Nadarajah, K.K. ROS homeostasis in abiotic stress tolerance in plants. *Int. J. Mol. Sci.* **2020**, *21*, 5208. [CrossRef]
6. Bechtold, U.; Field, B. Molecular mechanisms controlling plant growth during abiotic stress. *J. Exp. Bot.* **2018**, *69*, 2753–2758. [CrossRef]
7. Kollist, H.; Zandalinas, S.I.; Sengupta, S.; Nuhkat, M.; Kangasjärvi, J.; Mittler, R. Rapid responses to abiotic stress: Priming the landscape for the signal transduction network. *Trends Plant Sci.* **2019**, *24*, 25–37. [CrossRef]
8. De Storme, N.; Geelen, D. The impact of environmental stress on male reproductive development in plants: Biological processes and molecular mechanisms. *Plant Cell Environ.* **2013**, *37*, 1–18. [CrossRef]
9. Hasanuzzaman, M.; Nahar, K.; Alam, M.; Roychowdhury, R.; Fujita, M. Physiological, biochemical, and molecular mechanisms of heat stress tolerance in plants. *Int. J. Mol. Sci.* **2013**, *14*, 9643–9684. [CrossRef] [PubMed]
10. Chirivì, D.; Betti, C. Molecular links between flowering and abiotic stress response: A focus on *Poaceae*. *Plants* **2023**, *12*, 331. [CrossRef] [PubMed]
11. Riboni, M.; Galbiati, M.; Tonelli, C.; Conti, L. *GIGANTEA* enables drought escape response via abscisic acid-dependent activation of the florigens and *SUPPRESSOR OF OVEREXPRESSION OF CONSTANS1*. *Plant Physiol.* **2013**, *162*, 1706–1719. [CrossRef]
12. Kazan, K.; Lyons, R. The link between flowering time and stress tolerance. *J. Exp. Bot.* **2016**, *67*, 47–60. [CrossRef]
13. Ye, L.X.; Wu, Y.M.; Zhang, J.X.; Zhang, J.X.; Zhou, H.; Zeng, R.F.; Zheng, W.X.; Qiu, M.Q.; Zhou, J.J.; Xie, Z.Z.; et al. A bZIP transcription factor (*CiFD*) regulates drought- and low-temperature-induced flowering by alternative splicing in citrus. *J. Integr. Plant Biol.* **2022**, *65*, 674–691. [CrossRef]
14. Xu, Y.Y.; Zeng, R.F.; Zhou, H.; Qiu, M.Q.; Gan, Z.M.; Yang, Y.L.; Hu, S.F.; Zhou, J.J.; Hu, C.G.; Zhang, J.Z. Citrus FRIGIDA cooperates with its interaction partner dehydrin to regulate drought tolerance. *Plant J.* **2022**, *111*, 164–182. [CrossRef]
15. Osnato, M. The floral transition and adaptation to a changing environment: From model species to cereal crops. *Plant Cell* **2022**, *34*, E2. [CrossRef]
16. Forestan, C.; Farinati, S.; Zambelli, F.; Pavesi, G.; Rossi, V.; Varotto, S. Epigenetic signatures of stress adaptation and flowering regulation in response to extended drought and recovery in *Zea mays*. *Plant Cell Environ.* **2020**, *43*, 55–75. [CrossRef] [PubMed]
17. Ying, S.; Scheible, W.R.; Lundquist, P.K. A stress-inducible protein regulates drought tolerance and flowering time in *Brachypodium* and *Arabidopsis*. *Plant Physiol.* **2023**, *191*, 643–659. [CrossRef] [PubMed]
18. Chong, L.; Xu, R.; Huang, P.C.; Guo, P.C.; Zhu, M.K.; Du, H.; Sun, X.L.; Ku, L.X.; Zhu, J.K.; Zhu, Y.F. The tomato OST1-VOZ1 module regulates drought-mediated flowering. *Plant Cell* **2022**, *34*, 2001–2018. [CrossRef] [PubMed]
19. Fletcher, R.S.; Mullen, J.L.; Heiliger, A.; McKay, J.K. QTL analysis of root morphology, flowering time, and yield reveals trade-offs in response to drought in *Brassica napus*. *J. Exp. Bot.* **2015**, *66*, 245–256. [CrossRef] [PubMed]
20. Takeno, K. Stress-induced flowering: The third category of flowering response. *J. Exp. Bot.* **2016**, *67*, 4925–4934. [CrossRef] [PubMed]
21. Bastow, R.; Mylne, J.S.; Lister, C.; Lippman, Z.; Martienssen, R.A.; Dean, C. Vernalization requires epigenetic silencing of *FLC* by histone methylation. *Nature* **2004**, *427*, 164–167. [CrossRef]
22. Wu, Z.; Fang, X.; Zhu, D.; Dean, C. Autonomous pathway: *FLOWERING LOCUS C* repression through an antisense-mediated chromatin-silencing mechanism. *Plant Physiol.* **2020**, *182*, 27–37. [CrossRef]
23. Blázquez, M.A.; Ahn, J.H.; Weigel, D. A thermosensory pathway controlling flowering time in *Arabidopsis thaliana*. *Nat. Genet.* **2003**, *33*, 168–171. [CrossRef] [PubMed]
24. Osnato, M.; Castillejo, C.; Matías-Hernández, L.; Pelaz, S. *TEMPRANILLO* genes link photoperiod and gibberellin pathways to control flowering in *Arabidopsis*. *Nat. Commun.* **2012**, *3*, 808. [CrossRef]
25. Wang, J.-W. Regulation of flowering time by the miR156-mediated age pathway. *J. Exp. Bot.* **2014**, *65*, 4723–4730. [CrossRef] [PubMed]
26. Khan, M.R.G.; Ai, X.Y.; Zhang, J.Z. Genetic regulation of flowering time in annual and perennial plants. *Wiley Interdiscip. Rev. RNA* **2014**, *5*, 347–359. [CrossRef]
27. Corbesier, L.; Vincent, C.; Jang, S.H.; Fornara, F.; Fan, Q.Z.; Searle, I.; Giakountis, A.; Farrona, S.; Gissot, L.; Turnbull, C.; et al. FT protein movement contributes to long-distance signaling in floral induction of *Arabidopsis*. *Science* **2007**, *316*, 1030–1033. [CrossRef] [PubMed]
28. Lin, M.-K.; Belanger, H.; Lee, Y.-J.; Varkonyi-Gasic, E.; Taoka, K.-I.; Miura, E.; Xoconostle-Cázares, B.; Gendler, K.; Jorgensen, R.A.; Phinney, B.; et al. FLOWERING LOCUS T protein may act as the long-distance florigenic signal in the Cucurbits. *Plant Cell* **2007**, *19*, 1488–1506. [CrossRef] [PubMed]
29. Martínez, C.; Pons, E.; Prats, G.; León, J. Salicylic acid regulates flowering time and links defence responses and reproductive development. *Plant J.* **2004**, *37*, 209–217. [CrossRef]
30. Yamada, M.; Takeno, K. Stress and salicylic acid induce the expression of *PnFT2* in the regulation of the stress-induced flowering of *Pharbitis nil*. *J. Plant Physiol.* **2014**, *171*, 205–212. [CrossRef] [PubMed]
31. Wada, K.C.; Yamada, M.; Shiraya, T.; Takeno, K. Salicylic acid and the flowering gene *FLOWERING LOCUS T* homolog are involved in poor-nutrition stress-induced flowering of *Pharbitis nil*. *J. Plant Physiol.* **2010**, *167*, 447–452. [CrossRef] [PubMed]
32. Kim, W.-Y.; Ali, Z.; Park, H.J.; Park, S.J.; Cha, J.-Y.; Perez-Hormaeche, J.; Quintero, F.J.; Shin, G.; Kim, M.R.; Qiang, Z.; et al. Release of SOS2 kinase from sequestration with GIGANTEA determines salt tolerance in *Arabidopsis*. *Nat. Commun.* **2013**, *4*, 1352. [CrossRef] [PubMed]

33. Kurokura, T.; Mimida, N.; Battey, N.H.; Hytönen, T. The regulation of seasonal flowering in the *Rosaceae*. *J. Exp. Bot.* **2013**, *64*, 4131–4141. [CrossRef]
34. Mesejo, C.; Marzal, A.; Martinez-Fuentes, A.; Reig, C.; de Lucas, M.; Iglesias, D.J.; Primo-Millo, E.; Blazquez, M.A.; Agusti, M. Reversion of fruit-dependent inhibition of flowering in Citrus requires sprouting of buds with epigenetically silenced *CcMADS19*. *N. Phytol.* **2022**, *233*, 526–533. [CrossRef]
35. Agusti, M.; Mesejo, C.; Munoz-Fambuena, N.; Vera-Sirera, F.; de Lucas, M.; Martinez-Fuentes, A.; Reig, C.; Iglesias, D.J.; Primo-Millo, E.; Blazquez, M.A. Fruit-dependent epigenetic regulation of flowering in Citrus. *N. Phytol.* **2020**, *225*, 376–384. [CrossRef] [PubMed]
36. Yang, M.; Wu, Y.; Jin, S.; Hou, J.; Mao, Y.; Liu, W.; Shen, Y.; Wu, L. Flower bud transcriptome analysis of *Sapium sebiferum* (Linn.) Roxb. and primary investigation of drought induced flowering: Pathway construction and G-Quadruplex prediction based on transcriptome. *PLoS ONE* **2015**, *10*, e0118479. [CrossRef]
37. Agustí, M.; Reig, C.; Martínez-Fuentes, A.; Mesejo, C. Advances in citrus flowering: A review. *Front. Plant Sci.* **2022**, *13*, 868831. [CrossRef]
38. Abdelrahman, M.; Jogaiah, S.; Burritt, D.J.; Tran, L.-S.P. Legume genetic resources and transcriptome dynamics under abiotic stress conditions. *Plant Cell Environ.* **2018**, *41*, 1972–1983. [CrossRef]
39. Nishikawa, F. Regulation of floral induction in citrus. *J. Jpn. Soc. Hortic. Sci.* **2013**, *82*, 283–292. [CrossRef]
40. Su, Z.; Ma, X.; Guo, H.H.; Sukiran, N.L.; Guo, B.; Assmann, S.M.; Ma, H. Flower development under drought stress: Morphological and transcriptomic analyses reveal acute responses and long-term acclimation in *Arabidopsis*. *Plant Cell* **2013**, *25*, 3785–3807. [CrossRef]
41. Qi, J.; Song, C.-P.; Wang, B.; Zhou, J.; Kangasjärvi, J.; Zhu, J.-K.; Gong, Z. Reactive oxygen species signaling and stomatal movement in plant responses to drought stress and pathogen attack. *J. Integr. Plant Biol.* **2018**, *60*, 805–826. [CrossRef] [PubMed]
42. Ogura, T.; Goeschl, C.; Filiault, D.; Mirea, M.; Slovak, R.; Wolhrab, B.; Satbhai, S.B.; Busch, W. Root system depth in *Arabidopsis* is shaped by *EXOCYST70A3* via the dynamic modulation of auxin transport. *Cell* **2019**, *178*, 400–412. [CrossRef] [PubMed]
43. Pennisi, E. Plant genetics: Getting to the root of drought responses. *Science* **2008**, *320*, 173. [CrossRef] [PubMed]
44. Duan, H.L.; de Dios, V.R.; Wang, D.F.; Zhao, N.; Huang, G.M.; Liu, W.F.; Wu, J.P.; Zhou, S.X.; Choat, B.; Tissue, D.T. Testing the limits of plant drought stress and subsequent recovery in four provenances of a widely distributed subtropical tree species. *Plant Cell Environ.* **2022**, *45*, 1187–1203. [CrossRef] [PubMed]
45. Basu, S.; Ramegowda, V.; Kumar, A.; Pereira, A. Plant adaptation to drought stress. *F1000Research* **2016**, *5*, 1554. [CrossRef]
46. Riboni, M.; Test, A.R.; Galbiati, M.; Tonelli, C.; Conti, L. ABA-dependent control of *GIGANTEA* signalling enables drought escape via up-regulation of *FLOWERING LOCUS T* in *Arabidopsis thaliana*. *J. Exp. Bot.* **2016**, *67*, 6309–6322. [CrossRef] [PubMed]
47. Verslues, P.E.; Bailey-Serres, J.; Brodersen, C.; Buckley, T.N.; Conti, L.; Christmann, A.; Dinneny, J.R.; Grill, E.; Hayes, S.; Heckman, R.W.; et al. Burning questions for a warming and changing world: 15 unknowns in plant abiotic stress. *Plant Cell* **2023**, *35*, 67–108. [CrossRef]
48. Kooyers, N.J. The evolution of drought escape and avoidance in natural herbaceous populations. *Plant Sci.* **2015**, *234*, 155–162. [CrossRef]
49. Liang, M.T.; Xiao, S.M.; Cai, J.J.; Ow, D.W. OXIDATIVE STRESS 3 regulates drought-induced flowering through *APETALA 1*. *Biochem. Biophys. Res. Commun.* **2019**, *519*, 585–590. [CrossRef]
50. Pajoro, A.; Biewers, S.; Dougali, E.; Leal Valentim, F.; Mendes, M.A.; Porri, A.; Coupland, G.; Van de Peer, Y.; van Dijk, A.D.J.; Colombo, L.; et al. The evolution of gene regulatory networks controlling *Arabidopsis* plant reproduction: A two-decade history. *J. Exp. Bot.* **2014**, *65*, 4731–4745. [CrossRef]
51. Cui, Y.; Ouyang, S.; Zhao, Y.; Tie, L.; Shao, C.; Duan, H. Plant responses to high temperature and drought: A bibliometrics analysis. *Front. Plant Sci.* **2022**, *13*. [CrossRef]
52. Yang, Y.J.; Bi, M.H.; Nie, Z.F.; Jiang, H.; Liu, X.D.; Fang, X.W.; Brodribb, T.J. Evolution of stomatal closure to optimize water-use efficiency in response to dehydration in ferns and seed plants. *N. Phytol.* **2021**, *230*, 2001–2010. [CrossRef] [PubMed]
53. Marchin, R.M.; Backes, D.; Ossola, A.; Leishman, M.R.; Tjoelker, M.G.; Ellsworth, D.S. Extreme heat increases stomatal conductance and drought-induced mortality risk in vulnerable plant species. *Glob. Chang. Biol.* **2022**, *28*, 1133–1146. [CrossRef]
54. Groen, S.C.; Joly-Lopez, Z.; Platts, A.E.; Natividad, M.; Fresquez, Z.; Mauck, W.M.; Quintana, M.R.; Cabral, C.L.U.; Torres, R.O.; Satija, R.; et al. Evolutionary systems biology reveals patterns of rice adaptation to drought-prone agro-ecosystems. *Plant Cell* **2022**, *34*, 759–783. [CrossRef] [PubMed]
55. Juenger, T.E.; Verslues, P.E. Time for a drought experiment: Do you know your plants' water status? *Plant Cell* **2023**, *35*, 10–23. [CrossRef]
56. Ahanger, M.A.; Bhat, J.A.; Siddiqui, M.H.; Rinklebe, J.; Ahmad, P. Integration of silicon and secondary metabolites in plants: A significant association in stress tolerance. *J. Exp. Bot.* **2020**, *71*, 6758–6774. [CrossRef] [PubMed]
57. Shao, C.C.; Duan, H.L.; Ding, G.J.; Luo, X.Y.; Fu, Y.H.; Lou, Q. Physiological and biochemical dynamics of *Pinus massoniana* Lamb. seedlings under extreme drought stress and during recovery. *Forests* **2022**, *13*, 65. [CrossRef]
58. Li, Y.; Xu, Y.; Chen, Y.; Ling, L.; Jiang, Y.; Duan, H.; Liu, J. Effects of drought regimes on growth and physiological traits of a typical shrub species in subtropical China. *Glob. Ecol. Conserv.* **2020**, *24*, e01269. [CrossRef]
59. Hu, Y.; Yang, L.; Gao, C.; Liao, D.; Long, L.; Qiu, J.; Wei, H.; Deng, Q.; Zhou, Y. A comparative study on the leaf anatomical structure of *Camellia oleifera* in a low-hot valley area in Guizhou Province, China. *PLoS ONE* **2022**, *17*, e0262509. [CrossRef]

60. Gao, J.; Zhao, Y.; Zhao, Z.K.; Liu, W.; Jiang, C.H.; Li, J.J.; Zhang, Z.Y.; Zhang, H.L.; Zhang, Y.G.; Wang, X.N.; et al. *RRS1* shapes robust root system to enhance drought resistance in rice. *N. Phytol.* **2023**, *238*, 1146–1162. [CrossRef]
61. Scharwies, J.D.; Dinneny, J.R. Water transport, perception, and response in plants. *J. Plant Res.* **2019**, *132*, 311–324. [CrossRef] [PubMed]
62. Andrés, F.; Coupland, G. The genetic basis of flowering responses to seasonal cues. *Nat. Rev. Genet.* **2012**, *13*, 627–639. [CrossRef] [PubMed]
63. Su, P.; Sui, C.; Niu, Y.; Li, J.; Wang, S.; Sun, F.; Yan, J.; Guo, S. Comparative transcriptomic analysis and functional characterization reveals that the class III peroxidase gene *TaPRX-2A* regulates drought stress tolerance in transgenic wheat. *Front. Plant Sci.* **2023**, *14*, 1119162. [CrossRef] [PubMed]
64. Castro, B.; Citterico, M.; Kimura, S.; Stevens, D.M.; Wrzaczek, M.; Coaker, G. Stress-induced reactive oxygen species compartmentalization, perception and signalling. *Nat. Plants* **2021**, *7*, 403–412. [CrossRef] [PubMed]
65. Phua, S.Y.; De Smet, B.; Remacle, C.; Chan, K.X.; Van Breusegem, F. Reactive oxygen species and organellar signaling. *J. Exp. Bot.* **2021**, *72*, 5807–5824. [CrossRef] [PubMed]
66. Duan, Y.; Han, J.; Guo, B.; Zhao, W.; Zhou, S.; Zhou, C.; Zhang, L.; Li, X.; Han, D. *MbICE1* confers drought and cold tolerance through up-regulating antioxidant capacity and stress-resistant genes in *Arabidopsis thaliana*. *Int. J. Mol. Sci.* **2022**, *23*, 16072. [CrossRef] [PubMed]
67. Yang, Y.Q.; Guo, Y. Unraveling salt stress signaling in plants. *J. Integr. Plant Biol.* **2018**, *60*, 796–804. [CrossRef]
68. Jia, X.; Jia, X.; Li, T.; Wang, Y.; Sun, X.; Huo, L.; Wang, P.; Che, R.; Gong, X.; Ma, F. MdATG5a induces drought tolerance by improving the antioxidant defenses and promoting starch degradation in apple. *Plant Sci.* **2021**, *312*, 111052. [CrossRef]
69. Wei, T.L.; Wang, Y.; Xie, Z.Z.; Guo, D.Y.; Chen, C.W.; Fan, Q.J.; Deng, X.D.; Liu, J.H. Enhanced ROS scavenging and sugar accumulation contribute to drought tolerance of naturally occurring autotetraploids in *Poncirus trifoliata*. *Plant Biotechnol. J.* **2019**, *17*, 1394–1407. [CrossRef]
70. Zhao, S.; Gao, H.B.; Jia, X.M.; Wang, H.B.; Ke, M.; Ma, F.W. The HD-Zip I transcription factor *MdHB-7* regulates drought tolerance in transgenic apple (*Malus domestica*). *Environ. Exp. Bot.* **2020**, *180*, 104246. [CrossRef]
71. Srivastava, R.; Kobayashi, Y.; Koyama, H.; Sahoo, L. Cowpea NAC1/NAC2 transcription factors improve growth and tolerance to drought and heat in transgenic cowpea through combined activation of photosynthetic and antioxidant mechanisms. *J. Integr. Plant Biol.* **2022**, *65*, 25–44. [CrossRef] [PubMed]
72. Barth, C. The role of ascorbic acid in the control of flowering time and the onset of senescence. *J. Exp. Bot.* **2006**, *57*, 1657–1665. [CrossRef] [PubMed]
73. Shavrukov, Y.; Kurishbayev, A.; Jatayev, S.; Shvidchenko, V.; Zotova, L.; Koekemoer, F.; de Groot, S.; Soole, K.; Langridge, P. Early flowering as a drought escape mechanism in plants: How can it aid wheat production? *Front. Plant Sci.* **2017**, *8*, 1950. [CrossRef]
74. Feng, X.; Zhou, L.; Sheng, A.; Lin, L.; Liu, H. Comparative transcriptome analysis on drought stress-induced floral formation of *Curcuma kwangsiensis*. *Plant Signal. Behav.* **2022**, *17*, 2114642. [CrossRef]
75. Verslues, P.E.; Juenger, T.E. Drought, metabolites, and *Arabidopsis* natural variation: A promising combination for understanding adaptation to water-limited environments. *Curr. Opin. Plant Biol.* **2011**, *14*, 240–245. [CrossRef] [PubMed]
76. Bheemanahalli, R.; Sathishraj, R.; Manoharan, M.; Sumanth, H.N.; Muthurajan, R.; Ishimaru, T.; Krishna, J.S.V. Is early morning flowering an effective trait to minimize heat stress damage during flowering in rice? *Field Crops Res.* **2017**, *203*, 238–242. [CrossRef] [PubMed]
77. Xin, X.Y.; Su, T.B.; Li, P.R.; Wang, W.H.; Zhao, X.Y.; Yu, Y.J.; Zhang, D.S.; Yu, S.C.; Zhang, F.L. A histone H4 gene prevents drought-induced bolting in Chinese cabbage by attenuating the expression of flowering genes. *J. Exp. Bot.* **2021**, *72*, 623–635. [CrossRef]
78. Franks, S.J. Plasticity and evolution in drought avoidance and escape in the annual plant *Brassica rapa*. *N. Phytol.* **2011**, *190*, 249–257. [CrossRef]
79. Anbazhagan, K.; Bhatnagar-Mathur, P.; Sharma, K.K.; Baddam, R.; Kishor, P.B.K.; Vadez, V. Changes in timing of water uptake and phenology favours yield gain in terminal water stressed chickpea *At*DREB1A transgenics. *Funct. Plant Biol.* **2015**, *42*, 84–94. [CrossRef]
80. Chauhan, Y.; Allard, S.; Williams, R.; Williams, B.; Mundree, S.; Chenu, K.; Rachaputi, N.C. Characterisation of chickpea cropping systems in Australia for major abiotic production constraints. *Field Crops Res.* **2017**, *204*, 120–134. [CrossRef]
81. Yaish, M.W.; Colasanti, J.; Rothstein, S.J. The role of epigenetic processes in controlling flowering time in plants exposed to stress. *J. Exp. Bot.* **2011**, *62*, 3727–3735. [CrossRef]
82. Blanvillain, R.; Wei, S.; Wei, P.; Kim, J.H.; Ow, D.W. Stress tolerance to stress escape in plants: Role of the OXS2 zinc-finger transcription factor family. *EMBO J.* **2011**, *30*, 3812–3822. [CrossRef]
83. Riboni, M.; Robustelli Test, A.; Galbiati, M.; Tonelli, C.; Conti, L. Environmental stress and flowering time: The photoperiodic connection. *Plant Signal. Behav.* **2014**, *9*, e29036. [CrossRef]
84. Wang, Y.; Lu, Y.Y.; Guo, Z.Y.; Ding, Y.F.; Ding, C.Q. *RICE CENTRORADIALIS 1*, a *TFL1-like* gene, responses to drought stress and regulates rice flowering transition. *Rice* **2020**, *13*, 70. [CrossRef]
85. Galbiati, F.; Chiozzotto, R.; Locatelli, F.; Spada, A.; Genga, A.; Fornara, F. *Hd3a*, *RFT1* and *Ehd1* integrate photoperiodic and drought stress signals to delay the floral transition in rice. *Plant Cell Environ.* **2016**, *39*, 1982–1993. [CrossRef] [PubMed]

86. Bocco, R.; Lorieux, M.; Seck, P.A.; Futakuchi, K.; Manneh, B.; Baimey, H.; Ndjiondjop, M.N. Agro-morphological characterization of a population of introgression lines derived from crosses between IR 64 (*Oryza sativa* indica) and TOG 5681 (*Oryza glaberrima*) for drought tolerance. *Plant Sci.* **2012**, *183*, 65–76. [CrossRef] [PubMed]
87. Du, H.; Huang, F.; Wu, N.; Li, X.H.; Hu, H.H.; Xiong, L.H. Integrative regulation of drought escape through ABA-dependent and -independent pathways in rice. *Mol. Plant* **2018**, *11*, 584–597. [CrossRef] [PubMed]
88. Ghneim-Herrera, T.; Selvaraj, M.G.; Meynard, D.; Fabre, D.; Peña, A.; Ben Romdhane, W.; Ben Saad, R.; Ogawa, S.; Rebolledo, M.C.; Ishitani, M.; et al. Expression of the *Aeluropus littoralis AlSAP* gene enhances rice yield under field drought at the reproductive stage. *Front. Plant Sci.* **2017**, *8*, 994. [CrossRef] [PubMed]
89. Schmidt, J.; Tricker, P.J.; Eckermann, P.; Kalambettu, P.; Garcia, M.; Fleury, D. Novel alleles for combined drought and heat stress tolerance in wheat. *Front. Plant Sci.* **2020**, *10*, 1800. [CrossRef] [PubMed]
90. Li, J.-X.; Hou, X.-J.; Zhu, J.; Zhou, J.-J.; Huang, H.-B.; Yue, J.-Q.; Gao, J.-Y.; Du, Y.-X.; Hu, C.-X.; Hu, C.-G.; et al. Identification of genes associated with lemon floral transition and flower development during floral inductive water deficits: A hypothetical model. *Front. Plant Sci.* **2017**, *8*, 1013. [CrossRef]
91. Southwick, S.; Davenport, T. Characterization of water stress and low temperature effects on flower induction in citrus. *Plant Physiol.* **1986**, *81*, 26–29. [CrossRef] [PubMed]
92. Zhou, H.; Zeng, R.F.; Liu, T.J.; Ai, X.Y.; Ren, M.K.; Zhou, J.J.; Hu, C.G.; Zhang, J.Z. Drought and low temperature-induced *NF-YA1* activates *FT* expression to promote citrus flowering. *Plant Cell Environ.* **2022**, *45*, 3505–3522. [CrossRef] [PubMed]
93. Ma, X.W.; Su, Z.; Ma, H. Molecular genetic analyses of abiotic stress responses during plant reproductive development. *J. Exp. Bot.* **2020**, *71*, 2870–2885. [CrossRef]
94. Han, Y.; Zhang, X.; Wang, Y.; Ming, F. The suppression of *WRKY44* by *GIGANTEA-miR172* pathway is involved in drought response of *Arabidopsis thaliana*. *PLoS ONE* **2013**, *8*, e73541. [CrossRef]
95. Hwang, K.; Susila, H.; Nasim, Z.; Jung, J.Y.; Ahn, J.H. Arabidopsis ABF3 and ABF4 transcription factors act with the NF-YC complex to regulate *SOC1* expression and mediate drought-accelerated flowering. *Mol. Plant* **2019**, *12*, 489–505. [CrossRef] [PubMed]
96. Xu, M.Y.; Zhang, L.; Li, W.W.; Hu, X.L.; Wang, M.B.; Fan, Y.L.; Zhang, C.Y.; Wang, L. Stress-induced early flowering is mediated by miR169 in *Arabidopsis thaliana*. *J. Exp. Bot.* **2014**, *65*, 89–101. [CrossRef]
97. Wang, Z.; Wang, F.X.; Hong, Y.C.; Yao, J.J.; Ren, Z.Z.; Shi, H.Z.; Zhu, J.K. The flowering repressor SVP confers drought resistance in *Arabidopsis* by regulating abscisic acid catabolism. *Mol. Plant* **2018**, *11*, 1184–1197. [CrossRef]
98. Choi, K.; Kim, J.; Hwang, H.J.; Kim, S.; Park, C.; Kim, S.Y.; Lee, I. The FRIGIDA complex activates transcription of *FLC*, a strong flowering repressor in *Arabidopsis*, by recruiting chromatin modification factors. *Plant Cell* **2011**, *23*, 289–303. [CrossRef]
99. Sawa, M.; Kay, S.A. GIGANTEA directly activates *Flowering Locus T* in *Arabidopsis thaliana*. *Proc. Natl. Acad. Sci. USA* **2011**, *108*, 11698–11703. [CrossRef]
100. Fornara, F.; Montaigu, A.; Sánchez-Villarreal, A.; Takahashi, Y.; Ver Loren van Themaat, E.; Huettel, B.; Davis, S.J.; Coupland, G. The GI–CDF module of *Arabidopsis* affects freezing tolerance and growth as well as flowering. *Plant J.* **2015**, *81*, 695–706. [CrossRef]
101. Chen, L.; Hu, P.C.; Lu, Q.Q.; Zhang, F.; Su, Y.H.; Ding, Y. Vernalization attenuates dehydration tolerance in winter-annual *Arabidopsis*. *Plant Physiol.* **2022**, *190*, 732–744. [CrossRef] [PubMed]
102. Weng, X.; Wang, L.; Wang, J.; Hu, Y.; Du, H.; Xu, C.; Xing, Y.; Li, X.; Xiao, J.; Zhang, Q. *Grain number, plant height,* and *heading date7* is a central regulator of growth, development, and stress response. *Plant Physiol.* **2014**, *164*, 735–747. [CrossRef] [PubMed]
103. Xue, W.; Xing, Y.; Weng, X.; Zhao, Y.; Tang, W.; Wang, L.; Zhou, H.; Yu, S.; Xu, C.; Li, X.; et al. Natural variation in *Ghd7* is an important regulator of heading date and yield potential in rice. *Nat. Genet.* **2008**, *40*, 761–767. [CrossRef] [PubMed]
104. Kaneko-Suzuki, M.; Kurihara-Ishikawa, R.; Okushita-Terakawa, C.; Kojima, C.; Nagano-Fujiwara, M.; Ohki, I.; Tsuji, H.; Shimamoto, K.; Taoka, K.I. TFL1-Like proteins in rice antagonize rice FT-Like protein in inflorescence development by competition for complex formation with 14-3-3 and FD. *Plant Cell Physiol.* **2018**, *59*, 458–468. [CrossRef] [PubMed]
105. Xia, K.F.; Wang, R.; Ou, X.J.; Fang, Z.M.; Tian, C.G.; Duan, J.; Wang, Y.Q.; Zhang, M.Y. *OsTIR1* and *OsAFB2* downregulation via *OsmiR393* overexpression leads to more tillers, early flowering and less tolerance to salt and drought in rice. *PLoS ONE* **2012**, *7*, 364–373. [CrossRef]
106. Su, H.H.; Cao, Y.Y.; Ku, L.X.; Yao, W.; Cao, Y.Y.; Ren, Z.Z.; Dou, D.D.; Wang, H.T.; Ren, Z.B.; Liu, H.F.; et al. Dual functions of ZmNF-YA3 in photoperiod-dependent flowering and abiotic stress responses in maize. *J. Exp. Bot.* **2018**, *69*, 5177–5189. [CrossRef] [PubMed]
107. Swida-Barteczka, A.; Pacak, A.; Kruszka, K.; Nuc, P.; Karlowski, W.M.; Jarmolowski, A.; Szweykowska-Kulinska, Z. MicroRNA172b-5p/trehalose-6-phosphate synthase module stimulates trehalose synthesis and microRNA172b-3p/AP2-like module accelerates flowering in barley upon drought stress. *Front. Plant Sci.* **2023**, *14*, 1124785. [CrossRef] [PubMed]
108. Jin, S.; Kim, S.Y.; Susila, H.; Nasim, Z.; Youn, G.; Ahn, J.H. FLOWERING LOCUS M isoforms differentially affect the subcellular localization and stability of SHORT VEGETATIVE PHASE to regulate temperature-responsive flowering in *Arabidopsis*. *Mol. Plant* **2022**, *15*, 1696–1709. [CrossRef]
109. Bendahmane, M.; Airoldi, C.A.; McKay, M.; Davies, B. *MAF2* is regulated by temperature-dependent splicing and represses flowering at low temperatures in parallel with *FLM*. *PLoS ONE* **2015**, *10*, e0126516. [CrossRef]

110. Lyu, J.; Aiwaili, P.; Gu, Z.Y.; Xu, Y.J.; Zhang, Y.H.; Wang, Z.L.; Huang, H.F.; Zeng, R.H.; Ma, C.; Gao, J.P.; et al. Chrysanthemum MAF2 regulates flowering by repressing gibberellin biosynthesis in response to low temperature. *Plant J.* **2022**, *112*, 1159–1175. [CrossRef]
111. Jung, J.-H.; Seo, P.J.; Park, C.-M. The E3 ubiquitin ligase HOS1 regulates *Arabidopsis* flowering by mediating CONSTANS degradation under cold stress. *J. Biol. Chem.* **2012**, *287*, 43277–43287. [CrossRef]
112. Jung, J.-H.; Park, J.-H.; Lee, S.; To, T.K.; Kim, J.-M.; Seki, M.; Park, C.-M. The cold signaling attenuator HIGH EXPRESSION OF OSMOTICALLY RESPONSIVE GENE1 activates *FLOWERING LOCUS C* transcription via chromatin remodeling under short-term cold stress in *Arabidopsis*. *Plant Cell* **2013**, *25*, 4378–4390. [CrossRef]
113. Jung, J.-H.; Park, C.-M. HOS1-mediated activation of *FLC* via chromatin remodeling under cold stress. *Plant Signal. Behav.* **2014**, *8*, e27342. [CrossRef] [PubMed]
114. Ahn, G.; Park, H.J.; Jeong, S.Y.; Shin, G.-I.; Ji, M.G.; Cha, J.-Y.; Kim, J.; Kim, M.G.; Yun, D.-J.; Kim, W.-Y. HOS15 represses flowering by promoting GIGANTEA degradation in response to low temperature in *Arabidopsis*. *Plant Commun.* **2023**, *4*, 100570. [CrossRef] [PubMed]
115. Hsu, C.-Y.; Adams, J.P.; Kim, H.; No, K.; Ma, C.; Strauss, S.H.; Drnevich, J.; Vandervelde, L.; Ellis, J.D.; Rice, B.M.; et al. *FLOWERING LOCUS T* duplication coordinates reproductive and vegetative growth in perennial poplar. *Proc. Natl. Acad. Sci. USA* **2011**, *108*, 10756–10761. [CrossRef]
116. Chica, E.J.; Albrigo, L.G. Changes in *CsFT* transcript abundance at the onset of low-temperature floral induction in sweet orange. *J. Am. Soc. Hortic. Sci.* **2013**, *138*, 184–189. [CrossRef]
117. Chao, Y.; Yang, Q.; Kang, J.; Zhang, T.; Sun, Y. Expression of the alfalfa *FRIGIDA-like* Gene, *MsFRI-L* delays flowering time in transgenic *Arabidopsis thaliana*. *Mol. Biol. Rep.* **2012**, *40*, 2083–2090. [CrossRef]
118. Yan, L.; Fu, D.; Li, C.; Blechl, A.; Tranquilli, G.; Bonafede, M.; Sanchez, A.; Valarik, M.; Yasuda, S.; Dubcovsky, J. The wheat and barley vernalization gene *VRN3* is an orthologue of *FT*. *Proc. Natl. Acad. Sci. USA* **2006**, *103*, 19581–19586. [CrossRef]
119. Oliver, S.N.; Finnegan, E.J.; Dennis, E.S.; Peacock, W.J.; Trevaskis, B. Vernalization-induced flowering in cereals is associated with changes in histone methylation at the *VERNALIZATION1* gene. *Proc. Natl. Acad. Sci. USA* **2009**, *106*, 8386–8391. [CrossRef]
120. Trevaskis, B.; Chen, A.; Dubcovsky, J. Wheat TILLING mutants show that the vernalization gene *VRN1* down-regulates the flowering repressor *VRN2* in leaves but is not essential for flowering. *PLoS Genet.* **2012**, *8*, e1003134. [CrossRef]
121. Yang, F.X.; Zhu, G.F.; Wei, Y.L.; Gao, J.; Liang, G.; Peng, L.Y.; Lu, C.Q.; Jin, J.P. Low-temperature-induced changes in the transcriptome reveal a major role of *CgSVP* genes in regulating flowering of *Cymbidium goeringii*. *BMC Genom.* **2019**, *20*, 53. [CrossRef]
122. Vu, L.D.; Xu, X.Y.; Gevaert, K.; De Smet, I. Developmental plasticity at high temperature. *Plant Physiol.* **2019**, *181*, 399–411. [CrossRef] [PubMed]
123. Kumar, S.V.; Lucyshyn, D.; Jaeger, K.E.; Alos, E.; Alvey, E.; Harberd, N.P.; Wigge, P.A. Transcription factor PIF4 controls the thermosensory activation of flowering. *Nature* **2012**, *484*, 242–245. [CrossRef]
124. Thines, B.C.; Youn, Y.W.; Duarte, M.I.; Harmon, F.G. The time of day effects of warm temperature on flowering time involve PIF4 and PIF5. *J. Exp. Bot.* **2014**, *65*, 1141–1151. [CrossRef]
125. No, D.H.; Baek, D.; Lee, S.H.; Cheong, M.S.; Chun, H.J.; Park, M.S.; Cho, H.M.; Jin, B.J.; Lim, L.H.; Lee, Y.B.; et al. High-temperature conditions promote soybean flowering through the transcriptional reprograming of flowering genes in the photoperiod pathway. *Int. J. Mol. Sci.* **2021**, *22*, 1314. [CrossRef]
126. Hemming, M.N.; Walford, S.A.; Fieg, S.; Dennis, E.S.; Trevaskis, B. Identification of high-temperature-responsive genes in cereals. *Plant Physiol.* **2012**, *158*, 1439–1450. [CrossRef] [PubMed]
127. Zhang, B.Y.; Wu, S.H.; Zhang, Y.E.; Xu, T.; Guo, F.F.; Tang, H.S.; Li, X.; Wang, P.F.; Qian, W.F.; Xue, Y.B. A high temperature-dependent mitochondrial lipase EXTRA GLUME1 promotes floral phenotypic robustness against temperature fluctuation in rice (*Oryza sativa* L.). *PLoS Genet.* **2016**, *12*, e1006152. [CrossRef] [PubMed]
128. Nakano, Y.; Higuchi, Y.; Sumitomo, K.; Hisamatsu, T. Flowering retardation by high temperature in chrysanthemums: Involvement of *FLOWERING LOCUS T-like 3* gene repression. *J. Exp. Bot.* **2013**, *64*, 909–920. [CrossRef]
129. del Olmo, I.; Poza-Viejo, L.; Pineiro, M.; Jarillo, J.A.; Crevillen, P. High ambient temperature leads to reduced *FT* expression and delayed flowering in *Brassica rapa* via a mechanism associated with H2A.Z dynamics. *Plant J.* **2019**, *100*, 343–356. [CrossRef]
130. Taoka, K.; Ohki, I.; Tsuji, H.; Kojima, C.; Shimamoto, K. Structure and function of florigen and the receptor complex. *Trends Plant Sci.* **2013**, *18*, 287–294. [CrossRef]
131. Keller, S.R.; Levsen, N.; Ingvarsson, P.K.; Olson, M.S.; Tiffin, P. Local selection across a latitudinal gradient shapes nucleotide diversity in Balsam Poplar, *Populus balsamifera* L. *Genetics* **2011**, *188*, 941–952. [CrossRef]
132. Irwin, J.A.; Lister, C.; Soumpourou, E.; Zhang, Y.; Howell, E.C.; Teakle, G.; Dean, C. Functional alleles of the flowering time regulator *FRIGIDA* in the *Brassica oleracea* genome. *BMC Plant Biol.* **2012**, *12*, 21. [CrossRef] [PubMed]
133. Hyun, K.-g.; Oh, J.E.; Park, J.; Noh, Y.-S.; Song, J.-J. Structural analysis of FRIGIDA flowering-time regulator. *Mol. Plant* **2016**, *9*, 618–620. [CrossRef]
134. Renau-Morata, B.; Carrillo, L.; Dominguez-Figueroa, J.; Vicente-Carbajosa, J.; Molina, R.V.; Nebauer, S.G.; Medina, J. CDF transcription factors: Plant regulators to deal with extreme environmental conditions. *J. Exp. Bot.* **2020**, *71*, 3803–3815. [CrossRef] [PubMed]

135. Wu, K.; Zhang, L.; Zhou, C.; Yu, C.W.; Chaikam, V. HDA6 is required for jasmonate response, senescence and flowering in *Arabidopsis*. *J. Exp. Bot.* **2008**, *59*, 225–234. [CrossRef]
136. Kim, J.M.; To, T.K.; Matsui, A.; Tanoi, K.; Kobayashi, N.I.; Matsuda, F.; Habu, Y.; Ogawa, D.; Sakamoto, T.; Matsunaga, S.; et al. Acetate-mediated novel survival strategy against drought in plants. *Nat. Plants* **2017**, *3*, 17097. [CrossRef] [PubMed]
137. Yu, C.W.; Liu, X.C.; Luo, M.; Chen, C.Y.; Lin, X.D.; Tian, G.; Lu, Q.; Cui, Y.H.; Wu, K.Q. HISTONE DEACETYLASE6 interacts with FLOWERING LOCUS D and regulates flowering in *Arabidopsis*. *Plant Physiol.* **2011**, *156*, 173–184. [CrossRef]
138. Cui, L.G.; Shan, J.X.; Shi, M.; Gao, J.P.; Lin, H.X. The *miR156-SPL9-DFR* pathway coordinates the relationship between development and abiotic stress tolerance in plants. *Plant J.* **2014**, *80*, 1108–1117. [CrossRef]
139. Zheng, L.-L.; Qu, L.-H. Application of microRNA gene resources in the improvement of agronomic traits in rice. *Plant Biotechnol. J.* **2015**, *13*, 329–336. [CrossRef]
140. Giusti, L.; Mica, E.; Bertolini, E.; De Leonardis, A.M.; Faccioli, P.; Cattivelli, L.; Crosatti, C. microRNAs differentially modulated in response to heat and drought stress in durum wheat cultivars with contrasting water use efficiency. *Funct. Integr. Genom.* **2017**, *17*, 293–309. [CrossRef]
141. Pegler, J.; Grof, C.; Eamens, A. Profiling of the differential abundance of drought and salt stress-responsive microRNAs across grass crop and genetic model plant species. *Agronomy* **2018**, *8*, 118. [CrossRef]
142. Fang, L.; Wang, Y. MicroRNAs in woody plants. *Front. Plant Sci.* **2021**, *12*, 686831. [CrossRef] [PubMed]
143. Zhang, T.Q.; Wang, J.W.; Zhou, C.M. The role of miR156 in developmental transitions in *Nicotiana tabacum*. *Sci. China Life Sci.* **2015**, *58*, 253–260. [CrossRef] [PubMed]
144. Benlloch, R.; Lois, L.M. Sumoylation in plants: Mechanistic insights and its role in drought stress. *J. Exp. Bot.* **2018**, *69*, 4539–4554. [CrossRef] [PubMed]
145. Zhang, Z.; Hu, Q.; Gao, Z.; Zhu, Y.; Yin, M.; Shang, E.; Liu, G.; Liu, W.; Hu, R.; Cheng, H.; et al. Flowering repressor CmSVP recruits the TOPLESS corepressor to control flowering in chrysanthemum. *Plant Physiol.* **2023**, *193*, 2413–2429. [CrossRef] [PubMed]
146. Chang, P.; Hsieh, H.Y.; Tu, S.L. The U1 snRNP component RBP45d regulates temperature-responsive flowering in *Arabidopsis*. *Plant Cell* **2022**, *34*, 834–851. [CrossRef] [PubMed]
147. Woods, D.P.; McKeown, M.A.; Dong, Y.X.; Preston, J.C.; Amasino, R.M. Evolution of *VRN2/Ghd7-like* genes in vernalization-mediated repression of grass flowering. *Plant Physiol.* **2016**, *170*, 2124–2135. [CrossRef] [PubMed]
148. Ream, T.S.; Woods, D.P.; Schwartz, C.J.; Sanabria, C.P.; Mahoy, J.A.; Walters, E.M.; Kaeppler, H.F.; Amasino, R.M. Interaction of photoperiod and vernalization determines flowering time of *Brachypodium distachyon*. *Plant Physiol.* **2014**, *164*, 694–709. [CrossRef] [PubMed]
149. McKeown, M.; Schubert, M.; Marcussen, T.; Fjellheim, S.; Preston, J.C. Evidence for an early origin of vernalization responsiveness in temperate pooideae grasses. *Plant Physiol.* **2016**, *172*, 416–426. [CrossRef]
150. Kerbler, S.M.; Wigge, P.A. Temperature sensing in plants. *Annu. Rev. Plant Biol.* **2023**, *74*, 341–366. [CrossRef]
151. Li, G.Q.; Yu, M.; Fang, T.L.; Cao, S.H.; Carver, B.F.; Yan, L.L. Vernalization requirement duration in winter wheat is controlled by *Ta*VRN-A1 at the protein level. *Plant J.* **2013**, *76*, 742–753. [CrossRef]
152. Xie, L.; Zhang, Y.; Wang, K.; Luo, X.M.; Xu, D.A.; Tian, X.L.; Li, L.L.; Ye, X.G.; Xia, X.C.; Li, W.X.; et al. *TaVrt2*, an SVP-like gene, cooperates with *TaVrn1* to regulate vernalization-induced flowering in wheat. *N. Phytol.* **2021**, *231*, 834–848. [CrossRef]
153. Ochagavia, H.; Kiss, T.; Karsai, I.; Casas, A.M.; Igartua, E. Responses of barley to high ambient temperature are modulated by vernalization. *Front. Plant Sci.* **2022**, *12*, 776982. [CrossRef] [PubMed]
154. Xu, S.J.; Chong, K. Remembering winter through vernalisation. *Nat. Plants* **2018**, *4*, 997–1009. [CrossRef] [PubMed]
155. Zografos, B.R.; Sung, S.B. Vernalization-mediated chromatin changes. *J. Exp. Bot.* **2012**, *63*, 4343–4348. [CrossRef]
156. Zhou, J.X.; Liu, Z.W.; Li, Y.Q.; Li, L.; Wang, B.; Chen, S.; He, X.J. *Arabidopsis* PWWP domain proteins mediate H3K27 trimethylation on FLC and regulate flowering time. *J. Integr. Plant Biol.* **2018**, *60*, 1015. [CrossRef]
157. Shen, L.S.; Zhang, Y.; Sawettalake, N. A molecular switch for *FLOWERING LOCUS C* activation determines flowering time in *Arabidopsis*. *Plant Cell* **2022**, *34*, 818–833. [CrossRef]
158. Kong, X.X.; Luo, L.D.; Zhao, J.J.; Chen, Q.; Chang, G.X.; Huang, J.L.; Yang, Y.P.; Hu, X.Y. Expression of *FRIGIDA* in root inhibits flowering in *Arabidopsis thaliana*. *J. Exp. Bot.* **2019**, *70*, 5101–5114. [CrossRef]
159. Michaels, S.D.; Himelblau, E.; Kim, S.Y.; Schomburg, F.M.; Amasino, R.M. Integration of flowering signals in winter-annual *Arabidopsis*. *Plant Physiol.* **2005**, *137*, 149–156. [CrossRef] [PubMed]
160. Sheldon, C.C.; Finnegan, E.J.; Dennis, E.S.; Peacock, W.J. Quantitative effects of vernalization on *FLC* and *SOC1* expression. *Plant J.* **2006**, *45*, 871–883. [CrossRef] [PubMed]
161. Kim, D.H.; Doyle, M.R.; Sung, S.; Amasino, R.M. Vernalization: Winter and the timing of flowering in Plants. *Annu. Rev. Cell Dev. Biol.* **2009**, *25*, 277–299. [CrossRef]
162. Sheldon, C.C.; Rouse, D.T.; Finnegan, E.J.; Peacock, W.J.; Dennis, E.S. The molecular basis of vernalization: The central role of *FLOWERING LOCUS C (FLC)*. *Proc. Natl. Acad. Sci. USA* **2000**, *97*, 3753–3758. [CrossRef]
163. Li, Z.C.; Jiang, D.H.; He, Y.H. FRIGIDA establishes a local chromosomal environment for *FLOWERING LOCUS C* mRNA production. *Nat. Plants* **2018**, *4*, 836–846. [CrossRef]
164. Surkova, S.Y.; Samsonova, M.G. Mechanisms of vernalization-induced flowering in Legumes. *Int. J. Mol. Sci.* **2022**, *23*, 9889. [CrossRef] [PubMed]

165. Distelfeld, A.; Tranquilli, G.; Li, C.X.; Yan, L.L.; Dubcovsky, J. Genetic and molecular characterization of the *VRN2* loci in tetraploid wheat. *Plant Physiol.* **2009**, *149*, 245–257. [CrossRef]
166. Cui, G.Q.; Li, D.P.; Zhang, L.C.; Xia, C.; Kong, X.Y.; Liu, X. *GSK3* regulates *VRN1* to control flowering time in wheat. *J. Integr. Plant Biol.* **2023**, *65*, 1605–1608. [CrossRef]
167. Huan, Q.; Mao, Z.W.; Chong, K.; Zhang, J.Y. Global analysis of H3K4me3/H3K27me3 in *Brachypodium distachyon* reveals *VRN3* as critical epigenetic regulation point in vernalization and provides insights into epigenetic memory. *N. Phytol.* **2018**, *219*, 1373–1387. [CrossRef]
168. Milec, Z.; Strejcková, B.; Safár, J. Contemplation on wheat vernalization. *Front. Plant Sci.* **2023**, *13*, 1093792. [CrossRef] [PubMed]
169. Mulki, M.A.; von Korff, M. *CONSTANS* controls floral repression by up-regulating *VERNALIZATION2* (*VRN-H2*) in Barley. *Plant Physiol.* **2016**, *170*, 325–337. [CrossRef] [PubMed]
170. Taheripourfard, Z.S.; Izadi-Darbandi, A.; Ghazvini, H.; Ebrahimi, M.; Mortazavian, S.M.M. Characterization of specific DNA markers at *VRN-H1* and *VRN-H2* loci for growth habit of barley genotypes. *J. Genet.* **2018**, *97*, 87–95. [CrossRef]
171. Kennedy, A.; Geuten, K. The role of *FLOWERING LOCUS C* relatives in cereals. *Front. Plant Sci.* **2020**, *11*, 617340. [CrossRef]
172. Yan, L.L.; Loukoianov, A.; Blechl, A.; Tranquilli, G.; Ramakrishna, W.; SanMiguel, P.; Bennetzen, J.L.; Echenique, V.; Dubcovsky, J. The wheat *VRN2* gene is a flowering repressor down-regulated by vernalization. *Science* **2004**, *303*, 1640–1644. [CrossRef] [PubMed]
173. Thomashow, M.F. Plant cold acclimation: Freezing tolerance genes and regulatory mechanisms. *Annu. Rev. Plant Physiol. Plant Mol. Biol.* **1999**, *50*, 571–599. [CrossRef]
174. Li, Y.P.; Shi, Y.T.; Li, M.Z.; Fu, D.Y.; Wu, S.F.; Li, J.G.; Gong, Z.Z.; Liu, H.T.; Yang, S.H. The CRY2-COP1-HY5-BBX7/8 module regulates blue light-dependent cold acclimation in *Arabidopsis*. *Plant Cell* **2021**, *33*, 3555–3573. [CrossRef] [PubMed]
175. Cvetkovic, J.; Muller, K.; Baier, M. The effect of cold priming on the fitness of *Arabidopsis thaliana* accessions under natural and controlled conditions. *Sci. Rep.* **2017**, *7*, 44055. [CrossRef] [PubMed]
176. Zheng, G.W.; Li, L.X.; Li, W.Q. Glycerolipidome responses to freezing- and chilling-induced injuries: Examples in *Arabidopsis* and rice. *BMC Plant Biol.* **2016**, *16*, 70. [CrossRef]
177. Pareek, A.; Khurana, A.; Sharma, A.K.; Kumar, R. An overview of signaling regulons during cold stress tolerance in plants. *Curr. Genom.* **2017**, *18*, 498–511. [CrossRef]
178. Kerchev, P.I.; Van Breusegem, F. Improving oxidative stress resilience in plants. *Plant J.* **2021**, *109*, 359–372. [CrossRef]
179. Krasensky, J.; Jonak, C. Drought, salt, and temperature stress-induced metabolic rearrangements and regulatory networks. *J. Exp. Bot.* **2012**, *63*, 1593–1608. [CrossRef]
180. Theocharis, A.; Clément, C.; Barka, E.A. Physiological and molecular changes in plants grown at low temperatures. *Planta* **2012**, *235*, 1091–1105. [CrossRef]
181. Yadav, S.K. Cold stress tolerance mechanisms in plants. *Agron. Sustain. Dev.* **2010**, *30*, 515–527. [CrossRef]
182. Guy, C.L. Cold-acclimation and freezing stress tolerance-role of protein-metabolism. *Annu. Rev. Plant Physiol. Plant Mol. Biol.* **1990**, *41*, 187–223. [CrossRef]
183. Lyons, J.M. Chilling injury in plants. *Annu. Rev. Plant Physiol. Plant Mol. Biol.* **1973**, *24*, 445–466. [CrossRef]
184. Pan, T.; Deng, N.-M.; Guo, W.-X.; Wan, M.-Z.; Zheng, Y.-T.; Chen, S.-Y.; Liu, C.-L.; Li, H.-B.; Liang, S. *DnFCA* isoforms cooperatively regulate temperature-related flowering in *Dendrobium nobile*. *Biology* **2023**, *12*, 331. [CrossRef] [PubMed]
185. Li, D.; Liu, C.; Shen, L.; Wu, Y.; Chen, H.; Robertson, M.; Helliwell, C.A.; Ito, T.; Meyerowitz, E.; Yu, H. A repressor complex governs the integration of flowering signals in *Arabidopsis*. *Dev. Cell* **2008**, *15*, 110–120. [CrossRef]
186. Lee, J.H.; Yoo, S.J.; Park, S.H.; Hwang, I.; Lee, J.S.; Ahn, J.H. Role of *SVP* in the control of flowering time by ambient temperature in *Arabidopsis*. *Genes Dev.* **2007**, *21*, 397–402. [CrossRef]
187. Halliday, K.J.; Salter, M.G.; Thingnaes, E.; Whitelam, G.C. Phytochrome control of flowering is temperature sensitive and correlates with expression of the floral integrator *FT*. *Plant J.* **2003**, *33*, 875–885. [CrossRef]
188. Cao, S.; Ye, M.; Jiang, S. Involvement of *GIGANTEA* gene in the regulation of the cold stress response in *Arabidopsis*. *Plant Cell Rep.* **2005**, *24*, 683–690. [CrossRef]
189. Paltiel, J.; Amin, R.; Gover, A.; Ori, N.; Samach, A. Novel roles for GIGANTEA revealed under environmental conditions that modify its expression in *Arabidopsis* and *Medicago truncatula*. *Planta* **2006**, *224*, 1255–1268. [CrossRef]
190. Hatayama, T.; Takeno, K. The metabolic pathway of salicylic acid rather than of chlorogenic acid is involved in the stress-induced flowering of *Pharbitis nil*. *J. Plant Physiol.* **2003**, *160*, 461–467. [CrossRef]
191. Wilkie, J.D.; Sedgley, M.; Olesen, T. Regulation of floral initiation in horticultural trees. *J. Exp. Bot.* **2008**, *59*, 3215–3228. [CrossRef]
192. Zhang, J.Z.; Li, Z.M.; Mei, L.; Yao, J.L.; Hu, C.G. *PtFLC* homolog from trifoliate orange (*Poncirus trifoliata*) is regulated by alternative splicing and experiences seasonal fluctuation in expression level. *Planta* **2009**, *229*, 847–859. [CrossRef] [PubMed]
193. Nishikawa, F.; Endo, T.; Shimada, T.; Fujii, H.; Shimizu, T.; Omura, M.; Ikoma, Y. Increased *CiFT* abundance in the stem correlates with floral induction by low temperature in Satsuma mandarin (*Citrus unshiu* Marc.). *J. Exp. Bot.* **2007**, *58*, 3915–3927. [CrossRef] [PubMed]
194. Havko, N.E.; Das, M.R.; McClain, A.M.; Kapali, G.; Sharkey, T.D.; Howe, G.A. Insect herbivory antagonizes leaf cooling responses to elevated temperature in tomato. *Proc. Natl. Acad. Sci. USA* **2020**, *117*, 2211–2217. [CrossRef]
195. Allen, L.H.; Zhang, L.X.; Boote, K.J.; Hauser, B.A. Elevated temperature intensity, timing, and duration of exposure affect soybean internode elongation, mainstem node number, and pod number per plant. *Crop J.* **2018**, *6*, 148–161. [CrossRef]

196. Dixon, L.E.; Farré, A.; Finnegan, E.J.; Orford, S.; Griffiths, S.; Boden, S.A. Developmental responses of bread wheat to changes in ambient temperature following deletion of a locus that includes. *Plant Cell Environ.* **2018**, *41*, 1715–1725. [CrossRef] [PubMed]
197. Zhu, T.T.; De Lima, C.F.F.; De Smet, I. The heat is on: How crop growth, development, and yield respond to high temperature. *J. Exp. Bot.* **2021**, *72*, 7359–7373. [CrossRef] [PubMed]
198. Sato, S.; Peet, M.M.; Thomas, J.F. Determining critical pre- and post-anthesis periods and physiological processes in *Lycopersicon esculentum* Mill. exposed to moderately elevated temperatures. *J. Exp. Bot.* **2002**, *53*, 1187–1195. [CrossRef]
199. Sehgal, A.; Sita, K.; Siddique, K.H.M.; Kumar, R.; Bhogireddy, S.; Varshney, R.K.; HanumanthaRao, B.; Nair, R.M.; Prasad, P.V.V.; Nayyar, H. Drought or/and heat-stress effects on seed filling in food crops: Impacts on functional biochemistry, seed yields, and nutritional quality. *Front. Plant Sci.* **2018**, *9*, 1705. [CrossRef]
200. Draeger, T.; Martin, A.C.; Alabdullah, A.K.; Pendle, A.; Rey, M.D.; Shaw, P.; Moore, G. *Dmc1* is a candidate for temperature tolerance during wheat meiosis. *Theor. Appl. Genet.* **2020**, *133*, 809–828. [CrossRef]
201. Quint, M.; Delker, C.; Franklin, K.A.; Wigge, P.A.; Halliday, K.J.; van Zanten, M. Molecular and genetic control of plant thermomorphogenesis. *Nat. Plants* **2016**, *2*, 15190. [CrossRef] [PubMed]
202. Dusenge, M.E.; Duarte, A.G.; Way, D.A. Plant carbon metabolism and climate change: Elevated $CO_2$ and temperature impacts on photosynthesis, photorespiration and respiration. *N. Phytol.* **2019**, *221*, 32–49. [CrossRef] [PubMed]
203. Scafaro, A.P.; Fan, Y.Z.; Posch, B.C.; Garcia, A.; Coast, O.; Atkin, O.K. Responses of leaf respiration to heatwaves. *Plant Cell Environ.* **2021**, *44*, 2090–2101. [CrossRef]
204. Hieno, A.; Naznin, H.A.; Inaba-Hasegawa, K.; Yokogawa, T.; Hayami, N.; Nomoto, M.; Tada, Y.; Yokogawa, T.; Higuchi-Takeuchi, M.; Hanada, K.; et al. Transcriptome analysis and identification of a transcriptional regulatory network in the response to $H_2O_2$. *Plant Physiol.* **2019**, *180*, 1629–1646. [CrossRef]
205. Babbar, R.; Karpinska, B.; Grover, A.; Foyer, C.H. Heat-induced oxidation of the nuclei and cytosol. *Front. Plant Sci.* **2021**, *11*, 617779. [CrossRef] [PubMed]
206. Zhao, C.; Piao, S.L.; Wang, X.H.; Huang, Y.; Ciais, P.; Elliott, J.; Huang, M.T.; Janssens, I.A.; Li, T.; Lian, X.; et al. Plausible rice yield losses under future climate warming. *Nat. Plants* **2017**, *3*, 16202. [CrossRef] [PubMed]
207. Cossani, C.M.; Reynolds, M.P. Physiological traits for improving heat tolerance in wheat. *Plant Physiol.* **2012**, *160*, 1710–1718. [CrossRef]
208. Sadok, W.; Jagadish, S.V.K. The hidden costs of nighttime warming on yields. *Trends Plant Sci.* **2020**, *25*, 644–651. [CrossRef]
209. Kostaki, K.I.; Coupel-Ledru, A.; Bonnell, V.C.; Gustavsson, M.; Sun, P.; McLaughlin, F.J.; Fraser, D.P.; McLachlan, D.H.; Hetherington, A.M.; Dodd, A.N.; et al. Guard cells integrate light and temperature signals to control stomatal aperture. *Plant Physiol.* **2020**, *182*, 1404–1419. [CrossRef]
210. Bawa, G.; Feng, L.Y.; Chen, G.P.; Chen, H.; Hu, Y.; Pu, T.; Cheng, Y.J.; Shi, J.Y.; Xiao, T.; Zhou, W.G.; et al. Gibberellins and auxin regulate soybean hypocotyl elongation under low light and high-temperature interaction. *Physiol. Plant* **2020**, *170*, 345–356. [CrossRef]
211. Vu, L.D.; Xu, X.Y.; Zhu, T.T.; Pan, L.X.; van Zanten, M.; de Jong, D.; Wang, Y.W.; Vanremoortele, T.; Locke, A.M.; van de Cotte, B.; et al. The membrane-localized protein kinase MAP4K4/TOT3 regulates thermomorphogenesis. *Nat. Commun.* **2021**, *12*, 2842. [CrossRef]
212. Bellstaedt, J.; Trenner, J.; Lippmann, R.; Poeschl, Y.; Zhang, X.X.; Friml, J.; Quint, M.; Delker, C. A mobile auxin signal connects temperature sensing in cotyledons with growth responses in hypocotyls. *Plant Physiol.* **2019**, *180*, 757–766. [CrossRef] [PubMed]
213. Gol, L.; Tome, F.; von Korff, M. Floral transitions in wheat and barley: Interactions between photoperiod, abiotic stresses, and nutrient status. *J. Exp. Bot.* **2017**, *68*, 1399–1410. [CrossRef] [PubMed]
214. Jacott, C.N.; Boden, S.A. Feeling the heat: Developmental and molecular responses of wheat and barley to high ambient temperatures. *J. Exp. Bot.* **2020**, *71*, 5740–5751. [CrossRef] [PubMed]
215. Liu, L.-y.; Jia, M.-z.; Wang, S.-n.; Han, S.; Jiang, J.; Putterill, J. Identification and characterization of cotton PHYTOCHROME-INTERACTING FACTORS in temperature-dependent flowering. *J. Exp. Bot.* **2023**, *74*, 3765–3780. [CrossRef]
216. Chen, J.Y.; Zhang, H.W.; Zhang, H.L.; Ying, J.Z.; Ma, L.Y.; Zhuang, J.Y. Natural variation at *qHd1* affects heading date acceleration at high temperatures with pleiotropism for yield traits in rice. *BMC Plant Biol.* **2018**, *18*, 112. [CrossRef]
217. Wang, Y.Y.; Tao, H.B.; Tian, B.J.; Sheng, D.C.; Xu, C.C.; Zhou, H.M.; Huang, S.B.; Wang, P. Flowering dynamics, pollen, and pistil contribution to grain yield in response to high temperature during maize flowering. *Environ. Exp. Bot.* **2019**, *158*, 80–88. [CrossRef]
218. Jagadish, S.V.K.; Murty, M.V.R.; Quick, W.P. Rice responses to rising temperatures—Challenges, perspectives and future directions. *Plant Cell Environ.* **2015**, *38*, 1686–1698. [CrossRef]
219. Chiluwal, A.; Bheemanahalli, R.; Kanaganahalli, V.; Boyle, D.; Perumal, R.; Pokharel, M.; Oumarou, H.; Jagadish, S.V.K. Deterioration of ovary plays a key role in heat stress-induced spikelet sterility in sorghum. *Plant Cell Environ.* **2020**, *43*, 448–462. [CrossRef]
220. Sun, A.Q.; Somayanda, I.; Sebastian, S.V.; Singh, K.; Gill, K.; Prasad, P.V.V.; Jagadish, S.V.K. Heat stress during flowering affects time of day of flowering, seed set, and grain quality in spring wheat. *Crop Sci.* **2018**, *58*, 380–392. [CrossRef]
221. Noy-Porat, T.; Cohen, D.; Mathew, D.; Eshel, A.; Kamenetsky, R.; Flaishman, M.A. Turned on by heat: Differential expression of *FT* and *LFY-like* genes in *Narcissus tazetta* during floral transition. *J. Exp. Bot.* **2013**, *64*, 3273–3284. [CrossRef]

222. Noy-Porat, T.; Flaishman, M.A.; Eshel, A.; Sandler-Ziv, D.; Kamenetsky, R. Florogenesis of the Mediterranean geophyte *Narcissus tazetta* and temperature requirements for flower initiation and differentiation. *Sci. Hortic.* **2009**, *120*, 138–142. [CrossRef]
223. Oda, A.; Narumi, T.; Li, T.P.; Kando, T.; Higuchi, Y.; Sumitomo, K.; Fukai, S.; Hisamatsu, T. *CsFTL3*, a chrysanthemum *FLOWERING LOCUS T-like* gene, is a key regulator of photoperiodic flowering in chrysanthemums. *J. Exp. Bot.* **2012**, *63*, 1461–1477. [CrossRef] [PubMed]
224. Zheng, B.Y.; Chenu, K.; Chapman, S.C. Velocity of temperature and flowering time in wheat—Assisting breeders to keep pace with climate change. *Glob. Chang. Biol.* **2016**, *22*, 921–933. [CrossRef]
225. Majee, A.; Kumari, D.; Sane, V.A.; Singh, R.K. Novel roles of HSFs and HSPs, other than relating to heat stress, in temperature-mediated flowering. *Ann. Bot.* **2023**, *mcad*, 112. [CrossRef]
226. Guo, M.; Liu, J.H.; Ma, X.; Luo, D.X.; Gong, Z.H.; Lu, M.H. The plant heat stress transcription factors (HSFs): Structure, regulation, and function in response to abiotic stresses. *Front. Plant Sci.* **2016**, *7*, 114. [CrossRef] [PubMed]
227. Jeong, G.; Jeon, M.; Shin, J.; Lee, I. HEAT SHOCK TRANSCRIPTION FACTOR B2b acts as a transcriptional repressor of *VIN3*, a gene induced by long-term cold for flowering. *Sci. Rep.* **2022**, *12*, 10963. [CrossRef] [PubMed]
228. Wunderlich, M.; Gross-Hardt, R.; Schoffl, F. Heat shock factor HSFB2a involved in gametophyte development of *Arabidopsis thaliana* and its expression is controlled by a heat-inducible long non-coding antisense RNA. *Plant Mol. Biol.* **2014**, *85*, 541–550. [CrossRef]
229. Fragkostefanakis, S.; Mesihovic, A.; Simm, S.; Paupiere, M.J.; Hu, Y.J.; Paul, P.; Mishra, S.K.; Tschiersch, B.; Theres, K.; Bovy, A.; et al. HsfA2 controls the activity of developmentally and stress-regulated heat stress protection mechanisms in tomato male reproductive tissues. *Plant Physiol.* **2016**, *170*, 2461–2477. [CrossRef]
230. Kotak, S.; Larkindale, J.; Lee, U.; von Koskull-Döring, P.; Vierling, E.; Scharf, K.-D. Complexity of the heat stress response in plants. *Curr. Opin. Plant Biol.* **2007**, *10*, 310–316. [CrossRef]
231. Perez, D.E.; Hoyer, J.S.; Johnson, A.I.; Moody, Z.R.; Lopez, J.; Kaplinsky, N.J. BOBBER1 is a noncanonical *Arabidopsis* small heat shock protein required for both development and thermotolerance. *Plant Physiol.* **2009**, *151*, 241–252. [CrossRef] [PubMed]
232. Casal, J.J.; Balasubramanian, S. Thermomorphogenesis. *Annu. Rev. Plant Biol.* **2019**, *70*, 321–346. [CrossRef] [PubMed]
233. Kim, W.; Park, T.I.; Yoo, S.J.; Jun, A.R.; Ahn, J.H. Generation and analysis of a complete mutant set for the *Arabidopsis FT/TFL1* family shows specific effects on thermo-sensitive flowering regulation. *J. Exp. Bot.* **2013**, *64*, 1715–1729. [CrossRef]
234. Talbert, P.B.; Henikoff, S. Environmental responses mediated by histone variants. *Trends Cell Biol.* **2014**, *24*, 642–650. [CrossRef]
235. Arya, H.; Singh, M.B.; Bhalla, P.L. Genomic and molecular analysis of conserved and unique features of soybean *PIF4*. *Sci. Rep.* **2018**, *8*, 12569. [CrossRef] [PubMed]
236. Ford, B.; Deng, W.W.; Clausen, J.; Oliver, S.; Boden, S.; Hemming, M.; Trevaskis, B. Barley (*Hordeum vulgare*) circadian clock genes can respond rapidly to temperature in an *EARLY FLOWERING* 3-dependent manner. *J. Exp. Bot.* **2016**, *67*, 5517–5528. [CrossRef] [PubMed]
237. Ejaz, M.; von Korff, M. The genetic control of reproductive development under high ambient temperature. *Plant Physiol.* **2017**, *173*, 294–306. [CrossRef]
238. Prieto, P.; Ochagavia, H.; Savin, R.; Griffiths, S.; Slafer, G.A. Physiological determinants of fertile floret survival in wheat as affected by earliness *per se* genes under field conditions. *Eur. J. Agron.* **2018**, *99*, 206–213. [CrossRef]
239. Ochagavia, H.; Prieto, P.; Zikhali, M.; Griffiths, S.; Slafer, G.A. Earliness *Per Se* by temperature interaction on wheat development. *Sci. Rep.* **2019**, *9*, 2584. [CrossRef]
240. Penfield, S. Temperature perception and signal transduction in plants. *N. Phytol.* **2008**, *179*, 615–628. [CrossRef]
241. Takahashi, H.; Nishihara, M.; Yoshida, C.; Itoh, K. Gentian *FLOWERING LOCUS T* orthologs regulate phase transitions: Floral induction and endodormancy release. *Plant Physiol.* **2022**, *188*, 1887–1899. [CrossRef]
242. Barros, E.; Lezar, S.; Anttonen, M.J.; van Dijk, J.P.; Röhlig, R.M.; Kok, E.J.; Engel, K.H. Comparison of two GM maize varieties with a near-isogenic non-GM variety using transcriptomics, proteomics and metabolomics. *Plant Biotechnol. J.* **2010**, *8*, 436–451. [CrossRef]
243. Amiour, N.; Imbaud, S.; Clément, G.; Agier, N.; Zivy, M.; Valot, B.; Balliau, T.; Armengaud, P.; Quilleré, I.; Cañas, R.; et al. The use of metabolomics integrated with transcriptomic and proteomic studies for identifying key steps involved in the control of nitrogen metabolism in crops such as maize. *J. Exp. Bot.* **2012**, *63*, 5017–5033. [CrossRef]
244. Carrera, F.P.; Noceda, C.; Maridueña-Zavala, M.G.; Cevallos-Cevallos, J.M. Metabolomics, a powerful tool for understanding plant abiotic stress. *Agron. Basel* **2021**, *11*, 824. [CrossRef]
245. Ghatak, A.; Chaturvedi, P.; Weckwerth, W. Cereal crop proteomics: Systemic analysis of crop drought stress responses towards marker-assisted selection breeding. *Front. Plant Sci.* **2017**, *8*, 757. [CrossRef] [PubMed]
246. Guo, Q.; Li, X.; Niu, L.; Jameson, P.E.; Zhou, W. Transcription-associated metabolomic adjustments in maize occur during combined drought and cold stress. *Plant Physiol.* **2021**, *186*, 677–695. [CrossRef] [PubMed]
247. Kumar, M.; Kumar Patel, M.; Kumar, N.; Bajpai, A.B.; Siddique, K.H.M. Metabolomics and molecular approaches reveal drought stress tolerance in plants. *Int. J. Mol. Sci.* **2021**, *22*, 9108. [CrossRef] [PubMed]

agronomy

*Article*

# A Study on Petal Morphological and Physiological Characteristics of *Styrax japonicus* during the Flowering Period

**Chen Chen, Hong Chen, Ming Ni and Fangyuan Yu ***

Collaborative Innovation Centre of Sustainable Forestry in Southern China, College of Forest Science, Nanjing Forestry University, 159 Longpan Road, Nanjing 210037, China; cc0212@njfu.edu.cn (C.C.); hongchen@njfu.edu.cn (H.C.); haoyuw@njfu.edu.cn (M.N.)
* Correspondence: fyyu@njfu.edu.cn

**Abstract:** *Styrax japonicus* is a small ornamental tree with medicinal values, although its flowering period is short. To date, information about the morphological and physiological characteristics of the petals during the flowering period is limited. In this study, we observed the structure of the petals at the full flowering stage with a scanning electron microscope and detected the contents of nutrients, minerals, and endogenous hormones and the activities of enzymes at different flowering stages. The results showed that the content of soluble sugar exhibited an 'increase-decrease' trend, whereas the contents of soluble protein, nitrogen (N), phosphorus (P), and abscisic acid (ABA) showed a 'decrease-increase' pattern. The content of starch descended continuously, but the contents of potassium (K), gibberellic acid ($GA_3$), indoleacetic acid (IAA), and malondialdehyde (MDA) ascended continuously. The activities of peroxidase (POD) and superoxide dismutase (SOD) first rose and then declined during the flowering period. Higher contents of soluble sugar, N, K, and IAA promoted *S. japonicus* flowering; meanwhile, lower contents of starch, soluble protein, P, and $GA_3$ in addition to the lower activity of SOD might be some of the causes of the short flowering period. This work will serve as the foundation for a scientific technique to utilize the flowers and extend the flowering period in *S. japonicus*.

**Keywords:** *Styrax japonicus*; enzymes; hormones; minerals; nutrients; senescence

**Citation:** Chen, C.; Chen, H.; Ni, M.; Yu, F. A Study on Petal Morphological and Physiological Characteristics of *Styrax japonicus* during the Flowering Period. *Agronomy* **2021**, *11*, 1498. https://doi.org/10.3390/agronomy11081498

Academic Editors: Jinzhi Zhang and Avi Sadka

Received: 19 June 2021
Accepted: 26 July 2021
Published: 28 July 2021



## 1. Introduction

Controlling the flowering time and understanding the physiological mechanism of the induction of flowers are essential to the tourism, research, and flower industries [1]. In recent years, many researchers have studied the flowering process, including flower morphological structure, nutrients, mineral elements, enzymes, and relevant genes in flowers, so as to reveal the biological characteristics and regulate the flowering period of different plants [2–5]. Based on the study of flower morphology in the genus *Capsella*, Neuffer and Paetsch [6] specifically disclosed the petal size, pollen number, stigma length, and so on in *Capsella grandiflora*, *C. rubella*, and *C. bursa-pastoris*. Experiments were conducted to determine whether changing nutrient concentrations helped to induce *Eichhornia crassipes* flowering, and the result demonstrated that very low nutrient concentration conditions induced inflorescence formation [7], which reconfirmed the conclusion of Richards [8]. High soluble sugar content promoted *S. tonkinensis* flowering [9], while it was the opposite for olive [10]. Starch is one of the nutrients that may be converted into soluble sugar to meet the need for metabolic energy during the process of flower bud differentiation [9]. Moreover, soluble protein is one of the most important energy sources in plant bud initiation and the basis for flower organ morphogenesis [11]. Shahri et al. [12] pointed out that the soluble protein content in the *Helleborus orientalis* flower was similar from the tight bud stage to the half-open stage, but its content increased at the fully open stage and then decreased during senescence. Research on the changes of endogenous hormones is of great significance for the regulation of plant flowering. It was suggested that IAA content was directly related to flowering [13], ABA affected flowering through the regulation of the vegetative growth of plants [9], and GA controlled various development processes, such as flower development [14,15]. IAA may be an inhibitor of flower bud formation because the photoperiod induction of leaves was interfered with by IAA [16]. GA is also considered to inhibit flower bud growth, although it could contribute to the flowering of *Juglans regia* [17]. SOD, POD, and catalase (CAT) belong to the protective enzyme system and are considered to be key enzymes in removing reactive oxygen species (ROS). To investigate the activity of antioxidant enzymes in *Dahlia pinnata* petals, Kan et al. [18] divided its flowering period into six stages. They concluded that an 'increase-decrease' trend of CAT and SOD activity appeared with the development of flowers, but POD activity was the opposite. Li et al. [19] used RNA-seq technology to examine the expression of flowering-related genes during *S. japonicus* flower development. The experimental data showed that 31,471 differentially expressed unigenes were generated, and these genes were associated with four pathways, including phytohormone signaling, transcription factor, protein kinase, and circadian rhythms.

Numerous researchers have carried out studies about the regulation of the flowering period. Severe nutrient stress delayed *Arabidopsis thaliana* from flowering; however, flowering was accelerated when plants were grown with alternating 'high-low' nutrient levels [3]. Insufficient nutrients and cell membrane damage were the main causes of petal senescence and a short flowering period, which was proved by Liu et al. [20] in the research of *Nelumbo nucifera* flowering regulation. On the basis of mastering the law of plant growth and development, some technical measures could be adopted to regulate the flowering time or prolong the flowering period [21]. Exogenous $GA_3$ application could replace low temperature to enable some plants that need to undergo vernalization to bloom at room temperature and promote the flowering of wild *A. thaliana* under short-day conditions [22]. It was found that the exogenous application of $KNO_3$ advanced the flowering time of *Mangifera indica* and enhanced its flowering rate [23].

*Styrax japonicus*, a low-branched landscape tree species with slight fragrance [24], is widely distributed in Korea, Japan, and southern China [25]. Although *S. japonicus* is an ornamental species with a huge number of flowers, the flowering period is relatively short, which limits its production and ornamental value to some extent. Hence, artificial methods to prolong the flowering period and improve flowering quality are in demand. According to our observation of phenology, a single flower bloomed for 4–5 days then withered, and the flowering period for a group of flowers was about two weeks. Until now, little information has been available regarding the morphological and physiological characteristics of *S. japonicus* petals during the flowering period.

Therefore, we conducted experiments to investigate the structure of petals at the full flowering stage and the contents of nutrients, mineral elements, hormones, and the activities of antioxidant enzymes in petals at different flowering stages. We attempted to answer the following questions: (1) Which physiological indexes have the greatest effects on *S. japonicus* flowering? (2) What is the relationship between these indexes and flowering? We hope our work may contribute to a better understanding of the flowering physiological mechanism of *S. japonicus* and consequently facilitate effective methods to regulate its flowering period.

## 2. Materials and Methods

### 2.1. Plant Material

Plant material was collected and processed at the Jiangsu Guoxing Biotechnology Co. Ltd., located in Luhe District, Nanjing, China (32°54′ N, 118°50′ E) in 2019. The average temperature, length of the day, and rainfall per year were 15.3 °C, 2200 h, and 970 mm, respectively. *S. japonicus* blooms in late April and quickly enters the blooming stage with white flowers (Figure 1). Nine-year-old *S. japonicus* trees originating from Yichang, Hubei Province, were used in our experiment. These plants grew in natural conditions, and no fertilizer was used. Ten trees with similar height, growth, and development conditions were selected and tagged. Hence, each tree was regarded as one biological repetition. *S. japonicus* bloomed with stamens and pistil exposed at first, and then stigma exerted with bright anthers. Stigma, anthers, and petals were dry when flowers began to fall. Its flowering period was divided into three stages, including initial flowering (5% of the flowers on a flowering tree blossomed), full flowering (50% of the flowers on a flowering tree blossomed), and end flowering (95% of the flowers on a flowering tree blossomed).

**Figure 1.** General look of a flowering tree.

### 2.2. Sample Collection

To avoid the influence caused by different water, temperature, and light conditions, we harvested all the samples between 9:00–9:30 a.m. on a single day [26]. Thirty flowers were collected from the middle part of each tree at three stages. Then samples were mixed and taken back to the laboratory. Ten flowers from the full flowering stage were observed and photographed by scanning electron microscope (SEM) observation. Thirty flowers were chosen to measure the length and width of petals, calyx, and pedicel using a vernier caliper. The rest of the flowers at three flowering stages were used to determine several physiological indexes (including soluble sugar, starch, soluble protein, N, P, K, IAA, $GA_3$, ABA, SOD, POD, and MDA). Each index measurement contained three technical repetitions, then the average of the three values was calculated.

### 2.3. Determination Methods

SEM: Fresh petals were fixed in 4% glutaraldehyde in 0.2 M phosphate buffer and rinsed with 0.1 M phosphate buffer solution. Then these petals were fixed in 1% osmium tetroxide and rinsed with 0.1 M phosphate buffer again. After being dehydrated in ethanol, replaced with anhydrous alcohol and isoamyl ester, dried in a Hitachi HCP-2 critical point dryer (Hitachi, Tokyo, Japan) and coated with a mixture of gold/palladium in a Hitachi E-1010 sputter coater (Hitachi, Tokyo, Japan), fixed petals were observed and photographed by an FEI Quanta-200 SEM (FEI Company, Hillsboro, OR, USA) [27].

Soluble sugar and starch: A 0.2 g of sample in each replicate was ground and diluted to 10 mL. After two times of 30-min extraction with boiling water, they were diluted to 25 mL. Then 0.2 mL of extracting solution, 1.8 mL of distilled water, 0.5 mL of anthrone ethyl acetate, and 5 mL of 98% $H_2SO_4$ were added, respectively. Calculation of soluble sugar content was accomplished according to the method described by Li [28] and Wu et al. [29], and the OD value (at 630 nm) was read by a Beckman DU 800 UV-visible spectrophotometer

(Beckman Coulter, Inc., Brea, CA, USA; the same hereafter). The residues from extracting soluble sugar were transferred into test tubes and diluted to 10 mL. The tubes were placed in boiling water for full extraction for 15 min, then 2 mL of 9.2 mol/L perchloric acid were added, then the residues were extracted for another 15 min. The measurement steps were the same as above [9].

Soluble protein: A 0.2 g sample was ground and diluted to 5 mL. After centrifugation at 8000 r/min for 15 min at 4 °C (Allegra X-22R, F1010 Rotor, Beckman Coulter, Inc., USA), 1 mL of extraction and 5 mL of Coomassie brilliant blue G-250 were added to measure soluble protein content according to Li [28] and Bradford [30]. The OD value at 560 nm wavelength was read by a spectrophotometer.

Hormones: A fresh sample (0.3 g) was extracted with 80% cold methanol at 4 °C for 4 h. After purifying, vacuuming, and drying, the filtering samples were solved with PBS. $GA_3$, IAA, and ABA contents were measured using enzyme-linked immunosorbent assay (ELISA) according to Koshita et al. [31] and Weiler et al. [32].

N, P, and K: Petals were dried at 65 °C for 72 h, and then 0.1 g of petals were transferred into a Kjeldahl flask. A total of 9 mL of $H_2SO_4$ and 1 mL of $HClO_4$ were added, boiled, and filtered into a glass bottle for determination. Indigo colorimetry was adopted to estimate N using a spectrophotometer at 625 nm, molybdenum-antimony colorimetry to estimate P using a spectrophotometer at 700 nm, and flame atomic absorption spectrophotometry (AA900T, PerkinElmer, Waltham, MA, USA) to estimate K at 776 nm according to National Standards of the People's Republic of China.

POD, SOD, and MDA: A total of 0.2 g of petals were ground in 5 mL of phosphate buffer (pH = 7.8) using a pre-chilled mortar and pestle. The homogeneous liquid was made to determine POD, SOD, and MDA. The activities of POD and SOD were estimated based on the methods described by Beauchamp and Fridovich [33]. Specifically, 2.9 mL of the reaction solution (containing 28 μL of guaiacol and 19 μL of hydrogen peroxide) and 0.1 mL of the supernatant were mixed. POD activity was estimated with a spectrophotometer at 470 nm.

The homogenate was transferred to a centrifuge tube and centrifuged at 10,000 rpm for 20 min. The reaction system, including 1.5 mL of 50 mmol/L phosphate buffer, 0.3 mL of 130 mmol/L methionine, 0.3 mL of 750 μmol/L nitroblue tetrazolium, 0.3 mL of 100 μmol/L ethylene diamine tetraacetic acid, 0.3 mL of 20 μmol/L riboflavin, and 50 μL of phosphate buffer or supernatant, was made to estimate SOD activity using a spectrophotometer at 560 nm.

The homogenate was centrifuged at 3000 rpm for 20 min, 2 mL of supernatant and 2 mL of 0.6% thiobarbituric acid were added to the tube and placed in boiling water for 20 min. The OD values at 450 nm, 532 nm, and 600 nm were read, and the content of MDA was estimated according to the methods described by Hodges et al. [34] and Ennajeh et al. [35].

### 2.4. Data Analysis

Statistics were processed by Excel (Office 2013 Pro Plus, Microsoft Corporation, Redmond, WA, USA), and values were expressed as mean ± standard deviation for three replicates. One-way analysis of variance of assessments in *S. japonicus* petals at different flowering stages was performed by SPSS 22.0 (IBM, Armonk, NY, USA). The homogeneity of variance test was conducted and verified. The differences in Duncan's multiple comparisons at the significance level of 0.05 were established. $p$ values less than 0.05 were considered to indicate significance within groups.

## 3. Results

### 3.1. Structure Observation of S. japonicus Flowers

*S. japonicus* flowers were comprised of 5–6 white petals and looked like pendulous bells. Flowers at the full flowering stage were used as an example to exhibit the structural characteristics. Unfolded petals had a length of 18.504 ± 0.302 mm and a width of

8.184 ± 1.621 mm. Many stellate hairs were present on both sides of the petals, and the hairs were bigger and in greater number on the upper surface of the petals (Figure 2a). Hairs on the lower surface of the petals were sparser and darker (Figure 2b). The pistil was covered by several stamens, which consisted of yellow-narrow anthers (Figure 2c). Pollens were dense and oval-shaped (Figure 2d). The length of the calyx and pedicel was 5.02 ± 0.211 mm and 24.251 ± 2.472 mm, respectively.

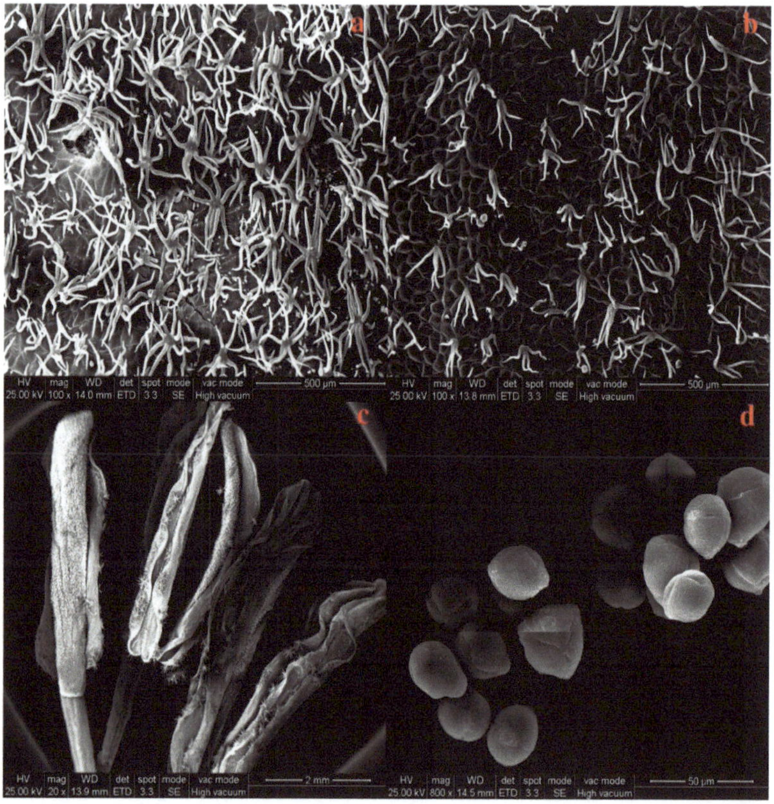

**Figure 2.** Structure of *S. japonicus* flowers at full flowering stage. (**a**) upper surface of petals; (**b**) lower surface of petals; (**c**) anthers; (**d**) pollens.

### *3.2. Comparison of Soluble Sugar, Starch, and Soluble Protein at Different Flowering Stages*

To analyze the contents of soluble sugar, starch, and soluble protein in *S. japonicus* petals at different flowering stages, we applied one-way analysis of variance and found an 'A' pattern in the change of soluble sugar content (Figure 3). The content at the full flowering stage (36.91 mg/g FW) was significantly higher than at the initial flowering and end flowering stages; however, the contents at these two stages were very close (30.77 mg/g FW and 29.54 mg/g FW, respectively). Starch content displayed a continuous decreasing trend from the initial flowering to end flowering stages (Figure 3). As shown in Figure 3, the content of soluble protein at the initial flowering stage was 2.33 mg/g FW and then decreased to 2.22 mg/g FW at the full flowering stage. Soluble protein content exhibited a 'decrease-increase' trend and reached the maximum at the end flowering stage (2.37 mg/g FW). No significant differences were found among the three stages.

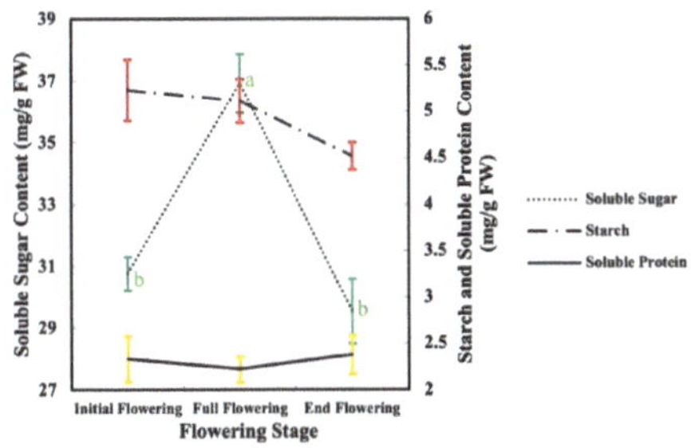


**Figure 3.** Changes of soluble sugar, starch, and soluble protein in *S. japonicus* petals during the flowering period. Data were shown as the average of three repetitions ± SD; different lowercase letters indicated significant differences at different stages.

### 3.3. Relationship between Mineral Elements and Flowering

The content of N at the initial flowering stage was significantly higher than at the full flowering and end flowering stages (Figure 4). With regard to P, the highest contents appeared at the initial flowering stage, and the lowest contents appeared at the full flowering stage, so the change trends of the content were 'decrease-increase' patterns (Figure 4). The contents of K at the initial flowering and full flowering stages were 0.30% DW and 0.31% DW, respectively, with no significant difference. Whereas the content at the end flowering stage was 0.39% DW, which was a significant difference from those at the other two stages (Figure 4).

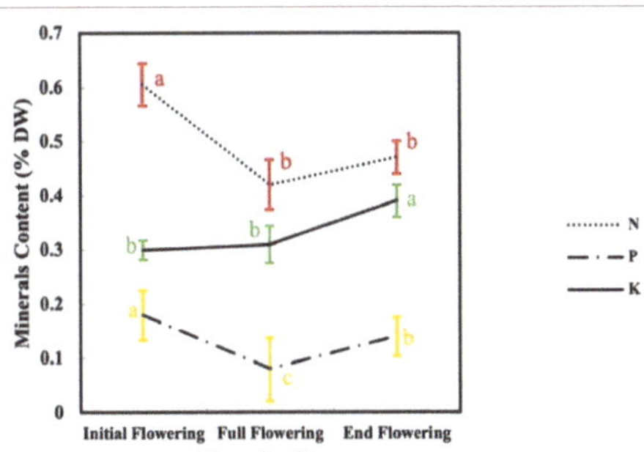


**Figure 4.** Changes of N, P, and K in *S. japonicus* petals during the flowering period. Data were shown as the average of three repetitions ± SD; different lowercase letters indicated significant differences at different stages.

### 3.4. Differences of the Contents of Hormones in S. japonicus Petals

$GA_3$ content increased continuously during the flowering period (Figure 5), and the difference between the full flowering and end flowering stages was significant (3.49 ng/g FW and 3.76 ng/g FW, respectively). The content of $GA_3$ increased slightly at the early stage and increased sharply at the late stage. The content of IAA increased continuously, and the difference among the three flowering stages was significant (Figure 5). The contents of ABA were more abundant at the initial flowering (54.50 ng/g FW) and end flowering (55.40 ng/g FW) stages than at the full flowering (51.90 ng/g FW) stage; however, no significant difference was observed among the three stages (Figure 5).

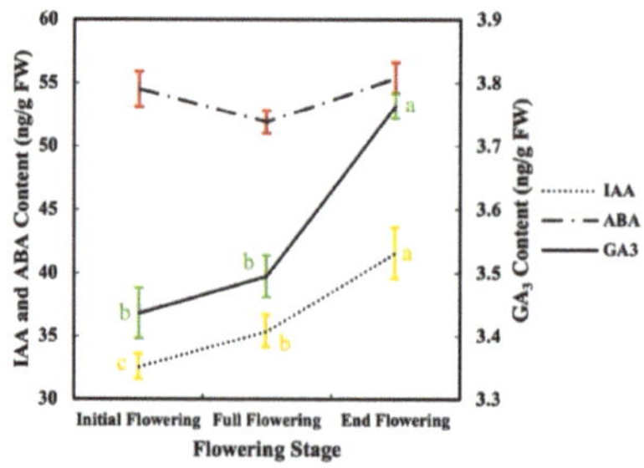


**Figure 5.** Changes of IAA, ABA, and $GA_3$ in *S. japonicus* petals during the flowering period. Data were shown as the average of three repetitions ± SD; different lowercase letters indicated significant differences at different stages.

### 3.5. Changes of Antioxidant Enzymes and MDA in S. japonicus Petals

The activity of POD ascended significantly from the initial flowering (472.17 U/g FW) to the full flowering (930.99 U/g FW) stage and thereafter descended significantly at the end flowering (666.16 U/g FW) stage (Figure 6a). The activity of POD at the full flowering stage was 97.2% higher than at the initial flowering stage. Figure 6a shows an 'increase-decrease' trend of the activity of SOD. Although there were no significant differences between the initial flowering and full flowering stages (23.43 U/g FW and 29.08 U/g FW, respectively), SOD activity at the end flowering (11.52 U/g FW) stage was significantly lower than those at the former two stages. MDA contents at the full flowering (2.61 μmol/mg FW) and end flowering (2.86 μmol/mg FW) stages were very close with no significant differences; however, they had significant differences with MDA content at the initial flowering (1.81 μmol/mg FW) stage. The changing pattern of the content of MDA rose continuously (Figure 6b).

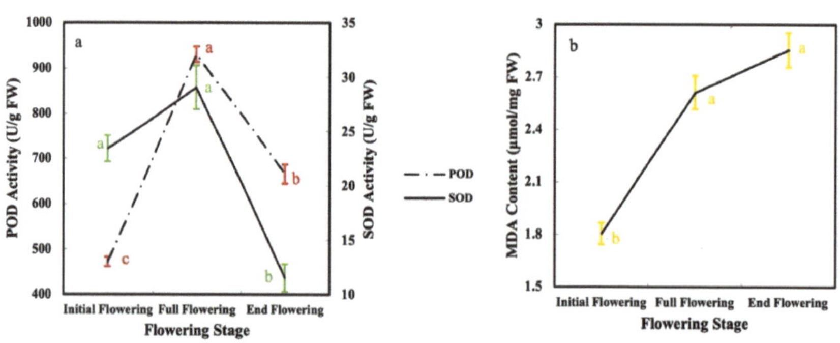


**Figure 6.** Changes of (**a**) POD, SOD, and (**b**) MDA in *S. japonicus* petals during the flowering period. Data were shown as the average of three repetitions ± SD; different lowercase letters indicated significant differences at different stages.

## 4. Discussion

### 4.1. Abundant Nutrients Promoted S. japonicus Flowering

Flowers are important reproductive organs of angiosperms, and the formation process of flowers consumes a large amount of energy. The dynamic changes of nutrients, for instance, soluble sugar, starch, and soluble protein, could reflect the growth and development of plants. Soluble sugar is a carrier of plant energy, as well as the main form of the transportation of carbohydrates, which could be directly utilized by plants, and its content change connects to the quality of flowers [36,37]. The present study indicated that soluble sugar content in petals at the initial flowering stage was 30.77 mg/g FW and then reached

36.91 mg/g FW at the full flowering stage; however, content declined to 29.54 mg/g FW at the end flowering stage. The possible reason behind this was the transition of the source-sink relationship caused by the energy demand during flowering. At the initial flowering stage, the flower organ was an important sink, which contained sufficient soluble sugar. The growth rate of *S. japonicus* flowers slowed down at the full flowering stage, resulting in the accumulation of soluble sugar [38]. Towards the end flowering stage, a decrease in soluble sugar content was presented because the flower as a sink was reduced. It was reported that soluble sugar was the main energy for flowers to sustain lifespan, and the decrease in soluble sugar in the late stage caused flower resistance to decline and accelerated the aging process [39]. Therefore, it was possible to enhance soluble sugar content so that the *S. japonicus* flowering period might be extended. Girdling would be an appropriate approach to accumulate soluble sugar content, thus facilitating bud induction and flowering [40].

Starch is a carbohydrate that has been reported as a resource for respiration and metabolic intermediates and structural components [41]. Numerous evidence confirmed that starch had great effects on plant flowering [42–44]. Wang et al. [44] explored starch mobilization in Oncidium 'Gower Ramsey' during the flowering process, and they indicated that little relationship was observed between the transition mechanism from vegetative to reproductive growth and the onset of starch mobilization. Obviously, their experimental result was not consistent with our common understanding that starch mobilization was helpful for floral induction [45]. Starch accumulation emerged as an 'S'-curve during sorghum flowering [46], whereas in the current experiment, starch content was much lower and decreased during the flowering period. We assumed that the starch in *S. japonicus* petals might be converted into soluble sugar to meet the need for metabolic energy during the flowering process.

Yu et al. [47] reported that soluble protein was one of the factors that impacted plant flowering, and our research proved their viewpoint. At the early stage, soluble protein was maintained at a relatively high level. Due to the flower's development, *S. japonicus* consumed many nutrients; thus, soluble protein content declined, then the petals absorbed and stored soluble protein from vegetative organs, leading to an increase in content.

### 4.2. Mineral Elements Affected *S. japonicus* Flowering Variously

In the present study, the average contents of N, P, and K were 0.5% DW, 0.13% DW, and 0.33% DW, respectively, and the change trends of these elements during the flowering stages were different. Yang et al. [48] analyzed the content of several minerals in *Rosa hybrida* petals and found that K was the most abundant mineral, followed by P, which was consistent with our results. Li et al. [49] discussed the link between the amount of minerals and the color of the petals in *P. lactiflora* and stated that red or purple petals had a higher amount of minerals than those that were white. K was the highest element in *P. lactiflora* petals (1.26%), which was much higher than in *S. japonicus* petals. In light of the above evidence, we might adopt some artificial measures to enhance K content in petals in order to attempt to change flower color. Fertilization was a common mineral supplement that directly affected plant growth. A profound understanding of the appropriate time and type of fertilization could effectively alleviate the increase of membrane permeability of petals, improve the activity of SOD, and significantly increase the longevity of flowers [50]. Low phosphorus content in *Matricaria chamomilla* petals promoted the synthesis of anthocyanin, which played a crucial role in free radical scavenging [51]. Similarly, P content in our study was lower when compared to N and K, so it could be considered to apply P fertilization to investigate its effect on the flowering of *S. japonicus*.

### 4.3. Hormones Closely Correlated with *S. japonicus* Flowering Period

Flowering is influenced by a complex network of genes that integrate multiple environmental factors and endogenous signals, thus ensuring that flowering occurs at the right time. Hormone regulation is vital in this process, and its signal transduction pathways are involved in the formation of flowers [52,53]. A study conducted by Halevy and Mayak [54] revealed that ABA content correlated with flower longevity negatively. Muller et al. [55] reached the same conclusion. Although no significant difference existed in ABA content at the three flowering stages, the average ABA content (53.93 ng/g FW) was much higher when compared to the other two hormones. Hence, we assumed that high ABA content could shorten the flowering period in *S. japonicus*, and lowering ABA content was favored for flowering regulation [56].

GAs are crucial plant growth regulators that play an important role in flowering pathways and affect floral transition [14,57]. Chrispeels and Varner [58] expounded that the action of GAs and ABA was antagonistic in most physiological processes. As we mentioned above, high ABA content probably shortened the *S. japonicus* flowering period, so we speculated that more $GA_3$ was used in other physiological activities or transferred to other organs, leading to the low content of $GA_3$ when compared to ABA. With the development of *S. tonkinensis*, $GA_3$ content in petals displayed an increasing trend [39], which was identical to our experimental result. A high concentration of $GA_3$, especially 25 mg/kg, obviously postponed *M. indica* flowering, but the fruiting quality was improved [59]. Therefore, the impacts of $GA_3$ on flowering depended on the plant species and varieties [60].

Keeping IAA at a suitable level was beneficial for plant growth and development [61]. Additionally, high IAA content correlated with intense cell division [62]. Distribution or gradient of auxin within plant tissues was different because of the polar movement of IAA, and these changes were dynamic during the plant development process, such as flowering [62,63]. Xu et al. [39] determined IAA content in petals during *S. tonkinensis* flowering, and the result showed that IAA content first increased and then decreased; however, IAA content increased continuously in *S. japonicus* petals. It was concluded that IAA acted dissimilarly on flower development in different tree species, even in the same genus. Compared to CK, IAA extended the *Brassica napus* flowering period for five days, and Zhu et al. [64] attributed this phenomenon to the capacity of IAA in increasing antioxidants. We wondered whether exogenous IAA treatment could regulate the *S. japonicus* flowering time, and if yes, reasonable IAA concentration required further research.

### *4.4. Antioxidant Enzyme Activity and MDA Content Significantly Impacted S. japonicus Flowering*

MDA is produced by membrane lipid peroxidation of unsaturated fatty acids in membrane lipids and could be inactivated by binding with proteins and enzymes. Normally, MDA is recognized to evaluate the degree of membrane damage [34]. The content of MDA is extremely low in growing plants, but it increases during flowering. Generally, MDA content equals the degree of petal senescence. As described in our study, MDA content increased gradually, indicating that the degradation of cell inclusions was intensified. The function of the enzymatic defense system decreased, and the degree of membrane lipid peroxidation was enhanced, revealing that flowering is an aging process.

POD is a biochemical marker, which could characterize the induction phase of the flowering process [65,66]. In addition, POD is involved in the production of ethylene and the oxygenolysis of IAA. McCord and Fridovich [67] found that SOD was capable of removing various free radicals and ROS produced by aerobionts. Both POD and SOD can convert superoxide anions or hydrogen peroxide into less active substances, reducing or eliminating their attacks on membrane lipids, so those membrane lipids are protected. The activities of POD and SOD in *S. japonicus* petals had the same trend of rising first and then declining. We supposed that *S. japonicus* could eliminate excessive ROS by enhancing the activities of POD and SOD to maintain the balance of metabolism at the early flowering stage; nonetheless, due to the reduction of protein degradation or synthesis, the activities of these two enzymes and scavenging capacity of ROS decreased at the late flowering stage. Our result was in agreement with Bartoli et al. [68] that the activities of SOD and POD increased first and then decreased. To conclude, *S. japonicus* inevitably produced MDA and other aging-promoting substances during the flowering period, but at the same time, SOD,

POD, and other anti-aging substances were generated, illustrating that these substances were antagonistic.

## 5. Conclusions

The current study elaborated the change trends of nutrients, minerals, hormones, and antioxidant enzymes and established the relationship between them and *S. japonicus* flowering. These physiological indexes were closely related to plant flowering, and they could be regulated by using appropriate flowering control techniques to achieve early, delayed, or even prolonged plant flowering periods. In general, light, temperature, and hormone regulation could be manipulated to affect the plant's flowering period. In recent years, with the deepening of flowering mechanism study, genetic engineering technology has also been applied to regulate plant flowering. Combined with high-throughput techniques, such as genomics, transcriptomics, and proteomics, genes associated with flowering were isolated, and the mechanism of plant flowering was illuminated, thereby realizing accurate regulation of the flowering period and improvement of flowering quality.

**Author Contributions:** Conceptualization, C.C. and F.Y.; methodology, C.C.; software, C.C.; formal analysis, C.C. and F.Y.; investigation, C.C., H.C. and M.N.; writing—original draft preparation, C.C.; writing—review and editing, F.Y.; funding acquisition, F.Y. All authors have read and agreed to the published version of the manuscript.

**Funding:** This research was funded by the National Natural Science Foundation of China (3197140894), A Project Funded by the Priority Academic Program Development of Jiangsu Higher Education Institutions (PAPD), and Postgraduate Research & Practice Innovation Program of Jiangsu Province (SJKY19_0882).

**Data Availability Statement:** Not applicable.

**Conflicts of Interest:** The authors declare no conflict of interest.

## References

1. Wu, P.P.; Wu, C.B.; Zhou, B.Y. Drought stress induces flowering and enhances carbohydrate accumulation in *Averrhoa carambola*. *Hortic. Plant J.* **2017**, *3*, 60–66. [CrossRef]
2. Hossain, Z.; Mandal, A.K.A.; Datta, S.K.; Biswas, A.K. Decline in ascorbate peroxidase activity—A prerequisite factor for tepal senescence in gladiolus. *J. Plant Physiol.* **2006**, *163*, 186–194. [CrossRef] [PubMed]
3. Kolář, J.; Seňkova, J. Reduction of mineral nutrient availability accelerates flowering of *Arabidopsis thaliana*. *J. Plant Physiol.* **2007**, *165*, 1601–1609. [CrossRef] [PubMed]
4. Li, X.M.; Fan, T.; Zou, P.; Zhang, W.H.; Wu, X.J.; Zhang, Y.X.; Liao, J.P. Can the anatomy of abnormal flowers elucidate relationships of the androecial members in the ginger (*Zingiberaceae*)? *EvoDevo* **2020**, *11*, 12. [CrossRef]
5. Luo, H.H.; Wang, Q.; Zhang, J.K.; Wang, L.S.; Li, Y.B.; Yang, G.Z. Minimum fertilization at the appearance of the first flower benefits cotton nutrient utilization of nitrogen, phosphorus and potassium. *Sci. Rep.* **2020**, *10*, 6815. [CrossRef]
6. Neuffer, B.; Paetsch, M. Flower morphology and pollen germination in the genus *Capsella* (Brassicaceae). *Flora* **2013**, *208*, 626–640. [CrossRef]
7. Watson, M.A.; Brochier, J. The role of nutrient levels on flowering in water hyacinth. *Aquat. Bot.* **1988**, *31*, 367–372. [CrossRef]
8. Richards, J. Development potential of axillary buds of water hyacinth, *Eichhornia crassipes* Solme (Pontederiaceae). *Am. J. Bot.* **1982**, *69*, 615–622. [CrossRef]
9. Chen, C.; Cao, Y.Y.; Wang, X.J.; Wu, Q.K.; Yu, F.Y. Do stored reserves and endogenous hormones in overwintering twigs determine flower bud differentiation of summer blooming plant—*Styrax tonkinensis*. *Int. J. Agric. Biol.* **2019**, *22*, 815–820.
10. Stutte, G.W.; Martin, G.C. Effect of light intensity and carbohydrate reserves on flowering in olive. *J. Am. Soc. Hortic. Sci.* **1986**, *111*, 27–31.
11. Zhang, H.; Wei, Q.; Zhang, H. Contents of microscopic structure, nucleic acid and soluble protein during flower bud differentiation in *Crocus sativus*. *J. Northeast For. Univ.* **2017**, *45*, 33–36.
12. Shahri, W.; Tahir, I.; Islam, S.T. Physiological and biochemical changes associated with flower development and senescence in so far unexplored *Helleborus orientalis* Lam. cv. Olympicus. *Physiol. Mol. Biol. Plant* **2011**, *17*, 33–39. [CrossRef]
13. Kumar, P.; Mishra, Y. Biochemical changes associated with flowering in *Bambusa arundinacea* and *Bambusa nutans*. *J. For. Res.* **2018**, *5*, 1315–1320. [CrossRef]
14. MacMillan, J. Occurrence of gibberellins in vascular plants, fungi, and bacteria. *J. Plant. Growth Regul.* **2001**, *20*, 387–442. [CrossRef]
15. Yamaguchi, S. Gibberellin metabolism and its regulation. *Annu. Rev. Plant Biol.* **2008**, *59*, 225–251. [CrossRef]

16. Bernier, G. The control of floral evocation and morphogenesis. *Annu. Rev. Plant Biol.* **1988**, *39*, 175–219. [CrossRef]
17. Hassankhah, A.; Rahemi, M.; Mozafari, M.R.; Vahdati, K. Flower development in walnut: Altering the flowering pattern by gibberellic acid application. *Not. Bot. Horti Agrobot. Cluj-Napoca* **2018**, *46*, 700–706. [CrossRef]
18. Kan, X.Q.; Sun, Y.; Su, M.; Zhou, L.L.; Xu, W.F.; Tan, J.Z. Analysis of enzyme activity of protective system in *Dahlia pinnata* cav petals during senescence. *North. Hortic.* **2010**, *19*, 124–127.
19. Li, W.; Xu, Z.Z.; Zhang, C.P.; Jiang, W.Q.; Wang, K.L. Transcriptomic identification of floral transition and development-associated genes in *Styrax japonicus*. *Forests* **2020**, *11*, 10. [CrossRef]
20. Liu, Y.P.; Huang, Z.Y.; Liang, L.; Meng, Y.N.; Kong, D.Z. The influence of auxin IAA on the flowering regulation of lotus. *J. Henan Agric. Sci.* **2019**, *48*, 141–145.
21. Franklin, K.A. Light and temperature signal crosstalk in plant development. *Curr. Opin. Plant Biol.* **2009**, *12*, 63–68. [CrossRef]
22. Blazquez, M.A.; Green, R.; Nilsson, O.; Sussman, M.R.; Weigel, D. Gibberellins promote flowering of *Arabidopsis* by activating the *LEAFY* promoter. *Plant Cell* **1998**, *10*, 791–800. [CrossRef]
23. Patil, K.R.; Burondkar, M.M.; Bhave, S.G. Post harvest chemical induction of vegetative growth and its physiological behavior in telation to regulation of flowering in 'alphonso' mango (*Mangifera indica* L.). *Acta Hortic.* **2013**, *992*, 193–200. [CrossRef]
24. Chen, C.; Chen, H.; Ni, M.; Yu, F.Y. Methyl jasmonate application and flowering stage affect scent emission of *Styrax japonicus*. *Flavour Fragr. J.* **2021**, *36*, 497–504. [CrossRef]
25. Li, W.; Zhang, C.P.; Jiang, X.Q.; Liu, Q.; Wang, K. De novo transcriptomic analysis and development of EST-SSRs for *Styrax japonicus*. *Forests* **2018**, *9*, 748. [CrossRef]
26. Yang, C.; Wei, H. Designing microarray and RNA-seq experiments for greater systems biology discovery in modern plant genomics. *Mol. Plant* **2015**, *8*, 196–206. [CrossRef]
27. Xu, L.P.; Yu, F.Y. Corolla structure and fragrance components in *Styrax tonkinensis*. *Trees* **2015**, *29*, 1127–1134. [CrossRef]
28. Li, H.S. *The Principles and Technologies for Plant Physiology and Biochemistry Experiments*; High Education Press: Beijing, China, 2006.
29. Wu, Q.K.; Cao, Y.Y.; Zhao, X.; Zhang, Z.H.; Yu, F.Y.; Guy, R.D. A comparative study of seed reserve accumulation in five *Styrax* species with potential for biofuel production. *Trees* **2020**, *34*, 891–902. [CrossRef]
30. Bradford, M.M. A rapid and sensitive method for the quantification of microgram quantities of protein utilizing the principle of protein dye binding. *Anal. Biochem.* **1976**, *72*, 248–254. [CrossRef]
31. Koshita, Y.T.; Ogata, T.T.; Goto, A. Involvement of endogenous plant hormones (IAA, ABA, $GA_S$) in leaves and flower bud formation of *Satsuma mandarin* (*Citrus unshiu* M.). *Sci. Hortic.* **1999**, *79*, 185–194. [CrossRef]
32. Weiler, E.W.; Jourdan, P.S.; Conrad, W. Levels of indole-3-acetic acid in intact and decapitated coleoptiles as determined by a specific and highly sensitive solid-phase enzyme immunoassay. *Planta* **1981**, *153*, 561–571. [CrossRef] [PubMed]
33. Beauchamp, C.; Fridovich, I. Superoxide dismutase: Improved assays and an assay applicable to acrylamide gels. *Anal. Biochem.* **1971**, *44*, 276–287. [CrossRef]
34. Hodges, D.M.; Delong, J.M.; Forney, C.F.; Prange, R.K. Improving the thiobarbituric acid-reactive-substances assay for estimating lipid peroxidation in plant tissues containing anthocyanin and other interfering compounds. *Planta* **1999**, *207*, 604–611. [CrossRef]
35. Ennajeh, M.; Vadel, A.M.; Khemira, H. Osmoregulation and osmoprotection in the leaf cells of two olive cultivars subjected to severe water deficit. *Acta Physiol. Plant.* **2009**, *31*, 711–721. [CrossRef]
36. Ma, Y.Y.; Zhang, Y.L.; Lu, J. Roles of plant soluble sugars and their response to plant cold stress. *Afr. J. Biotechnol.* **2009**, *8*, 2004–2010.
37. Boriboonkaset, T.; Theerawitaya, C.; Yamada, N.; Pichakum, A.; Supaibulwatana, K.; Cha-um, S.; Takabe, T.; Kirdmanee, C. Regulation of some carbohydrate metabolism-related genes, starch and soluble sugar contents, photosynthetic activities and yield attributes of two contrasting rice genotypes subjected to salt stress. *Protoplasma* **2013**, *250*, 1157–1167. [CrossRef]
38. Liu, Y.M.; Pan, S.Y.; Zhou, T.H.; Tian, F.Z. Changes of nutrients and endogenous hormones during flowering of Paeonia. *J. Heze Univ.* **2020**, *42*, 95–99.
39. Xu, L.P.; Liu, J.B.; Yu, F.Y. Physiological and biochemical changes in petals before and after pollination in *Styrax tonkinensis*. *Acta Agric. Univ. Jiangxiensis* **2019**, *41*, 464–475.
40. Samacha, A.; Smith, H.M. Constraints to obtaining consistent annual yields in perennials. II: Environment and fruit load affect induction of flowering. *Plant Sci.* **2013**, *207*, 168–176. [CrossRef]
41. Sheen, J.; Zhou, L.; Jang, J.C. Sugars as signaling molecules. *Curr. Opin. Plant Biol.* **1999**, *2*, 410–418. [CrossRef]
42. Menzel, C.M.; Rasmussen, T.S.; Simpson, D.R. Carbohydrate reserves in lychee trees (*Litchi chinensis* Sonn.). *J. Hortic. Sci.* **1995**, *70*, 245–255. [CrossRef]
43. Eshghi, S.; Tafazoli, E.; Dokhani, S.; Rahemi, M.; Emam, Y. Changes in carbohydrate contents in shoot tips, leaves and roots of strawberry (*Fragaria* × *ananassa* Duch.) during flower-bud differentiation. *Sci. Hortic.* **2007**, *113*, 255–260. [CrossRef]
44. Wang, C.Y.; Chiou, C.Y.; Wang, H.L.; Krishnamurthy, R.; Venkatagiri, S.; Tan, J.; Yeh, K.W. Carbohydrate mobilization and gene regulatory profile in the pseudobulb of *Oncidium* orchid during the flowering process. *Planta* **2008**, *227*, 1063–1077. [CrossRef]
45. Corbesier, L.; Lejeune, P.; Bernier, G. The role of carbohydrate in the induction of flowering in *Arabidopsis thaliana*: Comparison between the wild type and a starchless mutant. *Planta* **1998**, *206*, 131–137. [CrossRef]
46. Yi, B.; Zhou, Y.F.; Gao, M.Y.; Zhang, Z.; Han, Y.; Yang, G.D.; Xu, W.J.; Huang, R.D. Effect of drought stress during flowering stage on starch accumulation and starch synthesis enzymes in sorghum grains. *J. Integr. Agric.* **2014**, *13*, 2399–2406. [CrossRef]

47. Yu, P.T.; Pu, D.H.; Zhou, W.M. Analysis of soluble protein during flowering of *Petunia hybrida*. *J. Plant Physiol. Mol. Biol.* **2004**, *30*, 179–182.
48. Yang, M.; Cho, E.; Ha, J. Chemical composition of rose petals (*Rosa hybrida* L.) as a food material. *J. Korean Soc. Food Sci. Nutr.* **2002**, *31*, 539–542.
49. Li, C.Z.; Sun, Y.; Zhao, D.Q.; Tao, J.; Feng, L. Relationship between major mineral nutrient elements contents and flower colors of *Herbaceous peony* (*Paeonia lactiflora*). *Adv. J. Food Sci. Technol.* **2015**, *7*, 374–382. [CrossRef]
50. Wang, Y.X.; Wang, H.Y.; Zhou, Q.; Liu, L.L.; Yan, H.; Sun, J.J.; Ni, J.Y. Research of nitrogen nutrient on the preservation of carnation cut flowers. *Guangdong Agric. Sci.* **2009**, *8*, 81–84.
51. Kováčik, J.; Klejdus, B.; Bačkor, M.; Repčák, M. Phenylalanine ammonia-lyase activity and phenolic compounds accumulation in nitrogen-deficient *Matricaria chamomilla* leaf rosettes. *Plant Sci.* **2007**, *172*, 393–399. [CrossRef]
52. Xiao, F.; Wang, X.; Zhao, Y.; He, H. Flowering related comparative transcriptomics between Jatropha curcas and *Jatropha nigroviensrugosus*. *Intl. J. Agric. Biol.* **2018**, *20*, 1523–1532.
53. Campos-Rivero, G.; Osorio-Montalvo, P.; Sánchez-Borges, R.; Us-Camas, R.; Duarte-Aké, F.; De-la-Peña, C. Plant hormone signaling in flowering: An epigenetic point of view. *J. Plant Physiol.* **2017**, *214*, 16–27. [CrossRef]
54. Halevy, A.H.; Mayak, S. Interrelationship of several phytohormones in the regulation of rose petal senescence. *Acta Hortic.* **1975**, *41*, 103–116. [CrossRef]
55. Muller, R.; Andersen, A.S.; Serek, M. Differences in display life of miniature potted roses. *Sci. Hortic.* **1999**, *76*, 59–61. [CrossRef]
56. Saxena, S.N.; Kaushik, N.; Sharma, R. Effect of abscisic acid and proline on in vitro flowering in *Vigna aconitifolia*. *Biol. Plant* **2008**, *52*, 181–183. [CrossRef]
57. Teotia, S.; Tang, G. To bloom or not to bloom: Role of microRNAs in plant flowering. *Mol. Plant* **2015**, *8*, 359–377. [CrossRef]
58. Chrispeels, M.J.; Varner, J.E. Inhibition of gibberellic acid induced formation of α-amylase by abscisin II. *Nature* **1966**, *212*, 1066–1067. [CrossRef]
59. Li, G.L.; Pan, H.B.; Du, B.; Liu, B.; Zhou, W.J. Regulating effect of gibberellin on flowering characteristics of *Mangifera indica* cv. Keitt. *Hubei Agric. Sci.* **2011**, *50*, 2893–2896.
60. Chen, C.; Wang, X.J.; Cao, Y.Y.; Wu, Q.K.; Yu, F.Y. Time course morphological and histological changes in differentiating floral buds of *Styrax tonkinensis*. *Int. J. Agric. Biol.* **2019**, *22*, 1491–1496.
61. Davies, P.J. Regulatory factors in hormone action: Level, location and signal transduction. In *Plant Hormones: Biosynthesis, Signal Transduction, Action*; Davies, P.J., Ed.; Springer: Amsterdam, The Netherlands, 2010; pp. 16–35.
62. Tanaka, H.; Dhonukshe, P.; Brewer, P.B.; Friml, J. Spatiotemporal asymmetric auxin distribution: A means to coordinate plant development. *Cell. Mol. Life Sci.* **2006**, *63*, 2738–2754. [CrossRef]
63. Cheng, Y.; Zhao, Y. A role for auxin in flower development. *J. Integr. Plant Biol.* **2007**, *49*, 99–104. [CrossRef]
64. Zhu, L.R.; Liu, B.L.; Xiao, M.L.; Zhou, Q.H.; Fu, D.H. A study on chemical regulation of flowering in *Brassica napus*. *Acta Agric. Univ. Jiangxiensis* **2017**, *39*, 1057–1066.
65. Gasper, T.; Penel, C.; Castillo, F.J.; Greppin, H. A two-step control of basic and acidic peroxidases and its significance for growth and development. *Physiol. Plant* **1985**, *64*, 418–423. [CrossRef]
66. Bouazza, A.; Rambour, S.; Gaspar, T.; Legrand, B. Peroxidases during the course of callusing and organ differentiation from root explants of *Cichorium intybus*. *Biol. Plant* **1993**, *35*, 481–489. [CrossRef]
67. McCord, J.M.; Fridovich, I. Superoxide dismutase: Anenzumic function for erythrocuprein (hemocuprein). *Biol. Chem.* **1969**, *244*, 6049–6055. [CrossRef]
68. Bartoli, C.G.; Simontacchi, M.; Guiamet, J.J.; Montaldi, E.; Puntarulo, S. Antioxidant enzymes and lipid peroxidation during aging of *Chrysanthemum morifolium* RAM petals. *Plant Sci.* **1995**, *104*, 161–168. [CrossRef]

agronomy

MDPI

Article

# Morpho-Physiological and Transcriptome Analyses Provide Insights into the Wizened Bud Formation in Pear Trees

**Wei Du [1], Chunmei Shi [2], Syed Bilal Hussain [3], Mingqiu Li [1,2], Jing Fan [1], Qiliang Chen [1], Jingguo Zhang [1], Yongzhong Liu [2], Xiaoping Yang [1,*] and Hongju Hu [1,4,*]**

1 Research Institute of Fruit and Tea, Hubei Academy of Agricultural Science, Wuhan 430064, China; dw@hbaas.com (W.D.); lmqlushichuanshuo@163.com (M.L.); fanjing2013pear@163.com (J.F.); cql99@sina.com (Q.C.); zhjg@webmail.hzau.edu.cn (J.Z.)
2 Key Laboratory of Horticultural Plant Biology, Huazhong Agricultural University, Wuhan 430070, China; shichunmei@mail.hzau.edu.cn (C.S.); liuyongzhong@mail.hzau.edu.cn (Y.L.)
3 Citrus Research and Education Center, University of Florida, 700 Experiment Station Rd., Lake Alfred, FL 33850, USA; syedbilalhussain@ufl.edu
4 Hubei Hongshan Laboratory, Wuhan 430070, China
* Correspondence: yangxiaoping1981@163.com (X.Y.); hongjuhu@sina.com (H.H.)

**Abstract:** Wizened buds are frequently observed in pear (*Pyrus* spp.) trees, which greatly reduces the yield. However, little is known about the mechanism of wizened bud formation. Here, we analyzed physiological and transcriptomic differences between normal buds and wizened buds of '710' pear trees. The results indicated that the sorbitol and boron (B) contents, during bud differentiation, were significantly reduced in wizened buds. The microscopic observation and transcriptome analysis revealed that the collapse of the organ structure and cell wall loosening process may have a close relation with wizened bud formation. Moreover, reduced transcript levels of *PpyMYB39.1* and its downstream genes (*PpyHT1*, *PpyHT2*, *PpyPMEI1* and *PpyPMEI2*) were found in wizened buds. However, the transcript levels of pentose and glucuronate interconversion pathway genes (*PpyPME3*, *PpyPL18.1*, *PpyPL18.2*, *PpyPG1* and *PpyPG2*) and the concentration of pectin-degradation-related enzymes were increased in wizened buds. Correspondingly, the pectin concentration was significantly reduced in wizened buds. Taken together, *PpyMYB39.1* may promote pectin degradation and decrease carbohydrate transport by regulating its downstream genes and is supposed to play a vital role in the wizened bud formation resulting from the cell wall loosening process. Our study provides fundamental insights into wizened bud formation and strategies to reduce the wizened bud occurrence in pear trees.

**Keywords:** pear; wizened bud; cell wall loosening; pectin degradation; carbohydrate transport

**Citation:** Du, W.; Shi, C.; Hussain, S.B.; Li, M.; Fan, J.; Chen, Q.; Zhang, J.; Liu, Y.; Yang, X.; Hu, H. Morpho-Physiological and Transcriptome Analyses Provide Insights into the Wizened Bud Formation in Pear Trees. *Agronomy* **2022**, *12*, 484. https://doi.org/10.3390/agronomy12020484

Academic Editor: Sara Serra

Received: 13 January 2022
Accepted: 12 February 2022
Published: 15 February 2022



## 1. Introduction

Pear (*Pyrus* spp.) is one of the major temperate fruit crops [1], which has grown widely with the global production of over 23.1 million tons in the year 2020 [2]. In China, pear is an economical fruit crop, and its yield is affected by developmental processes, especially bud germination [3,4]. However, the bud differentiation is frequently hindered in many traditional pear species, such as 'Yuluxiang' (*P. bretschneideri.*), showing the occurrence of wizened bud, and results in a great loss of pear production [5,6]. Moreover, previous studies demonstrated that wizened bud formation is closely related to management practices and the nutritional status of buds [6,7].

The occurrence of wizened buds has been reported in many Rosaceae fruit trees, including pear [5], apple [8] and peach [9]. Typically, wizened buds have external scales loosening, internal bud organs browning and bud bursting and the presence of wizened buds at the later stage of bud differentiation [5]. Hence, the wizened bud formation is due to the impediment of bud differentiation. Mineral nutrients and carbohydrate utilization in buds are essential for the differentiation process [10,11]. For example, boron (B) is involved in the cell wall development of bud meristem, and sugar alcohols act as the carbon source for the cell division of bud organs during bud differentiation [12–14]. Sugar alcohols serve as major photo-assimilates and can be distributed to various tissues [15,16]. In pear, sorbitol is the dominant photosynthetic product that is synthesized in leaves [17]. Once the sorbitol is transported from the leaves to sink tissues, it is primarily converted into fructose by sorbitol dehydrogenase (SDH) for plant utilization [18,19]. Therefore, sorbitol, in addition to the energy source, plays an important role in reproductive development by regulating the expression of SDH genes [20,21].

Several genes are involved in the bud differentiation process, among which transcription factors play regulatory roles in the interaction with downstream target genes [22,23]. The MYB subfamily, which consists of MYB genes with conserved DNA-binding domains, play a vital role in bud development. For example, *CmMYB2* determines the bud differentiation time in chrysanthemum by regulating the gibberellin metabolism [24]. From Japanese morning glory (*Ipomoea nil*), *InMYB1* is active in the floral bud developmental stage and functions from young flower buds to bloomed flowers [25]. The triple knockdown mutant of Arabidopsis (*AtMYB37*, *AtMYB38* and *AtMYB84*) displayed a defect in axillary bud formation, showing the importance of MYBs in meristem bud formation [26]. DhMYB1 is involved in the morphogenesis of epidermal cells in *Dendrobium hybrida* and is therefore supposed to be required for floral bud development [27]. In addition to MYB transcription factors, the pectin methylesterase (PME) and pectin methylesterase inhibitor (PMEI) encoded genes have also been reported to play key roles in the bud growth of different species [28–30]. However, little is known about the molecular mechanisms of wizened bud formation in pears.

Here, we chose '710' sand pear plants, with typical wizened occurrence in the bud, as research material. Microscopic observations and metabolite quantification analyses were conducted to determine the physiological changes during the wizened bud formation. Moreover, RNA-seq analyses were used to conduct the transcriptomic profiling of normal buds and wizened buds. The aim of this study was to investigate the possible mechanisms of wizened bud formation and provide new insights to overcome wizened bud occurrence in pear.

## 2. Materials and Methods

### 2.1. Plant Materials and Sampling

The experiment was conducted from May 2021 to November 2021 in the Wuchang Sand Pear Germplasm Collection, Wuhan, China. For this experiment, 7-year-old '710' sand pear (*Pyrus Pyrifolia*, 'An Nong No. 1 × Cui Guan') grafted on Callery pear (*P. calleryana*) plants were selected. There were a total of 30 representative trees with 10 individual trees as a replication. The samples of wizened buds, normal buds and leaves were collected from the 1-year-old stem of selected trees in September 2020. Each bud sample consisted of more than 20 buds and the leaf sample consisted of more than 10 leaves, which were randomly collected from stems. Rhizosphere soil samples were collected from a depth of 30 cm around the selected trees.

### 2.2. Phloem Sap Extraction

Phloem sap was collected as described by Du et al. [31]. Briefly, wizened bud stems were washed with ultra-pure water twice. The bark was peeled off, and the xylem residues were removed by scalpel. Subsequently, the clean epidermis was cut into small pieces of 1 $cm^2$ and transferred into a 0.5 mL centrifuge tube punctured at the bottom. Then, a 0.5 mL centrifuged tube was inserted into a 2 mL centrifuge tube for centrifugation (12,000 rpm, 15 min, 20 °C), and about 200 μL phloem sap of each tube was collected for further analysis.

### 2.3. Determination of Nutrient Concentration

A fresh bud sample weighing around 0.1 g was used to measure the N concentration, by using the Vario MACRO Cube Analyzer (Elementar, Germany), following the protocol of Tian et al. [32]. Moreover, the concentrations of P, K, Ca, Mg, S, Fe, Mn, B, Cu and Zn were determined by taking a 0.3 g dried bud sample. The dried sample was ashed in a muffle furnace at 500 °C for 6 h; afterwards, the sample was dissolved in 30 mL of $HNO_3$ (1%) solution. The solution was used to determine the above-mentioned nutrient concentration by an inductively coupled plasma-optical emission spectrometer (ICP-OES, Finnigan MAT, Element I, Germany).

### 2.4. X-ray Micro-Computed (μCT) Tomography

X-ray micro-computed (μCT) tomography was performed as described by Buss et al. [33] with some modifications. Virtual image stacks of all bud specimens were detected by μCT (Versa 510, Carl Zeiss, Germany). Briefly, the samples were stored in dry tubes during the μCT-scanning process. Then, the samples were scanned with a Versa 510 X-ray microscope with the following parameters: 30–40 kV, 5–8 W, 2001 projections, 10–22 s acquisition time, objective: ×0.4, and voxel size: 5.0 μm.

### 2.5. RNA-seq, Data Processing and Gene Annotation

Transcriptome libraries were constructed from samples of normal buds and wizened buds in the early stage of wizened bud occurrence (September 2020) with 3 biological replications. Total RNA was isolated by using a TaKaRa Plant RNA Extraction Kit (TaKaRa, Dalian, China), and the transcriptome profiles were obtained by using NR604-VAHTS® Fast RNA-seq Library Prep Kit for Illumina (Vazyme, Nanjing, China), following the manufacturer's protocol. The original data have been submitted to NCBI (BioProject ID: PRJNA804671). Clean reads (high-quality reads) of 6 transcriptome libraries were filtered from raw reads by removing low-quality reads with ambiguous nucleotides (read length: 300 bp, total reads: 72,777,662 to 84,315,630) and adaptor sequences, and Tophat2 was used to map clean reads to the reference genome *Pyrus pyrifolia* in GDR (https://www.rosaceae.org/Analysis/9597119, 1 May 2021). Gene expression levels were calculated by the FPKM (Fragments Per Kilobase of transcript per Million mapped reads) method using the RSEM tool. A differential expression analysis was performed using the DESeq R package (1.10.1). The resulting *p* values were adjusted using Benjamini and Hochberg's approach for controlling the false discovery rate. Genes with a *p*-value < 0.05 found by DESeq were assigned as differentially expressed genes. The overview of RNA-seq statistics and the FPKM of differentially expressed genes (DEGs) are mentioned in Tables S3 and S4.

Gene annotation was conducted using the Blastp search against the nr database in NCBI, Swiss-Prot database and Pfam database. BlastKOALA was used to annotate the KOs of the KEGG ORTHOLOGY database. The protein sequences of genes were aligned against the GO database and KEGG pathway database using KOBAS 3.0 (http://kobas.cbi.pku.edu.cn/, 1 May 2021) to perform the enrichment analysis. A corrected *p*-value < 0.01 was set as a cut-off for GO enrichment 2.6. Quantitative Real-Time PCR (qRT-PCR) Analysis

The qRT-PCR experiment was conducted following the previous protocol [34]. Briefly, the total RNA of each sample was extracted by an OminiPlant RNA Kit (CWBIO, Beijing, China). One microgram of high-quality total RNA was used for the first-strand cDNA synthesis by TransScript One-step gDNA Removal and cDNA Synthesis SuperMix (TransGen Biotech, Beijing, China). The qRT-PCR was conducted by 3 biological replicates, and each biological replicate had 3 technical replicates. qRT-PCR was performed in a 10 μL reaction volume using Luna® Universal qPCR Master Mix (NEB) on a qRT-PCR system (CFX real-time PCR system, BioRad, Hercules, CA, USA) following the manufacturer's protocol. The reaction was started by an initial incubation at 94 °C for 2 min and then subjected to 34 cycles of 94 °C for 30 s, 57 °C for 30 s and 72 °C for 80 s, and then ended with 72 °C for 3 min. The primers used in this study are listed in Table S2. *PpyPP2A* (sequence ID: *Pbr020268.1*) was used as an internal control to normalize the expression level of the target genes between samples. The Livak method was employed to calculate relative expression values [35].

### 2.6. Determination of Pectin, Lignin, Cellulose, Pectin Methylesterase, Pectate Lyase, Polygalacturonase and Sorbitol Concentration

Two decigrams of fine powder of each bud sample were used to determine the pectin (by a plant pectin ELISA Kit), lignin (by a plant lignin ELISA Kit), cellulose (by a plant cellulose ELISA Kit), pectin methylesterase (by a plant PME ELISA Kit), pectate lyase (by a plant PL ELISA Kit) and polygalacturonase (by a plant PG ELISA Kit) concentration following the instructions of the kits (Comin Biotechnology Co., Ltd., Suzhou, China). Moreover, 1 g of fine powder of leaves, 1 g of buds and 0.5 mL of phloem sap were used to determine the sorbitol concentration using a gas chromatography-mass spectrometer. The detailed method for sorbitol detection was mentioned in the Supplementary Methods.

### 2.7. Statistical Analysis

Unless specially stated, each value was expressed as the means ± standard deviation (SD) of three replications. Data analysis was performed by using an independent-samples *t*-test ($p < 0.05$) in SPSS Statistics 19.0 (SPSS Inc., Chicago, IL, USA).

## 3. Results

### 3.1. Phenotype and Structure Comparison between Normal Buds and Wizened Buds

At the time of flower bud differentiation, wizened buds were observed on '710' sand pear plants. Phenotypically, the bud scales were wrapped tightly around normal buds with a hub shape; however, the bud scales were loosely surrounded by wizened buds with irregular shape and appeared like the bud burst from the inside to the outside (Figure 1A). Compared with normal buds, more than half of the total buds were wizened buds with significantly lower fresh weight (Figure 1B). Moreover, the ratio of leaf to bud in normal bud stems was 139:114, which was 2.5 times that of the wizened bud stems (54:112).

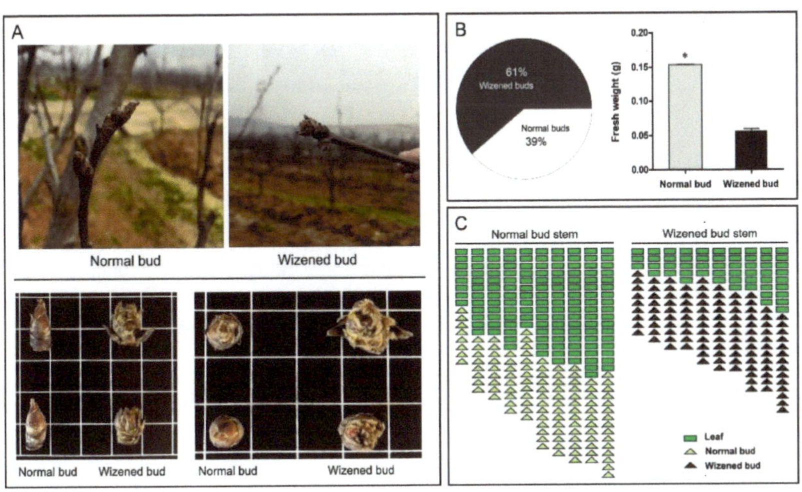


**Figure 1.** Phenotypic differences between normal buds and wizened buds, and their characteristics. The phenotypic illustration of normal buds and wizened buds (**A**). The proportion of wizened buds and normal buds and the comparison of fresh weight between normal buds and wizened buds (**B**) The comparison of leaf to bud ratio between normal bud stems and wizened bud stems (**C**). Values are the mean ± SD. The asterisk indicates a significant difference between normal bud and wizened bud (*t*-test, $p < 0.05$). In Figure 1A, the side length of each square indicates 1 cm.

According to the microscopic observations, the shape of inflorescence in the normal buds was trigonal pyramidal with a tight arrangement above the bud axis (Figure 2A); however, the inflorescence of wizened buds presented an irregular shape with a loose arrangement on the bud axis (Figure 2D). Internally, all inflorescences in normal buds were healthy with an integral structure (Figure 2B); by contrast, all inflorescences in the wizened bud were abnormal or even dead, with a collapsed structure and a shriveled and cracked appearance (Figure 2E). Furthermore, the structure of the flower parts in the normal bud was distinct and can be completely differentiated (Figure 2C); on the contrary, the structure of the wizened bud was indistinct, with an incomplete differentiation of stamen and pistil, and a lack of petal differentiation (Figure 2F).

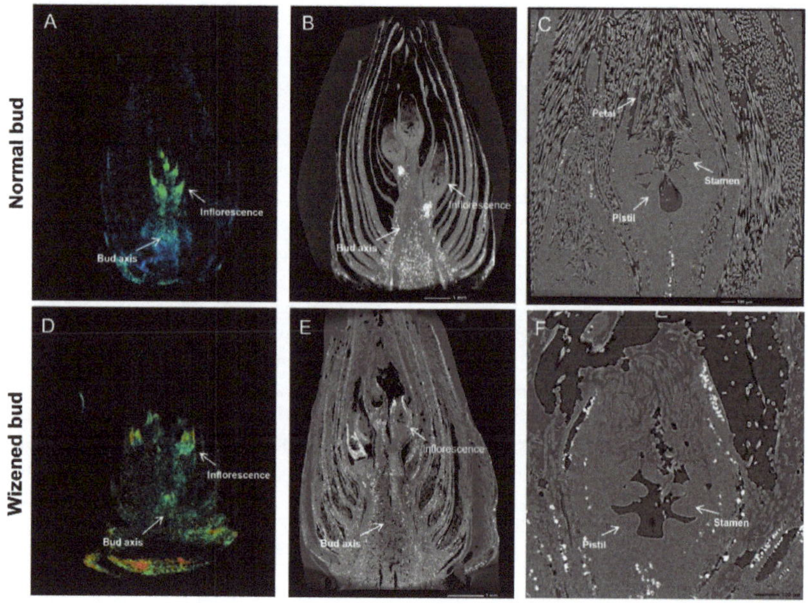


**Figure 2.** Microscopic illustration of normal buds and wizened buds. Appearance and inflorescence (**A**,**D**); longitudinal section (**B**,**E**); floral components (**C**,**F**) of normal buds and wizened buds, respectively.

### 3.2. *Comparison of B Concentration between Normal Buds and Wizened Buds*

Leaf nutrient analysis indicated that the concentration of K, Mg and B was significantly lower in wizened buds than in normal buds (Table S1). Moreover, the Mg concentration in orchard soil was higher than the normal threshold, while the B concentration in orchard soil was lower than the normal threshold (Table S1). According to Figure 3, wizened bud occurrence was started after 22 July and, before that, the B concentration in both types of bud tissues was the same. After 22 July, the B concentration started to reduce quickly in the wizened bud, but normal bud presented an increasing profile of B concentration; moreover, the B concentration of wizened buds was significantly lower than that of normal buds (Figure 3).

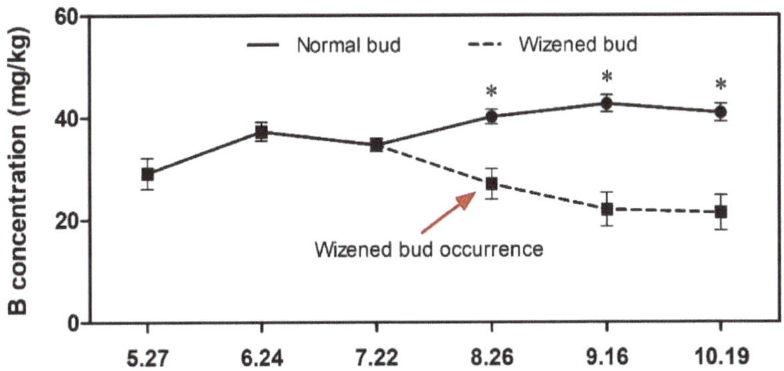


**Figure 3.** Changes of B concentration profile during flower bud differentiation starting from 27 May to 19 October. Values are the mean ± SD. The asterisk indicates a significant difference between normal buds and wizened buds at a given time point (*t*-test, $p < 0.05$).

### *3.3. Functional Enrichment and Coexpression Network of Differentially Expressed Genes*

To further explore the differences in the molecular regulation of wizened bud formation, a comparative transcriptome analysis was performed between normal buds and wizened buds. In total, 546 differentially expressed genes (DEGs) were identified; of them, 462 genes were down-regulated in wizened bud samples (Figure S1). The DEGs were enriched in 'response to wounding', 'suberin biosynthetic process', 'regulation of defense response', 'syncytium formation', 'nitrate transport' and 'plant-type cell wall loosening' processes (Figure 4A). Subsequently, a KEGG enrichment analysis revealed 10 typical pathways, and the 'pentose and glucuronate interconversions' pathway was related to the 'plant-type cell wall loosening' process (Figure 4B). To identify the key regulatory genes, we constructed gene networks via WGCNA and Cytoscape. Among all DEGs, 54 DEGs encoded transcription factors (TFs) and the highest edges of *PpyMYB39* were observed in the network analysis (Figure 4C).

### *3.4. Expression Levels of Differentially Expressed Genes and Candidate Target Genes of PpyMYB39*

To verify the authenticity of RNA-seq data, the DEGs belonging to the 'pentose and glucuronate interconversions' pathway and key TFs of the coexpression network were selected. The qRT-PCR results showed strong coherence with the finding of the RNA-seq results. Briefly, the expression level of *PpyPME3*, *PpyPL18.1*, *PpyPL18.2*, *PpyPG1* and *PpyPG2* were significantly higher in wizened buds than those of normal buds (Figure 5A). However, the expression level of *PpySDH* and all selected TFs were significantly lower in wizened buds than in normal buds (Figure 5A,B).

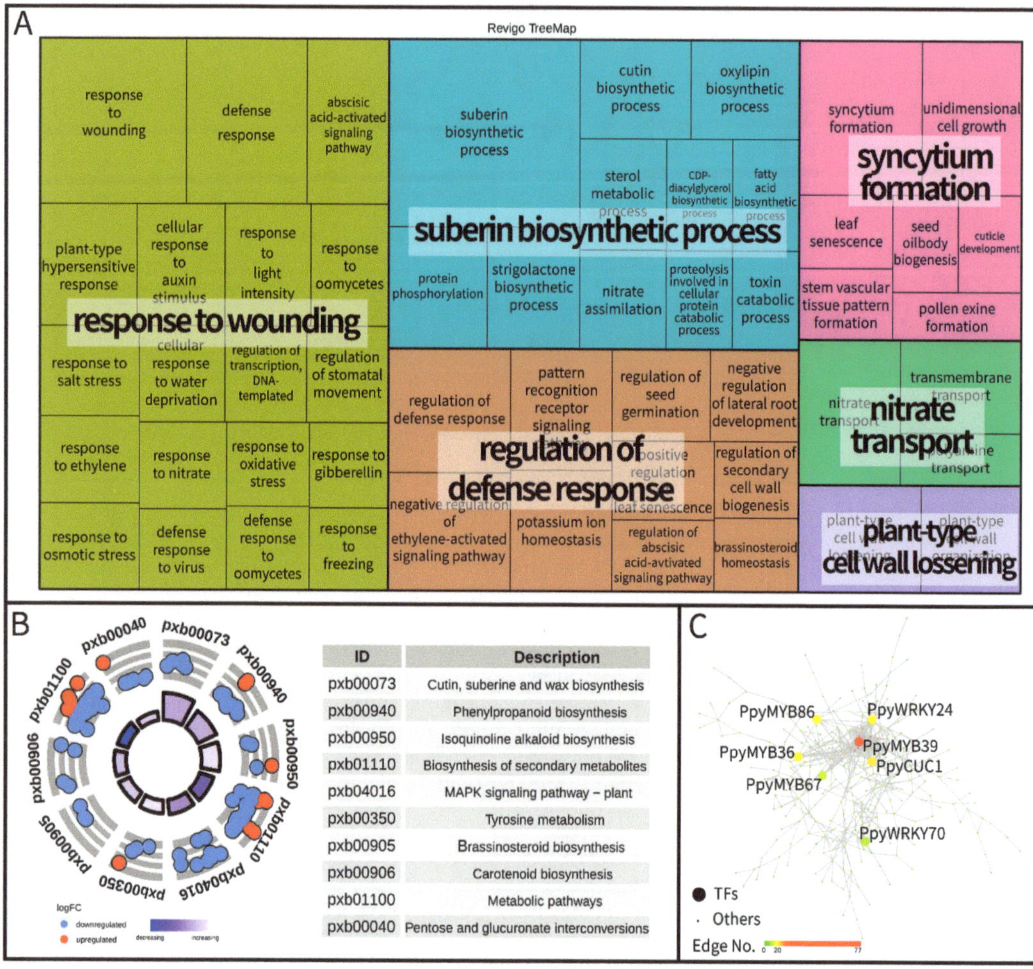


**Figure 4.** Go term enrichment analysis (**A**), KEGG pathway enrichment analysis (**B**) and correlation network of differentially transcription factors (**C**) between normal buds and wizened buds. Each rectangle of graph A is a single cluster, joined into 'superclusters' of loosely related terms, and the size of the rectangles is adjusted to reflect the *P*-value. The term logFC means log2 (FPKM_Wizened bud/FPKM_Normal bud) and the color of a bar from dark blue to light blue represents a hint regarding the pathway that is more likely to be decreased or increased in graph B. The transcription factors are shown by large circles in graph C, and the color from green to red represented the number of edge genes related to the transcription factor.

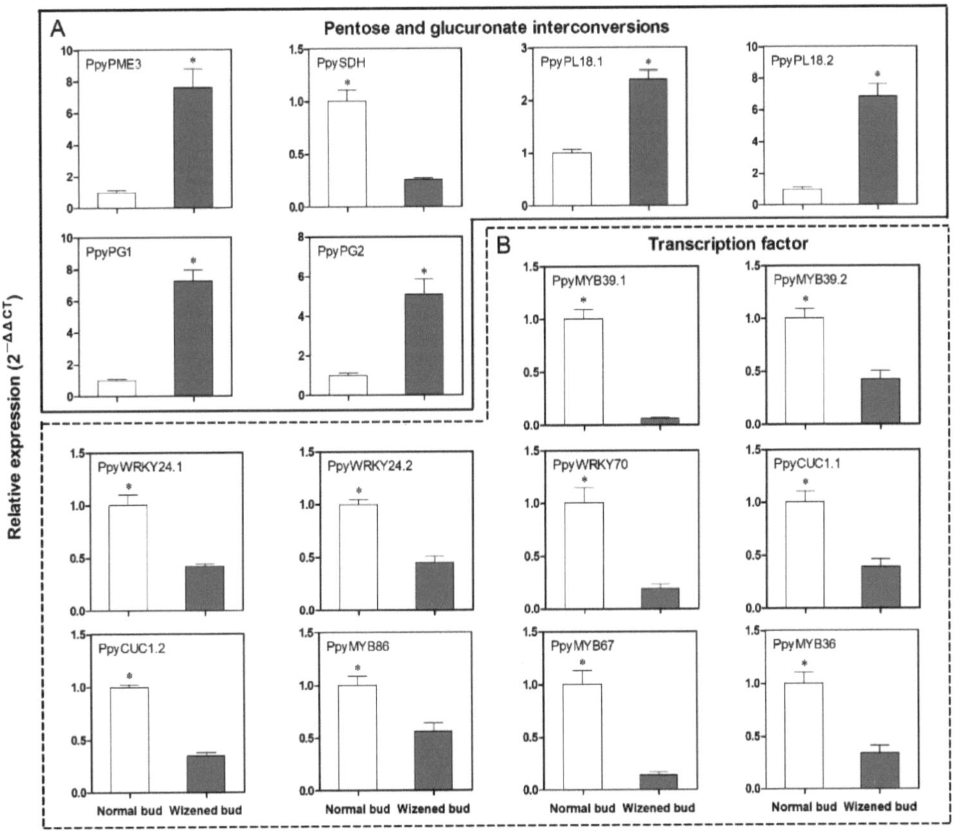


**Figure 5.** Transcript analysis of differential genes between normal buds and wizened buds. (**A**) refers to the transcript analysis of the pentose and glucuronate interconversions pathway-related genes, and (**B**) refers to the expression analysis of transcription factors. Values are the mean ± SD. PME: Pectin methylesterase. SDH: Sorbitol dehydrogenase, PL: Pectate lyase. PG: Polygalacturonase. The asterisk indicates a significant difference between normal bud and wizened bud (*t*-test, $p < 0.05$).

MdMYB39L plays a key role in sorbitol-modulated stamen development by regulating its downstream target genes, which are involved in hexose uptake, cell wall formation and modification, and microsporogenesis [21]. Based on the findings of this research, 8 candidate target genes of *PpyMYB39* were screened for qRT-PCR analysis between normal buds and wizened buds. These genes were related to bud development and derived from 2 functional categories, including hexose transporters (*PpyHT1* to *PpyHT4*) and pectin methylesterase inhibitors (*PpyPMEI1* to *PpyPMEI4*) (Figure 6A,B). The expression levels of *PpyHT1*, *PpyHT2*, *PpyPMEI1* and *PpyPMEI2* in wizened buds were significantly lower than those in normal buds; however, no significant difference was observed in the expression levels of *PpyHT3*, *PpyHT4*, *PpyPMEI3* and *PpyPMEI4* between normal buds and wizened buds (Figure 6A,B).

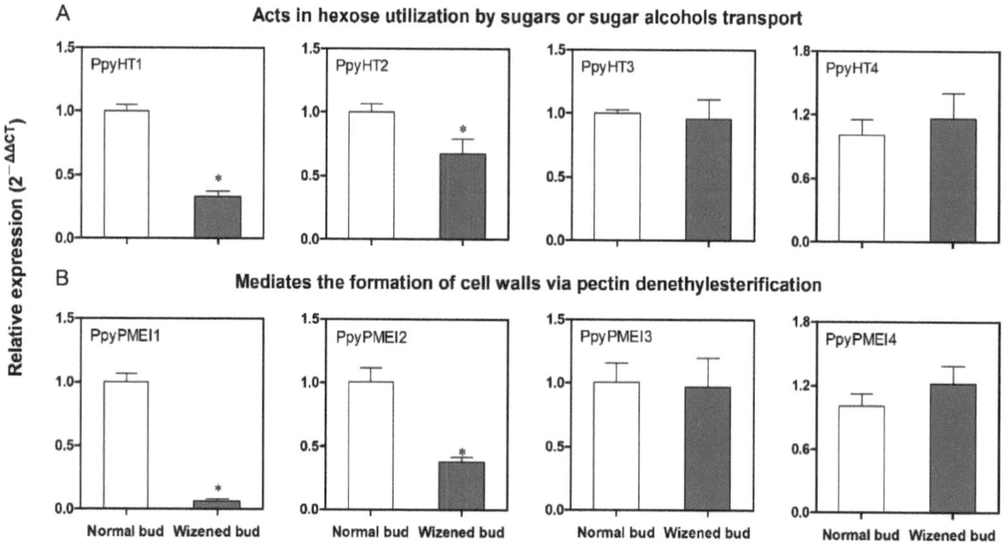


**Figure 6.** Transcript analysis of the comparison of *PpyMYB39* mRNA responsive genes between normal buds and wizened buds. (**A**) refers to the expression analysis of *Ppy*HT-encoded genes, and (**B**) refers to the expression analysis of *Ppy*PMEI-encoded genes. HT: hexose transporters. PMEI: pectin methylesterase inhibitors. Values are the mean ± SD. The asterisk indicates a significant difference between normal buds and wizened buds (*t*-test, $p < 0.05$).

### *3.5. Comparison of Pectin, Lignin, Cellulose and Pectin Methylesterase, Pectate Lyase Polygalacturonase Content Related to Pectin Degradation*

According to the KEGG enrichment analysis, the pectin degradation process was mainly induced after wizened bud occurrence. The quantitative analysis further confirmed that the pectin concentration in wizened buds was significantly lower than in normal buds (Figure 7A). However, no significant change was observed for the lignin and cellulose concentration between the normal buds and wizened buds (Figure 7B,C). Moreover, the content of pectin methylesterase, pectate lyase and polygalacturonase were all significantly higher in wizened buds than in normal buds (Figure 7D–F).

### *3.6. Changes in Sorbitol Content in the Buds, Leaves and Phloem Sap*

The determination of sorbitol content in different parts of the plant was conducted to compare the translocation and utilization of sorbitol content between normal bud stems and wizened bus stems. The results indicated that the sorbitol concentration in wizened buds and stem phloem sap was significantly lower than that in normal buds and stem phloem sap (Figure 8A,C); however, no difference was observed in leaf samples taken from normal bud stems and wizened bud stems (Figure 8B).

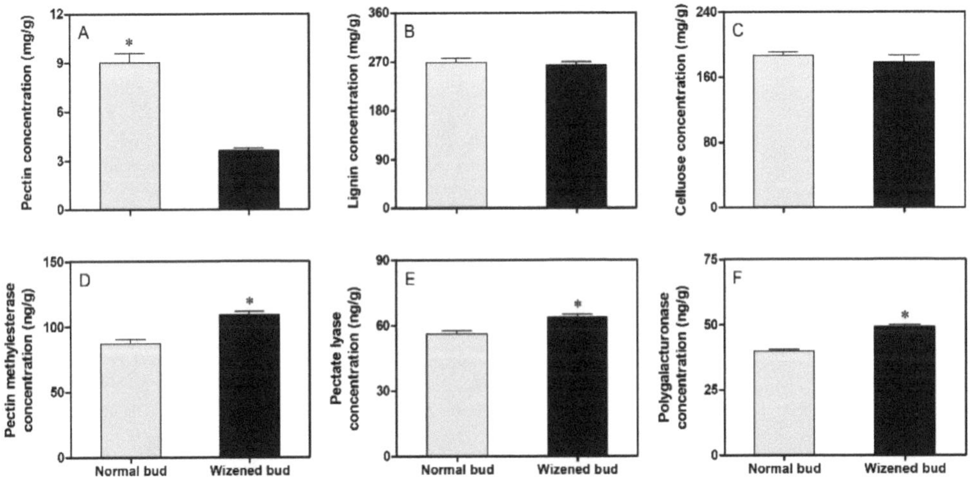


**Figure 7.** Comparison of pectin (**A**), lignin (**B**), cellulose (**C**), pectin methylesterase (**D**), pectate lyase (**E**) and polygalacturonase (**F**) concentration between normal buds and wizened buds. The values are the mean ± SD. The asterisk indicates a significant difference between normal buds and wizened buds (*t*-test, $p < 0.05$).

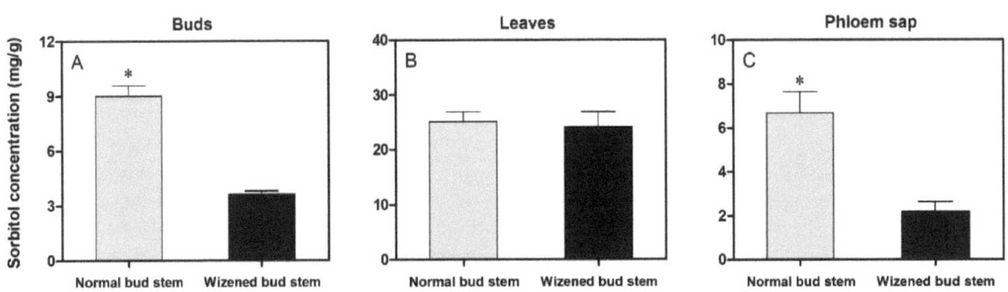


**Figure 8.** Comparison of sorbitol concentration in the buds (**A**), leaves (**B**), phloem sap (**C**) between normal bud stems and wizened bud stems. Values are the mean ± SD. The asterisk indicates a significant difference between normal bud stems and wizened bud stems (*t*-test, $p < 0.05$).

## 4. Discussion

In pear orchards, wizened bud occurrence is frequently observed, which impedes flower bud growth, resulting in a great loss of yield [6]. According to the present study, more than 60% of buds presented a wizened performance in '710' sand pear trees at the time of flower bud differentiation (Figure 1A,B), indicating that '710' sand pear trees have a typical character of wizened bud formation and could be used as interesting material to investigate the underlying mechanism. Mineral nutrients and carbohydrates play key roles in floral bud differentiation [36,37], and these are dominantly supplied from leaves by the source-sink transport mechanism [38,39]. In pear, sorbitol is the main transportable photo-assimilate, which was confirmed to contribute to flower bud differentiation [40,41]. The present study found that the number of source organs or the sorbitol concentration of buds and phloem sap was significantly lower in wizened bud stems than in normal bud stems (Figures 1C and 8A,C), thereby suggesting that wizened bud occurrence is related to the insufficient supply of sorbitol from source (leaves) to sink (bud) organs, and sorbitol may play a key role in the process of wizened bud formation.

According to the previous reports, B is crucial for floral bud differentiation by participating in cell wall development and membrane maintenance [42,43]; furthermore, its translocation via phloem is assisted with the presence of sorbitol in Rosaceae family plants, including pear [12,44]. Our results also found that B was typically decreased in wizened buds than in normal buds (Table S1). Temporal analysis indicated that the B concentration was the same in all buds during the early stages of floral bud differentiation; however, when buds showed wizened performance, the B concentration was rapidly decreased and was significantly lower than that of normal buds (Figure 3). These results supposed that the decrease of sorbitol transport from the leaves reduced the B concentration in buds, leading to abnormal bud differentiation.

A microscopic study showed that wizened buds might occur at or before the sepal differentiation phase in pear [5]. However, little is known about the inner structure of wizened buds and the microstructure of inflorescence in wizened buds. In the present study, the wizened buds showed loose and collapsed structures when compared with normal buds (Figure 2A,B,D,E). Moreover, the differentiation of stamen and pistil was initiated, but the formation of stamen and pistil was incomplete with indistinct structures in wizened buds (Figure 2C,F), indicating that the occurrence of wizened buds is likely due to the collapse of the organ structure, resulting in burst bud appearance.

To understand the molecular mechanism of wizened bud occurrence in pear, the samples with typical wizened bud occurrence (during the wizening process) were collected, and an RNA-seq experiment was conducted between normal buds and wizened buds. From 546 DEGs, many function-specific genes were identified between normal buds and wizened buds, which may contribute to the wizened bud formation process. It has been reported that cell wall modification can affect the bud formation process in woody plants [45,46]. In our study, the DEGs were enriched in different biological processes (Figure 4A); among them, the 'plant-type cell wall loosening' process was considered to be directly related to the collapse of the organ structure in buds (Figure 2). Then, the top 10 most enriched pathways and the DEGs enriched in these pathways were visualized using the GO plot package [47] (Figure 4B). Among these pathways, the 'pentose and glucuronate interconversions' pathway, which contained 5 DEGs associated with the pectin degradation process, was particularly focused on (Figure 4A,B). The expression levels of these genes, *PpyPME3* related to pectin methyl esterification, *PpyPL18.1* and *PpyPL18.2* related to pectin lysis, and *PpyPG1* and *PpyPG2* related to polygalacturonate were significantly higher in wizened buds than in normal buds (Figure 4A). Moreover, the concentration of pectin (except for lignin and cellulose) and the enzymes corresponding to these genes were significantly lower in wizened buds than in normal buds (Figure 7). These results suggested that the pectin degradation process was initiated after wizened bud occurrence. Previous studies demonstrated that pectin is one of the major components of the cell wall [48,49], and the degradation of pectin can trigger the cell wall loosening process either in reproductive or vegetative organs [50–52]. Taken together, we supposed that the occurrence of wizened buds has resulted from the cell wall loosening process induced by pectin degradation.

Gene network analyses identified some TFs, which were supposed to play vital roles in the regulation of wizened bud formation (Figure 4C). The qRT-PCR analysis indicated that 10 TFs genes related to the wizened bud formation process had a down-regulated expression in the wizened bud when compared with those of normal buds (Figure 5B). Most importantly, one MYB39-like transcription factor, *PpyMYB39.1*, showed a significantly lower expression in wizened buds (Figure 5B). As MYB39 genes can negatively regulate the flower bud growth either during the differentiation or the blooming stage [21,53,54], the lower expression of *PpyMYB39.1* in wizened buds suggested that the function of *PpyMYB39.1* is closely related to wizened bud occurrence. Moreover, its putative target genes based on the previous study [21], *PpyHT1*, *PpyHT2* and *PpyPMEI1*, and *PpyPMEI2* genes presented the same expression trend as *PpyMYB39.1* in wizened buds relative to normal buds (Figures 5B and 6). Taken together, *PpyMYB39.1* could be a key regulator of the wizened bud formation process.

Previous studies revealed that pectin methylesterases (PMEs) were involved in bud development by modifying cell wall stability [55–57], and PMEIs were able to inhibit PME activities [58,59]. In this study, lower expression levels of *PpyPMEI1* and *PpyPMEI2* genes and the higher expression level of *PpyPME3* suggested that both *PpyPMEI* genes were involved in bud development in pear by mediating PMEs that induced wizened bud occurrence (Figure 6B). On the other hand, *PpyHT1* and *PpyHT2* showed lower expression levels in wizened buds (Figure 6A). It has been reported that the hexose transporters contribute to bud formation and development by taking up hexoses for adequate carbohydrate utilization in the bud [60,61]. Our findings suggested that the lower expression levels of *PpyHT1* and *PpyHT2* in wizened buds suppressed the transport of hexose sugars into buds, thereby hindering bud formation and leading to wizened bud occurrence in pear plants (Figures 5A, 6A and 8A).

## 5. Conclusions

At the time of wizened bud occurrence, wizened buds were found to be deficient in sorbitol and B as compared to normal buds. The RNA-seq and metabolic analysis revealed that the cell wall loosening process induced by pectin degradation might be the main factor of wizened bud occurrence. The co-expression network of TFs further suggested that *PpyMYB39.1* might play a vital role in bud differentiation by regulating the pectin degradation and carbohydrate transport-related genes (*PpyPMEI1*, *PpyPMEI2*, *PpyHT1* and *PpyHT2*).

**Supplementary Materials:** The following supporting information can be downloaded at: https://www.mdpi.com/article/10.3390/agronomy12020484/s1, Figure S1: The number of differential expression genes by comparison of normal buds and wizened buds, Table S1: Comparison of nutrient concentration between normal bud and wizened bud or orchard soil and normal range, Table S2: The list of gene names, their IDs and primers used in this study. Table S3: The overview of RNA-seq statistics. Table S4: The FPKM of differentially expressed genes (DEGs) of RNA-seq data.

**Author Contributions:** Conceptualization, W.D. and H.H.; methodology, W.D., X.Y., C.S. and M.L.; formal analysis, W.D., C.S. and S.B.H.; investigation, W.D., J.Z., H.H., Q.C., J.F. and X.Y.; writing-original draft preparation, W.D.; writing-review and editing, W.D., S.B.H., Y.L., X.Y. and H.H.; supervision, J.F. and X.Y.; project administration, W.D.; funding acquisition, Y.L. and H.H. All authors have read and agreed to the published version of the manuscript.

**Funding:** This research was funded by the Major Program of Hubei Agricultural Science and Technology Innovation Center (2020-620-000-002-05), the National Key Research and Development Program (2020YFD1000202) and the China Agriculture Research System of MOF and MARA (CARS-28).

**Institutional Review Board Statement:** Not applicable.

**Informed Consent Statement:** Not applicable.

**Data Availability Statement:** Not applicable.

**Acknowledgments:** The authors sincerely thank Gui-Zhi Feng (Nanjing Agricultural University, Nanjing, China) and Wang Jia (Shanghai Bioprofile Technologe Company Ltd., Shanghai, China) for their help in data analysis.

**Conflicts of Interest:** The authors declare no conflict of interest.

## References

1. Zhang, M.-Y.; Xue, C.; Hu, H.; Li, J.; Xue, Y.; Wang, R.; Fan, J.; Zou, C.; Tao, S.; Qin, M.; et al. Genome-wide association studies provide insights into the genetic determination of fruit traits of pear. *Nat. Commun.* **2021**, *12*, 1144. [CrossRef] [PubMed]
2. FAOSTAT. *World Food and Agriculture Statistical Yearbook*; FAO-Food & Agriculture Organization of the United Nation: Rome, Italy, 2020.
3. Lee, S.-H.; Kim, W.-S.; Han, T.-H. Effects of post-harvest foliar boron and calcium applications on subsequent season's pollen germination and pollen tube growth of pear (*Pyrus pyrifolia*). *Sci. Hortic.* **2009**, *122*, 77–82. [CrossRef]

4. Yang, Q.; Niu, Q.; Tang, Y.; Ma, Y.; Yan, X.; Li, J.; Tian, J.; Bai, S.; Teng, Y. *PpyGAST1* is potentially involved in bud dormancy release by integrating the GA biosynthesis and ABA signaling in 'Suli' pear (*Pyrus pyrifolia* White Pear Group). *Environ. Exp. Bot.* **2019**, *162*, 302–312. [CrossRef]
5. Yang, S.; Bai, M.-D.; Gao, P.; Hao, G.-W.; Zhang, X.-W.; Guo, H.-P.; Li, L.-L. Relationship between change in endogenous hormone content and occurrence of wizened bud in Chinese pear. In Proceedings of the II Asian Horticultural Congress, Chengdu, China, 30 August 2018; pp. 235–244. [CrossRef]
6. Liu, Y.; Zhang, H.P.; Gu, C.; Tao, S.T.; Wang, D.S.; Guo, X.P.; Qi, K.J.; Zhang, S.L. Transcriptome profiling reveals differentially expressed genes associated with wizened flower bud formation in Chinese pear (*Pyrus bretschneideri* Rehd.). *J. Hortic. Sci. Biotechnol.* **2016**, *91*, 227–235. [CrossRef]
7. Zuo, X.; Zhang, D.; Wang, S.; Xing, L.; Li, Y.; Fan, S.; Zhang, L.; Ma, J.; Zhao, C.; Shah, K.; et al. Expression of genes in the potential regulatory pathways controlling alternate bearing in 'Fuji' (*Malus domestica* Borkh.) apple trees during flower induction. *Plant Physiol. Biochem.* **2018**, *132*, 579–589. [CrossRef] [PubMed]
8. Landsberg, J.J. Apple Fruit Bud Development and Growth; Analysis and an Empirical Model. *Ann. Bot.* **1974**, *38*, 1013–1023. [CrossRef]
9. Okie, W.R.; Werner, D.J. Genetic Influence on Flower Bud Density in Peach and Nectarine Exceeds That of Environment. *HortScience* **1996**, *31*, 1010–1012. [CrossRef]
10. Ruan, Y.-L. Sucrose Metabolism: Gateway to Diverse Carbon Use and Sugar Signaling. *Annu. Rev. Plant Biol.* **2014**, *65*, 33–67. [CrossRef]
11. Ramírez, F.; Davenport, T.L. Mango (*Mangifera indica* L.) flowering physiology. *Sci. Hortic.* **2010**, *126*, 65–72. [CrossRef]
12. Brown, P.H.; Bellaloui, N.; Hu, H.; Dandekar, A. Transgenically enhanced sorbitol synthesis facilitates phloem boron transport and increases tolerance of tobacco to boron deficiency. *Plant Physiol.* **1999**, *119*, 17–20. [CrossRef]
13. Chatterjee, M.; Tabi, Z.; Galli, M.; Malcomber, S.; Buck, A.; Muszynski, M.; Gallavotti, A. The boron efflux transporter ROTTEN EAR is required for maize inflorescence development and fertility. *Plant Cell* **2014**, *26*, 2962–2977. [CrossRef] [PubMed]
14. Patrick, J.W.; Botha, F.C.; Birch, R.G. Metabolic engineering of sugars and simple sugar derivatives in plants. *Plant Biotechnol. J.* **2013**, *11*, 142–156. [CrossRef] [PubMed]
15. Bieleski, R. Sugar alcohols. In *Plant Carbohydrates I*; Springer: Berlin/Heidelberg, Germany, 1982; pp. 158–192.
16. Reidel, E.J.; Rennie, E.A.; Amiard, V.; Cheng, L.; Turgeon, R. Phloem loading strategies in three plant species that transport sugar alcohols. *Plant Physiol.* **2009**, *149*, 1601–1608. [CrossRef] [PubMed]
17. Shen, C.; Wang, J.; Jin, X.; Liu, N.; Fan, X.; Dong, C.; Shen, Q.; Xu, Y. Potassium enhances the sugar assimilation in leaves and fruit by regulating the expression of key genes involved in sugar metabolism of Asian pears. *Plant Growth Regul.* **2017**, *83*, 287–300. [CrossRef]
18. Nosarzewski, M.; Downie, A.B.; Wu, B.; Archbold, D.D. The role of sorbitol dehydrogenase in *Arabidopsis thaliana*. *Funct. Plant Biol.* **2012**, *39*, 462–470. [CrossRef]
19. Park, S.W.; Song, K.J.; Kim, M.Y.; Hwang, J.-H.; Shin, Y.U.; Kim, W.-C.; Chung, W.-I. Molecular cloning and characterization of four cDNAs encoding the isoforms of NAD-dependent sorbitol dehydrogenase from the Fuji apple. *Plant Sci.* **2002**, *162*, 513–519. [CrossRef]
20. Zhou, R.; Cheng, L.; Dandekar, A.M. Down-regulation of sorbitol dehydrogenase and up-regulation of sucrose synthase in shoot tips of the transgenic apple trees with decreased sorbitol synthesis. *J. Exp. Bot.* **2006**, *57*, 3647–3657. [CrossRef]
21. Meng, D.; He, M.; Bai, Y.; Xu, H.; Dandekar, A.M.; Fei, Z.; Cheng, L. Decreased sorbitol synthesis leads to abnormal stamen development and reduced pollen tube growth via an MYB transcription factor, *MdMYB39L*, in apple (*Malus domestica*). *New Phytol.* **2018**, *217*, 641–656. [CrossRef]
22. He, W.; Chen, Y.; Gao, M.; Zhao, Y.; Xu, Z.; Cao, P.; Zhang, Q.; Jiao, Y.; Li, H.; Wu, L. Transcriptome analysis of *Litsea cubeba* floral buds reveals the role of hormones and transcription factors in the differentiation process. *G3 Genes Genomes Genet.* **2018**, *8*, 1103–1114. [CrossRef]
23. Fan, L.; Chen, M.; Dong, B.; Wang, N.; Yu, Q.; Wang, X.; Xuan, L.; Wang, Y.; Zhang, S.; Shen, Y. Transcriptomic analysis of flower bud differentiation in *Magnolia sinostellata*. *Genes* **2018**, *9*, 212. [CrossRef]
24. Zhu, L.; Guan, Y.; Liu, Y.; Zhang, Z.; Jaffar, M.A.; Song, A.; Chen, S.; Jiang, J.; Chen, F. Regulation of flowering time in chrysanthemum by the R2R3 MYB transcription factor *CmMYB2* is associated with changes in gibberellin metabolism. *Hortic. Res.* **2020**, *7*, 96. [CrossRef] [PubMed]
25. Azuma, M.; Morimoto, R.; Hirose, M.; Morita, Y.; Hoshino, A.; Iida, S.; Oshima, Y.; Mitsuda, N.; Ohme-Takagi, M.; Shiratake, K. A petal-specific InMYB1 promoter from Japanese morning glory: A useful tool for molecular breeding of floricultural crops. *Plant Biotechnol. J.* **2016**, *14*, 354–363. [CrossRef]
26. Müller, D.R.; Schmitz, G.; Theres, K. Blind homologous R2R3 Myb genes control the pattern of lateral meristem initiation in Arabidopsis. *Plant Cell* **2006**, *18*, 586–597.
27. Lau, S.-E.; Schwarzacher, T.; Othman, R.Y.; Harikrishna, J.A. dsRNA silencing of an R2R3-MYB transcription factor affects flower cell shape in a *Dendrobium* hybrid. *BMC Plant Biol.* **2015**, *15*, 194. [CrossRef] [PubMed]
28. Tan, C.; Liu, Z.; Huang, S.; Li, C.; Ren, J.; Tang, X.; Liu, W.; Peng, S.; Feng, H. Pectin methylesterase inhibitor (PMEI) family can be related to male sterility in Chinese cabbage (*Brassica rapa ssp. Pekinensis*). *Mol. Genet. Genom.* **2018**, *293*, 343–357. [CrossRef] [PubMed]

29. Chen, C.; Liu, S.; Hao, X.; Chen, G.; Cao, B.; Chen, Q.; Lei, J. Characterization of a Pectin Methylesterase Gene Homolog, *CaPME1*, Expressed in Anther Tissues of *Capsicum annuum* L. *Plant Mol. Biol. Rep.* **2012**, *30*, 403–412. [CrossRef]
30. David, P.-L.; Deyholos, M.K. Pectinmethylesterases (PME) and Pectinmethylesterase Inhibitors (PMEI) Enriched during Phloem Fiber Development in Flax (*Linum usitatissimum*). *PLoS ONE* **2014**, *9*, e105386. [CrossRef]
31. Du, W.; Pan, Z.-Y.; Hussain, S.B.; Han, Z.-X.; Peng, S.-A.; Liu, Y.-Z. Foliar Supplied Boron Can Be Transported to Roots as a Boron-Sucrose Complex via Phloem in Citrus Trees. *Front. Plant Sci.* **2020**, *11*, 250. [CrossRef]
32. Tian, J.; Lu, X.; Chen, Q.; Kuang, X.; Liang, C.; Deng, L.-S.; Lin, D.; Cai, K.; Tian, J. Phosphorus fertilization affects soybean rhizosphere phosphorus dynamics and the bacterial community in karst soils. *Plant Soil* **2020**, 1–16. [CrossRef]
33. Buss, D.J.; Reznikov, N.; McKee, M.D. Crossfibrillar mineral tessellation in normal and Hyp mouse bone as revealed by 3D FIB-SEM microscopy. *J. Struct. Biol.* **2020**, *212*, 107603. [CrossRef]
34. Shi, C.-Y.; Hussain, S.B.; Yang, H.; Bai, Y.-X.; Khan, M.A.; Liu, Y.-Z. *CsPH8*, a P-type proton pump gene, plays a key role in the diversity of citric acid accumulation in citrus fruits. *Plant Sci.* **2019**, *289*, 110288. [CrossRef] [PubMed]
35. Livak, K.J.; Schmittgen, T.D. Analysis of Relative Gene Expression Data Using Real-Time Quantitative PCR and the 2−ΔΔCT Method. *Methods* **2001**, *25*, 402–408. [CrossRef] [PubMed]
36. Hwang, H.S.; Jeong, H.W.; Lee, H.R.; Jo, H.G.; Kim, H.M.; Hwang, S.J. Acceleration of Flower Bud Differentiation of Runner Plants in "Maehyang" Strawberries Using Nutrient Solution Resupply during the Nursery Period. *Agronomy* **2020**, *10*, 1127. [CrossRef]
37. Luo, T.; Zhang, J.; Khan, M.N.; Liu, J.; Xu, Z.; Hu, L. Temperature variation caused by sowing dates significantly affects floral initiation and floral bud differentiation processes in rapeseed (*Brassica napus* L.). *Plant Sci.* **2018**, *271*, 40–51. [CrossRef]
38. Kebrom, T.H.; Mullet, J.E. Photosynthetic leaf area modulates tiller bud outgrowth in sorghum. *Plant Cell Environ.* **2015**, *38*, 1471–1478. [CrossRef]
39. Corot, A.; Roman, H.; Douillet, O.; Autret, H.; Perez-Garcia, M.-D.; Citerne, S.; Bertheloot, J.; Sakr, S.; Leduc, N.; Demotes-Mainard, S. Cytokinins and abscisic acid act antagonistically in the regulation of the bud outgrowth pattern by light intensity. *Front. Plant Sci.* **2017**, *8*, 1724. [CrossRef]
40. Zhang, Q.-J.; Tao, S.-T.; Li, M.; Qi, X.-X.; Wu, J.; Yin, H.; Deng, J.-L.; Zhang, S.-L. Identification of differentially expressed genes using digital gene expression profiles in *Pyrus pyrifolia* Nakai cv. Hosui bud release following early defoliation. *Tree Genet. Genomes* **2015**, *11*, 34. [CrossRef]
41. Dong, X.; Wang, Z.; Tian, L.; Zhang, Y.; Qi, D.; Huo, H.; Xu, J.; Li, Z.; Liao, R.; Shi, M. De novo assembly of a wild pear (*Pyrus betuleafolia*) genome. *Plant Biotechnol. J.* **2020**, *18*, 581–595. [CrossRef]
42. Wang, N.; Yang, C.; Pan, Z.; Liu, Y.; Peng, S.A. Boron deficiency in woody plants: Various responses and tolerance mechanisms. *Front. Plant Sci.* **2015**, *6*, 916. [CrossRef]
43. Musacchi, S.; Iglesias, I.; Neri, D. Training systems and sustainable orchard management for European pear (*Pyrus communis* L.) in the Mediterranean area: A review. *Agronomy* **2021**, *11*, 1765. [CrossRef]
44. Hu, H.; Penn, S.G.; Lebrilla, C.B.; Brown, P.H. Isolation and characterization of soluble boron complexes in higher plants (the mechanism of phloem mobility of boron). *Plant Physiol.* **1997**, *113*, 649–655. [CrossRef] [PubMed]
45. Gholizadeh, J.; Sadeghipour, H.R.; Abdolzadeh, A.; Hemmati, K.; Vahdati, K. Bud break accompanies with the enhanced activities of hemicellulase and pectinase and the mobilization of cell wall thickenings in Persian walnut bud scales. *Trees* **2021**, *35*, 1399–1410. [CrossRef]
46. Zhao, X.; Han, X.; Wang, Q.; Wang, X.; Chen, X.; Li, L.; Fu, X.; Gao, D. EARLY BUD BREAK 1 triggers bud break in peach trees by regulating hormone metabolism, the cell cycle, and cell wall modifications. *J. Exp. Bot.* **2020**, *71*, 3512–3523. [CrossRef] [PubMed]
47. Walter, W.; Sánchez-Cabo, F.; Ricote, M. GOplot: An R package for visually combining expression data with functional analysis. *Bioinformatics* **2015**, *31*, 2912–2914. [CrossRef]
48. Jung, N.U.; Giarola, V.; Chen, P.; Knox, J.P.; Bartels, D. Craterostigma plantagineum cell wall composition is remodelled during desiccation and the glycine-rich protein CpGRP1 interacts with pectins through clustered arginines. *Plant J.* **2019**, *100*, 661–676. [CrossRef]
49. Zhang, H.; Guo, Z.; Zhuang, Y.; Suo, Y.; Du, J.; Gao, Z.; Pan, J.; Li, L.; Wang, T.; Xiao, L. MicroRNA775 regulates intrinsic leaf size and reduces cell wall pectin levels by targeting a galactosyltransferase gene in Arabidopsis. *Plant Cell* **2021**, *33*, 581–602. [CrossRef]
50. Wang, D.; Yeats, T.H.; Uluisik, S.; Rose, J.K.; Seymour, G.B. Fruit softening: Revisiting the role of pectin. *Trends Plant Sci.* **2018**, *23*, 302–310. [CrossRef]
51. Kim, J.-H.; Lee, Y.; Kim, E.-J.; Gu, S.; Sohn, E.J.; Seo, Y.S.; An, H.J.; Chang, Y.-S. Exposure of iron nanoparticles to *Arabidopsis thaliana* enhances root elongation by triggering cell wall loosening. *Environ. Sci. Technol.* **2014**, *48*, 3477–3485. [CrossRef]
52. Rose, J.K.; Saladié, M.; Catalá, C. The plot thickens: New perspectives of primary cell wall modification. *Curr. Opin. Plant Biol.* **2004**, *7*, 296–301. [CrossRef]
53. Chen, G.; Ye, X.; Zhang, S.; Zhu, S.; Yuan, L.; Hou, J.; Wang, C. Comparative transcriptome analysis between fertile and CMS flower buds in Wucai (*Brassica campestris* L.). *BMC Genom.* **2018**, *19*, 908. [CrossRef]
54. Feng, G.; Xu, X.; Xu, L.; Yang, Z.; Nie, G.; Ma, X.; Huang, L.; Zhang, X. Comparative Transcript Profiling Suggests Distinct Flowering Response of Early-and Late-Flowering Phenotypes in Forage Grass *Dactylis glomerata* L. *J. Plant Growth Regul.* **2021**, *40*, 2124–2138. [CrossRef]

55. Yap, Y.-M.; Loh, C.-S.; Ong, B.-L. Regulation of flower development in *Dendrobium crumenatum* by changes in carbohydrate contents, water status and cell wall metabolism. *Sci. Hortic.* **2008**, *119*, 59–66. [CrossRef]
56. Louvet, R.; Cavel, E.; Gutierrez, L.; Guénin, S.; Roger, D.; Gillet, F.; Guerineau, F.; Pelloux, J. Comprehensive expression profiling of the pectin methylesterase gene family during silique development in *Arabidopsis thaliana*. *Planta* **2006**, *224*, 782–791. [CrossRef]
57. Peaucelle, A.; Louvet, R.; Johansen, J.N.; Salsac, F.; Morin, H.; Fournet, F.; Belcram, K.; Gillet, F.; Höfte, H.; Laufs, P. The transcription factor BELLRINGER modulates phyllotaxis by regulating the expression of a pectin methylesterase in Arabidopsis. *Development* **2011**, *138*, 4733–4741. [CrossRef] [PubMed]
58. Di Matteo, A.; Giovane, A.; Raiola, A.; Camardella, L.; Bonivento, D.; De Lorenzo, G.; Cervone, F.; Bellincampi, D.; Tsernoglou, D. Structural basis for the interaction between pectin methylesterase and a specific inhibitor protein. *Plant Cell* **2005**, *17*, 849–858. [CrossRef]
59. Hocq, L.; Sénéchal, F.; Lefebvre, V.; Lehner, A.; Domon, J.-M.; Mollet, J.-C.; Dehors, J.; Pageau, K.; Marcelo, P.; Guérineau, F. Combined experimental and computational approaches reveal distinct pH dependence of pectin methylesterase inhibitors. *Plant Physiol.* **2017**, *173*, 1075–1093. [CrossRef]
60. Liao, S.; Wang, L.; Li, J.; Ruan, Y.-L. Cell wall invertase is essential for ovule development through sugar signaling rather than provision of carbon nutrients. *Plant Physiol.* **2020**, *183*, 1126–1144. [CrossRef] [PubMed]
61. Ko, H.-Y.; Ho, L.-H.; Neuhaus, H.E.; Guo, W.-J. Transporter *SlSWEET15* unloads sucrose from phloem and seed coat for fruit and seed development in tomato. *Plant Physiol.* **2021**, *187*, 2230–2245. [CrossRef] [PubMed]

agronomy

MDPI

Article

# Differentially Expressed Transcription Factors during Male and Female Cone Development in *Pinus halepensis*

**Gilad Reisfeld [1,2], Adi Faigenboim [1], Hagar Fox [1,3], Hanita Zemach [1], Leor Eshed Williams [2] and Rakefet David-Schwartz [1,*]**

[1] Institute of Plant Sciences, Agricultural Research Organization, Volcani Center, Rishon LeZion 7505101, Israel; gilad.reisfeld@mail.huji.ac.il (G.R.); adif@volcani.agri.gov.il (A.F.); hagar.fox@weizmann.ac.il (H.F.); hanita@volcani.agri.gov.il (H.Z.)

[2] The Robert H. Smith Institute of Plant Sciences and Genetics in Agriculture, The Hebrew University of Jerusalem, Rehovot 76100, Israel; leor.williams@mail.huji.ac.il

[3] Department of Plant and Environmental Sciences, Weizmann Institute of Science, Rehovot 76100, Israel

* Correspondence: rakefetd@agri.gov.il

**Abstract:** The molecular regulation of induction and development of reproductive organs is well established in angiosperms, while it is slowly accumulating in gymnosperms. Here, we conducted comparative transcriptomic profiles at different stages of male and female cone development in *Pinus halepensis*. Our aim was to reveal transcription factor encoding genes involved in reproductive induction and development. For accurate developmental stage identification, histological analysis preceded the molecular analysis. The flowering induction genes *FT/TFL* were expressed mainly at the late developmental stages of the male cone, suggesting involvement in vegetative bud dormancy instead of flowering induction. The male cone development was associated with the expression of the C-class *PhMADS2*, and the B-class genes *PhDAL11* and *PhDAL13*, while the female cone development was associated with the expression of the C-class *MADS1* and *DAL14* genes. This study adds valuable knowledge to the profile of transcription factors and MADS-box genes regulating cone development in gymnosperms.

**Keywords:** flowering induction; conifers; cone development; MADS-box; *Pinus halepensis*

**Citation:** Reisfeld, G.; Faigenboim, A.; Fox, H.; Zemach, H.; Eshed Williams, L.; David-Schwartz, R. Differentially Expressed Transcription Factors during Male and Female Cone Development in *Pinus halepensis*. *Agronomy* **2022**, *12*, 1588. https://doi.org/10.3390/agronomy12071588

Academic Editors: Anastasios Darras and Fengjie Sun

Received: 1 February 2022
Accepted: 28 June 2022
Published: 30 June 2022



## 1. Introduction

Reproduction in plants includes induction and organ differentiation stages. The molecular regulation of these processes is well characterized in angiosperms compared to gymnosperms [1–5]. Yet, the current genomic tools permit a slow bridging of this gap and provide valuable knowledge to understand cone development in conifers.

Flowering induction in angiosperms is regulated by transcription factors, including LEAFY (LFY), APETALA1 (AP1), CAULIFLOWER (CAL), and FRUITFUL (FUL) [1]. Other key regulators are genes belonging to the *FLOWERING LOCUS T (FT)/TERMINAL FLOWER 1 (TFL1)* family [6,7]. The FT protein (also known as florigen) is produced in the leaves, then transported to the meristem, where it forms a complex with the FLOWERING LOCUS D (FD) transcription factor, to induce flowering. In contrast, TFL1 is produced in the meristem and forms a complex with FD that inhibits the expression of the flowering-promoting genes *LFY* and the MADS-box gene, *APETALA1 (AP1)*, preventing flowering induction [8,9]. *SUPPRESSOR OF OVEREXPRESSION OF CO 1 (SOC1)*, encoding a MADS-box transcription factor, is another key gene in flowering induction [2,10,11]. The photoperiodic regulation of *SOC1* is mediated by *CONSTANS (CO)*, mainly through *FT*, which acts as an activator. SOC1 directly activates *LFY* and *AGL24*, a MADS-box transcription factor [2]. Genes encoding basic helix-loop-helix-type (bHLH) transcription factors are also related to flowering induction. These include *FLOWERING BHLH (FBH1-4)*, which together with *TEOSINTE BRANCHED 1/CYCLOIDEA/PROLIFERATING CELL NUCLEAR ANTIGEN FACTOR* (*TCP*) positively regulate *CO* [3].

Studies on induction of reproductive organs in gymnosperms, specifically in the *Pinaceae*, indicate similar and dissimilar mechanisms to angiosperms [12]. Contrary to the one copy of *LFY* in angiosperms, gymnosperms possess two copies; *LFY* and *Needly* (*NLY*) [13–15]. Both are expressed in vegetative buds, male, and female developing cones [14,16–18]. However, the *Picea abies LFY* (*PaLFY*) mRNA levels are slightly increased prior to the female cone setting in the early-cone-setting mutant *acrocona* [18]. Similarly, the *Pinus caribaea LFY* homolog, *PcLFY*, is found to be expressed dominantly in the female cone in early developmental stages [19]. Furthermore, both *LFY* and *NLY* from *Pinus radiata* as well as *LFY* from *P. caribaea* largely rescued the Arabidopsis *lfy* mutant phenotype, suggesting a role in transition to the reproductive stage [15,17,19].

The gymnosperm homologous of *FT*/*TFL1* was identified in *Pinaceae* [20,21]. The expression levels of the *P. abies* homologs *PaFTL1* and *PaFTL2* reached their maximum level before growth cessation, suggesting a role in dormancy regulation. Moreover, expression of *FTL1* and *FTL2* from *P. abies* and *Pinus tabuliformis* in Arabidopsis inhibited floral transition, demonstrating the functional homology of these genes to the *FT/TFL1* in angiosperms [21,22]. A *SOC1* homolog of *P. abies*, *DEFICIENCE-AGAMOUS*-LIKE 19 (*DAL19*) was highly expressed in the *acrocona* mutant prior to cone set, suggesting a role in flower induction similarly to *SOC1* [18].

Following flower induction, the development of flower organs is regulated mainly by transcription factors from the MADS-box family [23]. The ABC model, later expanded to the ABCDE model, describes the interactions between gene groups that regulate the different floral organs development [5,24]. In Arabidopsis, overlapping expression of E-class genes, which include *SEPALLATA*1,2,3,4 (*SEP*1,2,3,4) and *AGL6*-like, and A-class genes, which include *AP1* and *APETALA2* (*AP2*), induce sepal formation; E-class together with B-class, which include *APETALA3* (*AP3*) and *PISTILLATA* (*PI*), and A-class induce petal formation; E-class together with B- and C-class, which include *AGAMOUS* (*AG*) genes, form stamens; E- and C-class induce carpels, while E- together with C- and D-classes are responsible for ovule formation [5,25].

Many genes homologous to angiosperm floral organ determining genes were found in gymnosperms [5,21]. It appears that B-class homologous genes, including the *P. radiata PrDGL* and *DAL11-13* from *P. abies* and *P. tabuliformis*, are expressed specifically in developing male cones [21,26,27]. Gymnosperm genes homologous to angiosperm C-class genes include *PrMADS1-3* in *P. radiata*, as well as *DAL2* in *P. abies* and *P. tabuliformis* [21,28,29]. The *DAL2* homologs in other conifers such as *Podocarpus reichei* and *Taxus globose* exhibit female cone expression [30]. Interestingly, the *DAL10* MADS-box gene from *P. abies*, which do not show homology to any known angiosperm gene, exhibits a male specific function similar to that of B-class genes [31]. The overexpression of the *P. tabuliformis DAL10* homolog in Arabidopsis resulted in very early flowering induction, suggesting also a role in the transition to the reproductive phase in gymnosperms [21].

According to the gymnosperms model, overlapping C- and B-class gene expression results in male cone development, while the expression of only C-class genes causes female cone development. In both cases, the tetramers contain A class proteins [5]. A recent study revealed several physical interactions among MADS-box proteins in *P. tabuliformis*, supporting the hypothesis of the requirement for tetramerization of MADS-box proteins [21]. Furthermore, the large number of MADS-box homologous genes in conifers suggest complex regulatory interaction.

*P. halepensis* is a monoecious species (bearing male and female cones on the same tree), which begins its reproductive phase with the production of female cones approximately at the age of five years. Once the tree grows sufficiently in size, it begins to form male cones as well, usually one to two years later [32,33]. An early female cone development and a spatial distribution of the cones on a mature tree (female cones at the upper part and male cones at the middle and lower part, Supplementary Figure S1) facilitate outcrossing, ensuring high genetic diversity [32]. Female cone buds occur in a lateral position on a vegetative shoot bud. The female cone contains ovule-bearing scales (megasporophylls) attached to bracts and arranged spirally on a central axis, while the male cones are organized in clusters around a central axis (Figure 1). Vegetative to reproductive transition occurs in October–November depending on the provenance and environmental conditions, and cone development continues until March when pollination takes place [34].

**Figure 1.** Developmental stages of vegetative buds, and female and male cones in *P. halepensis*. (**A–C**) Vegetative buds at stages 1–3. (**A**) VS1, Early vegetative buds. (**B**) VS2, Dwarf shoots are visible. (**C**) VS3, Needles are visible. (**D–G**) Female cones. (**D**) FS1, Tightly enclosed bud. (**E**) FS2, Elongated and swelled bud. (**F**) FS3, Cone emergence through the top of the scales. (**G**) FS4, A visible cone. (**H–K**) Male cones. (**H**) MS1, Tightly enclosed bud. (**I**) MS2, Apparent buds. (**J**) MS3, Cone emergence from the scales. (**K**) MS4, Fully developed cones.

To gain knowledge on the molecular regulation of the reproductive phase of *P. halepensis*, we analyzed differentially expressed genes encoding for transcription factors at different stages of male and female cones development. The transcriptomic analysis was accompanied by histological analysis to achieve an accurate developmental stage identification. We identified several TF families that were dominantly expressed in either female or male cones, and also showed the cone specific expression of several MADS-box genes.

## 2. Materials and Methods

### 2.1. Plant Material

Plant material was collected from *Pinus halepensis* trees that grow at the Volcani Center (31.9872, 34.8212, Rishon LeZion, Israel). For transcriptome analysis, samples (13–18 shoot apices from 3–5 trees) were collected every week from December 2019 to March 2020 between 10:00 and 11:00 a.m. to minimize the effect of the circadian rhythm. Each shoot apex was immediately cut longitudinally into two parts. One part was frozen in liquid nitrogen and stored at −80 °C for mRNA extraction and sequencing, while the other half was fixated in formaldehyde-acetic acid–alcohol (FAA, 10:5:50 in double-distilled water) solution for histological analysis. The aim of the histological analysis was to determine the developmental stage and meristem identity (male/female/vegetative) of each collected sample.

For candidate gene analysis, samples were collected in a similar way from October 2020 to February 2021.

### 2.2. Histological Analysis

Samples in FAA were put in a vacuum chamber for 15 min and were stored overnight at 4 °C. Following dehydration in an ethanol series (70, 80, 90, and 100%, 30 min each), the samples went through a gradual Histoclear solution change up to 100% Histoclear. Overnight incubation at room temperature with Paraplast chips (Leica, Wetzlar, Germany, Paraplast Plus) was followed by several hours of incubation at 42 °C. Dissolved pure paraffin was changed twice a day for four days at 62 °C oven. The samples were embedded in blocks, sectioned into 10 µm width longitudinal sections (using Leica, RM2245 Microtome), and mounted on slides, which were incubated overnight on a 40 °C plate. The sections were then stained with Fast Green and Safranin [35], observed in a light microscope (Nikon Eclipse Ni-E), and images were taken using DS-Ri2 camera (Nikon, Tokyo, Japan).

### 2.3. Sample Selection for RNA Sequencing

Following histological analysis, terminal buds were chosen for mRNA sequencing according to their developmental stage [36] (Table 1). All developmental stages were identified, except for stage 1 for vegetative, male, and female, which was absent (probably due to late sampling). For each stage, we included three biological replicates.

**Table 1.** Developmental stages of vegetative shoot, and male and female buds.

| Meristem | Stage | Description |
|---|---|---|
| Vegetative | 1 * | Undifferentiated lateral meristems. |
| | 2 | Differentiated non-visible needles. |
| | 3 | Visible elongated needles. |
| Male | 1 * | Male flowering bud clearly identified. |
| | 2 | Differentiated microsporophyll. |
| | 3 | Developed pollen sacs with a tapetal layer on the edges, containing pollen mother cells. |
| Female | 1 * | Female flowering bud (ovulate cone) clearly identified. |
| | 2 | Bract differentiation, ovuliferous scales are not differentiated. |
| | 3 | Ovuliferous scales are at early differentiation. |
| | 4 | Ovuliferous scales are differentiated, ovules are visible. |

* Stages that were missing from the collected samples for RNA-seq.

### 2.4. RNA Isolation, Library Preparation and Sequencing

The selected buds were ground with liquid nitrogen, using a mortar and pestle. Total RNA was extracted with SpectrumTM Plant Total RNA Kit (Sigma, St Louis, MO, USA) following the manufacturer's instructions, and the RNA quantity was determined using a NanoDrop 1000 spectrophotometer (Thermo Fisher Scientific, Inc.). The RNA samples were sent to Macrogen Inc. (Amsterdam, the Netherlands) for the following sequencing procedure: RNA quality analysis using the 2200 TapeStation System (Agilent Technologies, Santa Clara, CA, USA), mRNA enrichment, library preparation using Illumina TruSeq RNA Library v2 kit, and 100 bp paired end sequencing by Illumina NovaSeq (Illumina, San Diego, CA, USA).

### 2.5. Transcriptome Analysis: Assembly and Differential Expression Analysis

A transcriptome database was de novo assembled using Trinity software (version: v2.1.1) [37] based on the sequences derived from the RNA sequencing. The transcript quantification from the RNA-Seq data was performed using the Bowtie2 aligner [38] and the expectation maximization method (RSEM) by estimating maximum likelihood expression levels [39] via the perl script align_and_estimate_abundance.pl with –est_method RSEM from Trinity protocol [40]. According to the sample correlation matrix (Supplementary Figure S2), three samples showed expression patterns which were not sufficiently correlated with their equivalent biological replicates. These samples (Male stage 3 replicate 1, Female stage 2 replicate 2, and Vegetative stage 1 replicate 1) were excluded from further analyses. Differential expression analyses were done using the edgeR R package [41].

### 2.6. Sequence Similarity and Functional Annotation

The merged transcriptome was used as a query term for a search of the NCBI non-redundant (nr) protein database that was carried out with the DIAMOND program [42]. Homologous sequences were also identified by searching the Swiss-Prot database with the BLASTx tool [43] and an E-value threshold of $10^{-5}$. The search results were imported into Blast2GO version 4.0 [44] for gene ontology (GO) assignments. Enzyme codes and KEGG pathway annotations were based on the mapping of GO terms to their enzyme codes. The KAAS tool (Kegg Automatic Annotation Server; http://www.genome.jp/tools/kaas/, accessed on 1 July 2021) was used for KEGG orthology and KEGG pathway assignments. Gene ontology enrichment analysis was carried out using the Blast2GO program based on Fisher's exact test [45] with multiple testing correction of false discovery rate (FDR) [46]. The threshold was set as FDR with a corrected *p*-value of less than 0.05. Gene ontology analysis was done by comparing the GO terms in the test sample to the GO terms in a background reference. Every comparison had hundreds to thousands of DE genes. In order to focus on transcription factors, the differentially expressed (DE) genes were blasted against the transcription factors database PlantFTDB (http://planttfdb.gao-lab.org/, accessed on 1 September 2021). In addition, the homologs of the MADS-box genes were isolated from the identified DE transcription factors.

### 2.7. Phylogenetic Analysis

A phylogenetic tree of the MADS-box transcription factors was constructed with MEGA X [47] using the maximum likelihood method. In silico translated amino acid sequences were used for this analysis. One isoform was selected from each gene, according to the highest protein annotation score. The *P. halepensis* proteins were aligned using the MUCSLE algorithm [48] with MADS-box protein sequences from Arabidopsis (retrieved from The Arabidopsis Information Resource, TAIR) and other conifer MADS-box proteins (retrieved from UniProt, https://www.uniprot.org/, accessed on 1 September 2021). The conserved MADS and K domains together with the linker between them were used for the alignment. A thousand replicates were used for bootstrapping. The evolutionary history was inferred by using the maximum likelihood method and JTT matrix-based model [49]. Initial tree(s) for the heuristic search were obtained automatically by applying Neighbor-Join and BioNJ algorithms to a matrix of pairwise distances estimated using the JTT model, and then selecting the topology with a superior log likelihood value.

### 2.8. Quantitative Polymerase Chain Reaction (qPCR) Analysis

Candidate genes were analyzed by qPCR in samples that were collected in the second year (see above). Primers were designed to amplify 70–200 bp segment of each candidate gene (Supplementary Table S1). Each type of bud at each developmental stage was represented by three to four biological replicates, with two technical replicates for each. Since sample collection for the qPCR analysis started earlier in the second year than in the first year, it included earlier developmental stages; 1, 3 and 4 for female, and 2 and 3 for male. RNA was extracted as described above from all the samples and diluted equally to reach a concentration of 1000 ng/μL. DNA was degraded using RQ1 RNase-Free DNase (Promega, Madison, WI, USA). cDNA was synthesized using the qScript cDNA Synthesis Kit (QIAGEN, Hilden, Germany). Finally, qPCR was done using ABsolute Blue QPCR Mix, SYBR Green, ROX (Thermo Scientific™, Waltham, MA, USA). The expression level of the chosen genes was normalized according to the expression levels of the *P. halepensis* homologs of *UPL6* (E3 ubiquitin-protein ligase) TRINITY_DN211436_c2_g2_i2. This gene showed little variation in expression levels between the different bud types, suggesting reliability for qPCR relative expression levels normalization. The relative gene expression was calculated according to the $2^{-\Delta\Delta CT}$ method [50], using the housekeeping gene CT values, and the average CT values of vegetative stage 1 tissues, defined as the control group. The derived values were converted to its logarithmic base 2 for stabilization of variances.

## 3. Results

### 3.1. Phenotypic Characterization of the Developmental Stages

To gain insights into the genes involved in reproduction in *P. halepensis*, we collected the three different bud types, vegetative, female and male at different developmental stages (Figure 1), for RNA-seq, and a year later for qPCR analysis as described in Material and Methods. The buds presented in Figure 1 and the histology analysis in Figure 2 represent all stages identified in this study.

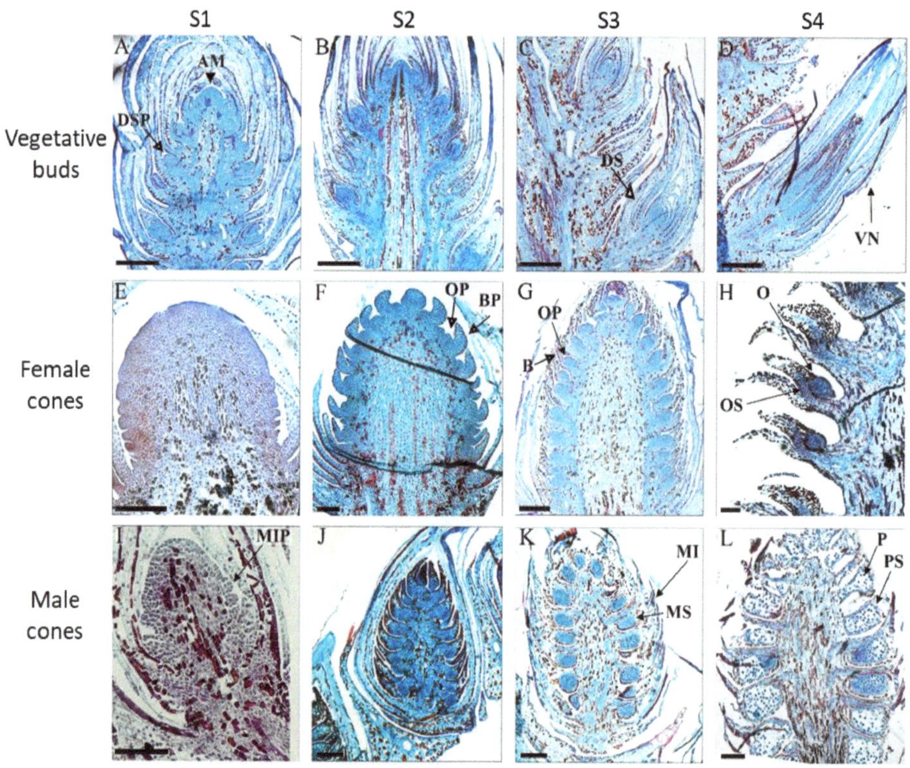


**Figure 2.** Longitudinal sections of vegetative, female and male *P. halepensis* buds at different developmental stages. **(A–D)** Vegetative buds developmental stages. **(A)** S1, Dwarf shoot (Brachyblasts) primordia (DSP) with apical meristem (AM). **(B)** (S2) and **(C)** (S3), Dwarf shoots (DS) emerge through the scales. **(D)** S4, Visible needles (VN). **(E–H)** Female cones developmental stages. **(E)** S1, Bulge-like structure enclosed within the scales with few lateral bract anlagen. **(F)** S2, bract primordia (BP) and ovuliferous primordia (OP). **(G)** S3, Bracts (B) are fully developed covering the developing ovuliferous scales. **(H)** S4, bract and ovuliferous scale (OS) complex with developing ovule (O). **(I–L)** Male cones developmental stages. **(I)** S1, Lateral microsporophyll primordia (MIP). **(J)** S2, Young cones hidden within the scales, contains developing microsporophyll. **(K)** S3, Cone emergence from bud scales and the microsporophyll (MI) contains microsporangium (MS). **(L)** S4, fully developed cones with pollen sacs (PS) and pollen mother cells (P). Scale Bar, **(A–E,I)** 500 μM, **(F–H,J–L)** 250 μM.

At their early developmental stage, the vegetative, female and male closed buds look similar (Figure 1A,D,H). Soon after the buds elongate and swell, their different morphology becomes clearly visible (Figure 1B,E,I). The vegetative shoot develops needles, which further elongate (Figure 1C), and the female cone emerges through the top of the scales (Figure 1F,G), while the many small male cones emerge from the bud scales to be ready for pollen release (Figure 1J,K).

### 3.2. Histological Analysis

Histological analysis that was performed on half of each sample (see Materials and Method) revealed the developmental stages of the three different bud types (Figure 2). In early developmental stage, the lateral meristems of the vegetative bud are still not fully developed and enclosed within the bud scales, this stage was called "vegetative stage 1" (VS1). At stage 2 (VS2), the lateral meristems are apparent, and the developing brachyblasts (dwarf shoots) are growing outside the bud scales. At stage 3 (VS3), the bud is fully developed and the elongated needles are visible.

At stage 1 of the female cone developmental (FS1), a huge swollen structure with a few lateral primordial bracts appear, enclosed within the bud scales. At stage 2 (FS2), the bud is enlarged in size, the bract primordia are elongated, and the ovuliferous scale anlagen becomes visible. During stage 3 (FS3), developed bracts cover the developing ovuliferous scales. At stage 4 (FS4), the cone is fully developed with ovules, ready to be pollinated. The developmental stages presented in Figure 2 for the female cone are consistent with the previously defined stages [51].

The first stage of the male cone development is characterized by lateral microsporophyll primordia. In the second stage (MS2), a swollen structure, hidden within the bud scales, contains developing microsporophyll. At MS3, the cones are emerged from the bud scales, and the microsporophyll contains a microsporangium (pollen sac), with pollen mother cells surrounded by a tapetal layer. In the final stage of male cone development (MS4), the cone contains fully developed pollen grains ready to be carried by wind for pollination.

### 3.3. Transcriptome Analysis of the Three Bud Types at Early and Late Differentiation Stages

Based on the developmental stages determined by the histology analysis in the first year, we realized that we started our sampling a little late. Thus, the selected buds for RNA-seq represented stages 2 and 3 for vegetative bud (VS2 and VS3) and male cone (MS2 and MS3), and stages two to four for the female cone (FS2-FS4). Twenty-one cDNA libraries yielded an average of 44 million 100 bp paired-end reads, which were assembled using Trinity as described in Material and Methods. The transcriptome catalog contained 546,815 transcripts corresponding to 414,456 genes. The average contig length was 735.42 bp, with an N50 size of 1487 bp corresponding to a total length of 402 Mbp. Annotating the transcriptome catalog resulted in 81,521 contigs. Differential gene expression analyses between the three buds, both at early and late developmental stages, revealed that the number of DE genes between male and vegetative types is the highest, whereas between the female and the vegetative types is the lowest (Table 2, full list of genes in Supplementary Table S2). Moreover, more genes were DE at late developmental stages than in early developmental stages (Table 2), reflecting the advanced differentiation of each organ. Comparison between stage 2 and stage 3 of each bud type revealed that the vegetative bud exhibits the smallest number of DE between the two stages, whereas the male cones exhibited the largest number of DE genes. A similar number of genes were up- and down-regulated between stage 2 and stage 3 in all three bud types (Table 2). Shared expressed genes between the three developmental stages of the female cone are presented in Figure 3. It appears that 649 genes were uniquely up-regulated in developmental FS3 compared to FS4, suggesting that at the latter stage, some processes were completed.

**Table 2.** Number of differentially expressed genes between the different bud types, and between the developmental stages in each bud type.

| | Comparison | Up-Regulated | Down-Regulated | Total |
|---|---|---|---|---|
| Early developmental Stage | FS2 vs. MS2 | 417 | 405 | 822 |
| | FS2 vs. VS2 | 92 | 88 | 180 |
| | MS2 vs. VS2 | 519 | 689 | 1208 |
| Late developmental Stage | FS3 vs. MS3 | 2382 | 2115 | 4497 |
| | FS3 vs. VS3 | 1020 | 1300 | 2320 |
| | MS3 vs. VS3 | 2042 | 2521 | 5563 |
| Vegetative bud development | VS2 vs. VS3 | 139 | 224 | 363 |
| Male cone development | MS2 vs. MS3 | 2200 | 2155 | 4355 |
| Female cone development | FS2 vs. FS3 | 558 | 738 | 1296 |
| | FS2 vs. FS4 | 1219 | 1023 | 2242 |
| | FS3 vs. FS4 | 1283 | 832 | 2115 |

S, developmental stage; F, female; M, male; V, vegetative; DE, differentially expressed.

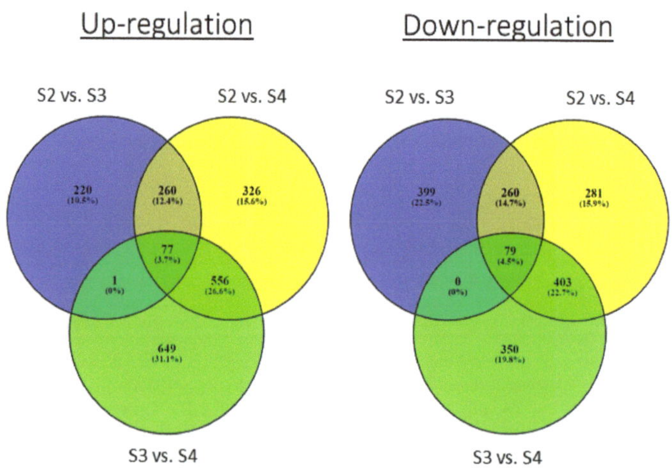


**Figure 3.** Venn diagram of up-regulated and down-regulated genes showing shared genes between the female cone three developmental stages. S2–S4, developmental stages.

### 3.4. *Transcription Factors Families Involved in Early and Late Stages of Cone Development*

To identify transcription factors related to cone development, we blasted the DE genes against the transcription factors database PlantFTDB (http://planttfdb.gao-lab.org/, accessed on 1 September 2021) and grouped the genes into TF families (Figure 4). Included in Figure 4 are TF families with at least three members differentially expressed in at least one comparison. The number of TFs that were DE during the male cone development was dramatically higher than DE TFs during the female cone development (Figure 4). For female cone development, more TFs were up-regulated in the late than in the early developmental stages. The most represented TF families (>10 members in both stages) in the male cone were bHLH, ERF, MYB-related, NAC, and MIKC-MADS (Figure 4). The MYB-related family was highly represented in the late female stages, while the ERF, bHLH, and MYB families were highly represented in the most advanced developmental stage, FS4. Interestingly, the HD-ZIP family was dominantly expressed at MS2 with 15 family members. In addition, male cone development exhibited TFs (GRF, HB-other, bZIP, ZF-HD, NF-YC, FAR1, and Nin-like) that were not DE during the female or vegetative bud development. MICK_MADS and M-type_MADS were enriched in the early and late stages of male and female cone development (Figure 4).

**a** Up -Regulated TFs at Stage 2

| TF Family | MS2/MS3 | FS2/FS3 | FS2/FS4 | VS2/VS3 |
|---|---|---|---|---|
| bHLH | 21 | 2 | 13 | 0 |
| ERF | 16 | 6 | 17 | 1 |
| MYB_related | 15 | 3 | 9 | 0 |
| HD-ZIP | 15 | 1 | 2 | 0 |
| NAC | 13 | 5 | 9 | 0 |
| MIKC_MADS | 10 | 4 | 9 | 1 |
| C2H2 | 9 | 4 | 6 | 1 |
| LBD | 9 | 2 | 8 | 1 |
| NF-YC | 8 | 0 | 2 | 0 |
| MYB | 8 | 2 | 12 | 1 |
| C3H | 8 | 1 | 3 | 0 |
| WRKY | 7 | 1 | 9 | 0 |
| Trihelix | 7 | 0 | 1 | 1 |
| G2-like | 7 | 2 | 3 | 0 |
| FAR1 | 6 | 1 | 2 | 0 |
| ARF | 5 | 1 | 1 | 0 |
| GeBP | 5 | 1 | 1 | 1 |
| * GRF | 5 | 0 | 0 | 0 |
| HSF | 4 | 2 | 1 | 0 |
| M-type_MADS | 4 | 1 | 7 | 0 |
| AP2 | 4 | 2 | 5 | 0 |
| * HB-other | 4 | 0 | 0 | 0 |
| * bZIP | 3 | 0 | 0 | 0 |
| WOX | 3 | 0 | 1 | 0 |
| * ZF-HD | 3 | 0 | 0 | 0 |
| B3 | 3 | 0 | 1 | 0 |
| YABBY | 3 | 0 | 4 | 0 |
| NF-YB | 3 | 0 | 2 | 0 |
| Total: | 208 | 41 | 128 | 7 |

**b** Up -Regulated TFs at Stage 3

| TF Family | MS3/MS2 | FS3/FS2 | FS4/FS2 | VS3/VS2 |
|---|---|---|---|---|
| ERF | 24 | 6 | 27 | 1 |
| bHLH | 24 | 7 | 10 | 1 |
| MYB_related | 22 | 12 | 14 | 2 |
| NAC | 18 | 5 | 8 | 0 |
| MYB | 16 | 6 | 12 | 0 |
| C2H2 | 16 | 3 | 4 | 1 |
| LBD | 15 | 4 | 5 | 0 |
| MIKC_MADS | 14 | 4 | 4 | 0 |
| Trihelix | 12 | 4 | 2 | 0 |
| M-type_MADS | 8 | 3 | 6 | 0 |
| C3H | 8 | 0 | 2 | 0 |
| WRKY | 7 | 4 | 7 | 1 |
| bZIP | 7 | 2 | 6 | 0 |
| GATA | 7 | 0 | 2 | 1 |
| * NF-YC | 6 | 0 | 0 | 0 |
| B3 | 5 | 2 | 1 | 0 |
| TCP | 4 | 1 | 1 | 0 |
| * FAR1 | 4 | 0 | 0 | 0 |
| ZF-HD | 4 | 4 | 2 | 1 |
| ARF | 3 | 0 | 2 | 1 |
| * Nin-like | 3 | 0 | 0 | 0 |
| HD-ZIP | 3 | 6 | 3 | 0 |
| YABBY | 3 | 0 | 1 | 0 |
| AP2 | 3 | 2 | 1 | 0 |
| CO-like | 3 | 1 | 1 | 0 |
| GRAS | 3 | 0 | 2 | 0 |
| NF-YB | 1 | 0 | 5 | 6 |
| Total: | 243 | 76 | 128 | 15 |

**Figure 4.** Heat-map of up-regulated transcription factors (TF) grouped into families involved in cone development. Included are families which showed at least three up-regulated members in at least one meristem type in the early (**a**) and late (**b**) stages of cone development. Color intensity of the heat-map reflects the number of genes. Common families between the early and late stages share the same color, while uncommon families are not colored. Asterisks indicate TF families found only in the male cone. MS2, Male Stage 2; MS3, Male Stage 3; FS2, Female Stage 2; FS3, Female Stage 3; FS4, Female Stage 4; VS2, Vegetative Stage 2; VS3, Vegetative Stage 3.

### *3.5. Differentially Expressed TFs Involved in Flowering Induction*

The *LFY* homolog gene (TRINITY_DN212345_c0_g2) exhibiting 97.5% similarity to the *P. radiata NLY* (Accession U76757) had higher expression level in MS2 compared to MS3 (log2FC = 2.48, Supplementary Table S2). Except for that, *LFY* homologs did not show differential expression between meristem types or developmental stages.

Two *FT/TFL1* homologous contigs (TRINITY_DN197941_c0_g2 and TRINITY_DN197941_c0_g3) were up-regulated in the male cones at S3, compared to vegetative buds at S3 (log2FC = 10.33 and 5.75, respectively), and also up-regulated in MS3 compared to MS2 (log2FC = 4.8 and 10.5, respectively). Both contigs exhibit high similarity (93% and 97%, respectably) to the *Pinus armandii* FT/TFL protein. TRINITY_DN197941_c0_g3 (*FT/TFL1*) was found to decline in female cones as development proceeded. In addition, a MOTHER OF FT and TFL1-like homolog (MFT) (TRINITY_DN203996_c0_g3) was highly expressed at S3 of the female cone (Supplementary Table S2, Figure S3).

From the bHLH family, 31 TFs were identified as DE among the six comparisons between different bud types at early and late developmental stages (Figure 5). DE *bHLH* TFs were more pronounced in the male than the female cones. Early differentiated female cones had one up-regulated gene, a MYC2–like homolog. Early differentiated male cones had seven up-regulated genes, including homologous to *HEC2*, *EAT1* and *bHLH63*. Three down-regulated genes in early differentiated male cones included homologs of *ABORTED MICROSPORES*, *bHLH112* and *bHLH16*. Late differentiated female cones had two up-regulated genes (*bHLH094*, *bHLH49* homologs) and one down-regulated gene (*BHLH140*). Like MADS-box genes, the most DE bHLH TFs were up-regulated in late differentiated male cones. Substantial down-regulation of the *bHLH041* homolog was observed in late differentiated male cones (Figure 5).

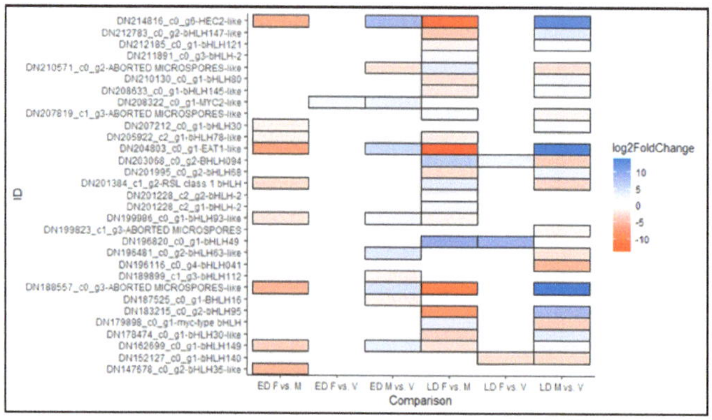

**Figure 5.** *bHLH* TFs differentially expressed in six comparisons. Gene ID is composed of the TRINITY accession ID, followed by NCBI BLASTP derived homolog. ED, Early development (Stage 2); LD, Late development (Stage 3); F, female; V, vegetative; M, male.

### 3.6. Differentially Expressed TF Involved in Female Cone Differentiation

Altogether, 25 MADS-box TFs were identified as DE in six comparisons (Figure 6). Three of them, homologs of *DAL3* were up-regulated in early differentiating female cones. Late differentiating female tissues had six up-regulated genes, including *DAL14*, *MADS1*, *MADS2* and *MADS5* homologous genes. Those tissues had five down-regulated genes as well, among them are *DAL13*, *AGL1* and *SOC1* homologous genes, together with two putative MADS-box genes.

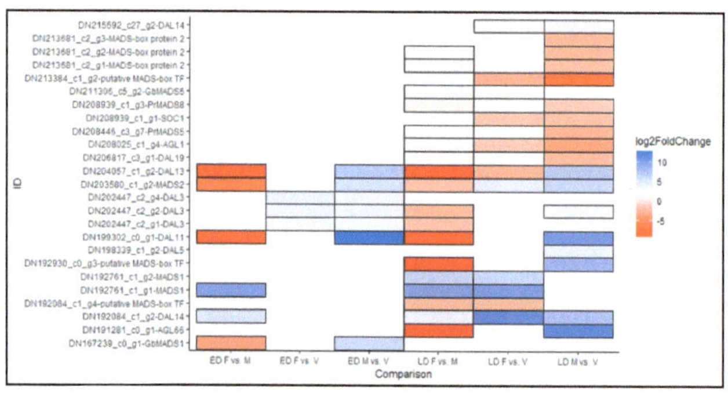

**Figure 6.** MADS-box TFs differentially expressed in six comparisons. Gene ID is composed of the TRINITY accession ID, followed by NCBI BLASTP derived homolog gene. ED, Early development (Stage 2); LD, Late development (Stage 3); F, female; V, vegetative; M, male.

### 3.7. Differentially Expressed TF Involved in Male Cone Differentiation

Seven genes, including homologs of *MADS1*, *MADS2*, *DAL3*, *DAL11* and *DAL13*, were up-regulated in early differentiating male cones. Late differentiated male cones were the most prominent bud type in terms of the magnitude of DE MADS-box genes, with nine up-regulated genes (*DAL3*, *DAL5*, *DAL11*, *DAL13*, *DAL14*, *MADS2*, *AGL66*, and a putative MADS-box homologous gene), and seven down-regulated genes (*MADS-box 2*, *MADS8*, *SOC1*, *MADS5*, *AGL1*, *DAL19*, and a putative MADS-box homologous gene).

### 3.8. Phylogenetic Analysis

The importance of MADS-box genes in flowering regulation as well as other developmental processes, led us to focus on those genes for further analysis. We found that most of MADS-box homologs had several isoforms. Thus, we generated a phylogenetic tree of the 25 identified MADS-box TFs together with homologs from Arabidopsis and representative gymnosperms, including *P. radiata*, *Ginkgo biloba*, and *P. abies*. The 25 TFs were divided into 14 different clades, while the clade containing the higher number of *P. halepensis* MADS-box proteins was the sister clade of the Arabidopsis *SOC1* TF, which included the *P. abies* proteins *DAL3* and *DAL19* as well (Supplementary Figure S4).

### 3.9. Differentially Expressed Gene Validation

Eight MADS-box genes were chosen for further characterization by qPCR. We chose them due to their potential role in cone development manifested by their up-regulation in at least one developmental stage (Figure 6; Supplementary Table S3). Each gene was represented by its most expressed isoform. The selected genes were named according to their closest coniferous homologous inferred from their closest NCBI BLASTP results, as well as the derived phylogenetic tree.

Homogeny of variances between the different bud types was determined by the Bartlett test. We found the variances to be unequal in *PhDAL11* and *PhMADS1*. Therefore, significant differences in *PhDAL11* and *PhMADS1* were determined by six different t-tests, under the assumption of unequal variances. Bonferroni correction was applied in *PhMADS1* and *PhDAL11* resulting in $\alpha$ values of 0.0083. The Tukey–Kramer HSD test supported significant differences found by the independent t-tests. Therefore, the differences in relative expression levels of all genes were determined using the Tukey–Kramer HSD test.

The qPCR showed that *PhDAL11* and *PhDAL13* were highly expressed across the male cone developmental stages (Figure 7). *PhDAL14*, *PhMADS1*, and *PhMADS2* were highly expressed across the female cone developmental stages; however, they were also highly expressed at late developmental stages of the male cone. The expression of *PhMADS1* was increasing gradually throughout the female cones development, while *PhMADS2* showed a weak correlation with cone development. The expression patterns of all genes described above, corresponded to the RNA-seq results (Figure 6, Supplementary Table S3). However, the expression patterns of *PhDAL3.1*, *PhDAL3.2*, and *PhMADS5* were not consistent with the RNA-seq results (Figure 7).

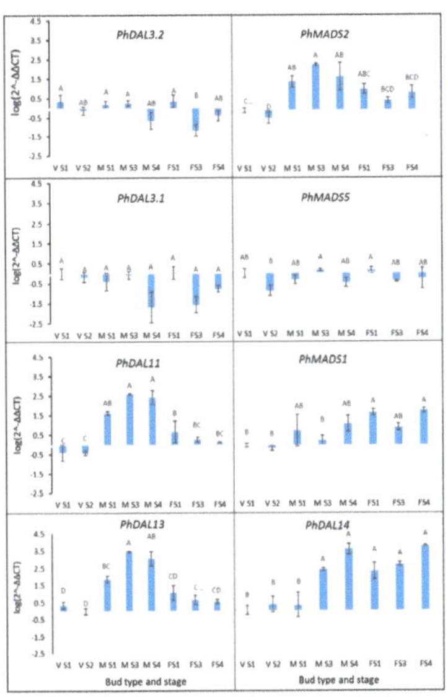


**Figure 7.** Relative expression levels of eight MADS-box genes in vegetative buds, male, and female cones during several developmental stages. Uppercase letters indicate significant differences ($p < 0.05$). Bars represent standard errors. VS1, Vegetative Stage 1; VS2, Vegetative Stage 2; MS1, Male Stage 1; MS3, Male Stage 3; FS1, Female Stage 1; FS3, Female Stage 3; FS4, Female Stage 4.

## 4. Discussion

This study describes the involvement of transcription factors in the induction and development of the male and female cones in *P. halepensis*. Identification of the different developmental stages through histology prior to the molecular analysis allowed us to identify differentially expressed TFs at early and late developmental stages. This study enriches our knowledge on the regulation of reproductive development in gymnosperms, and particularly in pines.

By analyzing the transcriptome of vegetative, male, and female tissues at different developmental stages, we found that the higher number of DE genes correlated with the male tissues (Table 2). That may be attributed to a more complexed regulation of male compared to female cone development [5]. Male cone development follows the BC model, according to which expression of both B- and C-class TFs are required, whereas female cone development requires solely C-class TFs [5]. DE genes were also prominent in all tissues in late compared to early developmental stages, suggesting that cone structure and tissue specification at late developmental stages are complex.

### 4.1. Reproductive Development Is Accompanied by Enriched TF Expression

TFs that were DE in early and late cone developmental stages were enriched in the five families bHLH, ERF, MYB-related, NAC, and MIKC-MADS (Figure 4). These TF families are essential in many processes throughout the plant's life cycle. In cone development, the bHLH family was prominent with more than 20 genes in early and late male cone development stages. This suggests that *bHLH* genes may have a significant regulatory role in reproductive organ development in conifers, similar to what was previously described in Arabidopsis [3]. A study on Arabidopsis and the liverwort *Marchantia polymorpha* showed that *bHLH* genes are crucial for the formation of generative cells that give rise to the male gametes, suggesting that *bHLH* may be key regulators of reproductive development across plant species [52]. Recently, it was shown that the *bHLH* expression in *P. tabuliformis* is age dependent, suggesting a role in the vegetative growth transition [53]. In light of these reports, the current study results, which suggest massive involvement of bHLH genes in the reproductive stage, encourage additional studies in conifers. Another noticeable TF family is the ERF, a subfamily within the AP2/ERF super-family [54]. We found that most *ERF* genes were up-regulated in late female cone development (Figure 4). *AP2* homologs, *AP2L1* and *AP2L2*, were identified in *P. abies* and *Pinus thunbergii*. *AP2L1* was dominantly expressed in the female cone, whereas *AP2L2* was expressed in both cone types in both species [55,56]. In addition, *PaAP2L2* was able to rescue the *ap2-1* Arabidopsis mutant phenotype [57]. These functional studies imply that ERF TFs in *P. halepensis* likely possess *AP2/ERF* homologs regulating late stages of female cone development.

### 4.2. Floral Induction Homologous Genes

Unexpectedly, we found only one *LFY* homolog, which was DE in the male cone tissue only. Its expression level was higher at stage 2 compared to stage 3. This result is inconsistent with previous studies that showed high expression of *LFY* and *NLY* in early stages of male and female cone development [58]. It has been shown that *NLY* is expressed in reproductive as well as in vegetative tissue in pines. However, its capacity to rescue the Arabidopsis *lfy* mutant implies that it also plays a role in flowering induction [17]. In the current study, the absence of *LFY* homologs in the DE genes may be due to late tissue collection for the RNA-seq analysis, after the expression peak of *LFY* and *NLY*.

The expression patterns of the two *FT/TFL1* genes imply a role in late male cone development (Figure S3A). One of those genes (TRINITY_DN197941_c0_g3) had the highest expression level at FS2 and then declined as development progressed (Figure S3B). The *FT/TFL* homologs exhibited a similar expression level in vegetative and female buds. Yet, an additional *FT/TFL* homolog (a MOTHER OF FT and *TFL1*-like, Table S2) was DE between the three developmental stages of the female cone, with highest expression level seen at stage 3, suggesting that *FT/TFL* may have a role in female cone development. The opposite expression pattern of the *FT/TFL1* homologs in female and male cones indicates a different function, which may be antagonistic. It has been shown that *FT/TFL1* homologs from conifers act similarly to *TFL1* in repressing instead of promoting flowering, and do not seem to have a major role in the reproductive development of pines [20,22]. A recent publication reporting on age-related gene expression in *Pinus tabuliformis* suggests that interactions between the MADS-box genes *MADS11* and *DAL1* mediate flowering induction in pines [53]. The same study showed that the over-expression of *PtMADS11* in Arabidopsis rescued the late flowering *ft* phenotype, implying a role in flowering induction [53]. Since in our study we did not include pre-cone differentiation stages in the RNA-seq analysis, we could not confirm that *MADS11* and *DAL1* are involved in the flowering induction in *P. halepensis*.

The *PhDAL3.1* and *PhDAL3.2* genes exhibited high expression level during early differentiation stages of both cone types. The closest homologous gene of *PhDAL3.1* and *PhDAL3.2* is *DAL3*, which belongs to an orthologous sister clade and thus to the entire angiosperm SOC1 clade [58]. Our phylogenetic analysis also classified the *P. halepensis* DAL3 homologs in a sister clade of the SOC1 clade, together with the *P. abies* DAL3 (Figure S4). The central role of SOC1 in the transition to flowering in Arabidopsis through multiple pathways [2] also implies a central role for *PhDAL3.1*, *PhDAL3.2* in pine, assuming functional similarity between angiosperms and gymnosperms. The qPCR results showed no difference in *PhDAL3.1*, *PhDAL3.2* expression between both cone types and the vegetative buds, in all developmental stages. The only significant difference observed was a lower expression of *PhDAL3.2* in FS3 cones. Therefore, our data suggest a role for *PhDAL3.1* and *PhDAL3.2* in the development of all bud types, including vegetative growth.

Although the RNA-seq data showed that *PhMADS5* is up-regulated in late developing female cones, the qPCR results found no difference in the expression levels of this gene in any specific tissue (Figure 7). The closest Arabidopsis homolog of *PhMADS5* is *SHORT VEGETATIVE PHASE* (*SVP*), involved in flowering time regulation and transition to flowering [59,60]. Like *SOC1* orthologs, *SVP* is a multiple copies family in gymnosperms [61], which indicates that like *SOC1*, different *SVP* paralogs may have a different role in developmental processes in conifers.

### 4.3. Genes Involved in Male and Female Cone Development

We found that the MADS-box genes *PhDAL11* and *PhDAL13* expressed dominantly in the Male cones. These genes were previously found to be phylogenetically related to B- class MADS-box genes, expressed specifically in male cones of *P. abies* and *P. tabuliformis* [58,62], and had similar phenotypic effects to endogenous Arabidopsis B-class proteins when expressed ectopically in Arabidopsis [27]. A new study published recently further demonstrated the male cone specific expression of *DAL11* and *DAL13* in *P. tabuliformis*, and showed physical interactions between these two proteins [21]. Together with these studies, our study supports the model proposing that B- class genes are essential for male identity determination [5]. An additional B-class gene examined in the current study, *PhMADS2*, was up-regulated in both cone types, yet most prominently in the male cone. Niu et al. [21] showed that PtMADS2 interacts with PtDAL11 and PtDAL13, supporting the involvement of PhMADS2 in the male cone development.

*PhMADS1* was up-regulated in female cones (Figure 7). The most compatible alignment of PhMADS1 in Arabidopsis was the AP1 protein, which is an A-class TF, important for both floral meristem identity transition and the development of petal and sepals [1,5]. *MADS1* in *P. radiata* was shown to be highly expressed in early cone development and was localized to the ovuliferousscale primordia [13]. MADS1 homologs from *P. tabuliformis*, *PtMADS1* and *PtMADS10* were also specifically expressed in the developing female cone [21]. They were shown to physically interact with other female-related MADS-box proteins, such as PtDAL14 and PtDAL1, to regulate the female cone development. It is likely that PhMADS1 plays a similar role in *P. halepensis*.

In our study, *PhDAL14* was highly expressed throughout the female cone development; however, it was also up-regulated in the male cone during the middle to late developmental stages (Figure 7). This expression pattern is in agreement with *PaDAL14* expression in *P. abies* and *P. tabuliformis* female and male cone development [21,58]. AGL6 is the most closely related Arabidopsis protein to PhDAL14. AGL6 has an E function of floral organ development, with additional functions including floral meristem regulation, ovule and seed development [63]. The function of AGL6 and other E class proteins in angiosperms seem similar to that of *PhDAL14* in conifers. This suggests that *PhDAL14* is an essential component for both female and male cone development [28,64]. Physical interactions between DAL14 and the male-specific proteins DAL11 and DAL13 in *P. tabuliformis* support its involvement in regulating both cone types [21].

Function of the genes presented in the current study is confirmed by heterologous complementation done in Arabidopsis mutants in similar studies, as mentioned above. However, gene silencing via the CRISPR/Cas9 approach offers a better understanding of gene function, assuming that transformation and regeneration platforms are available for coniferous species. Silencing key genes in the development of the female pine cone can result in sterile trees, an outcome desirable for both commercial and ecological purposes. Sterile trees might direct the energy to vegetative growth, offering a great economic potential for timber production. In particular, sterility in *P. halepensis* might prevent the negative environmental impacts caused by uncontrolled regeneration in forests, thereby increasing the efficiency of forest management.

## 5. Conclusions

Our study contributes to the accumulating evidence of the molecular regulation of the reproduction phase in conifers. The expression pattern of the *P. halepensis* MADS-box genes was very similar to that found in other conifers, as well as to the newly studied homologs from *P. tabuliformis* [21], suggesting highly conserved mechanisms in Pinaceae. Yet, the large number of transcription factors revealed in our study suggests that the reproductive phase in Pinaceae is much more complex. We believe that this study will urge further investigation into the regulation of reproductive development in conifers.

**Supplementary Materials:** The following supporting information can be downloaded at: https://www.mdpi.com/article/10.3390/agronomy12071588/s1, Figure S1: A *P. halepensis* tree with male and female cones; Figure S2: Heatmap of Pearson correlation analysis evaluating the similarity between samples; Figure S3: Differential expression of *FT/TFL1* homologs between different developmental stages of the male (A) and female (B) cones. Figure S4: Maximum likelihood phylogenetic analysis of 25 MADS-box proteins from *P. halepensis* related to MADS-box proteins from other species. Table S1: Primers used in qPCR; Table S2: Transcriptome analysis of *Pinus halepensis* cones development; Table S3: Selected *P. halepensis* MADS-box genes and their expression patterns.

**Author Contributions:** Conceptualization, R.D.-S. and G.R.; methodology, G.R.; validation, G.R., H.Z. and H.F.; formal analysis, A.F.; investigation, G.R.; data curation, G.R.; writing—original draft preparation, G.R.; writing—review and editing, R.D.-S., G.R. and L.E.W.; supervision, R.D.-S. and L.E.W.; project administration, R.D.-S.; funding acquisition, R.D.-S. All authors have read and agreed to the published version of the manuscript.

**Funding:** This research received no external funding.

**Informed Consent Statement:** Not applicable.

**Data Availability Statement:** Data archiving statement. Data of this project have been deposited with links to BioProject accession number SUB11014167 in The National Center for Biotechnology Information (NCBI). The raw data from the Illumina sequencing have been deposited in NCBI Sequence Read Archive (SRA)—Accession numbers will be provided later, and before publication.

**Acknowledgments:** Gilad Reisfeld acknowledges financial support from the Appleby Foundation.

**Conflicts of Interest:** The authors declare no conflict of interest.

## References


1. Ferrandiz, C.; Gu, Q.; Martienssen, R.; Yanofsky, M.F. Redundant regulation of meristem identity and plant architecture by FRUITFULL, APETALA1 and CAULIFLOWER. *Development* **2000**, *127*, 725–734. [CrossRef]
2. Lee, J.; Lee, I. Regulation and function of SOC1, a flowering pathway integrator. *J. Exp. Bot.* **2010**, *61*, 2247–2254. [CrossRef]
3. Ito, S.; Song, Y.H.; Josephson-Day, A.R.; Miller, R.J.; Breton, G.; Olmstead, R.G.; Imaizumi, T. FLOWERING BHLH transcriptional activators control expression of the photoperiodic flowering regulator *CONSTANS* in *Arabidopsis*. *Proc. Natl. Acad. Sci. USA* **2012**, *109*, 3582–3587. [CrossRef]
4. Silva, C.S.; Puranik, S.; Round, A.; Brennich, M.; Jourdain, A.; Parcy, F.; Hugouvieux, V.; Zubieta, C. Evolution of the Plant Reproduction Master Regulators LFY and the MADS Transcription Factors: The Role of Protein Structure in the Evolutionary Development of the Flower. *Front. Plant Sci.* **2016**, *6*, 1193. [CrossRef]
5. Theißen, G.; Melzer, R.; Rümpler, F. MADS-domain transcription factors and the floral quartet model of flower development: Linking plant development and evolution. *Development* **2016**, *143*, 3259–3271. [CrossRef]
6. Kobayashi, Y.; Kaya, H.; Goto, K.; Iwabuchi, M.; Araki, T. A Pair of Related Genes with Antagonistic Roles in Mediating Flowering Signals. *Science* **1999**, *286*, 1960–1962. [CrossRef]
7. Wickland, D.P.; Hanzawa, Y. The FLOWERING LOCUS T/TERMINAL FLOWER 1 Gene Family: Functional Evolution and Molecular Mechanisms. *Mol. Plant* **2015**, *8*, 983–997. [CrossRef]
8. Moraes, T.S.; Dornelas, M.C.; Martinelli, A.P. FT/TFL1: Calibrating Plant Architecture. *Front. Plant Sci.* **2019**, *10*, 97. [CrossRef]
9. Liljegren, S.J.; Gustafson-Brown, C.; Pinyopich, A.; Ditta, G.S.; Yanofsky, M.F. Interactions among APETALA1, LEAFY, and TERMINAL FLOWER1 specify meristem fate. *Plant Cell* **1999**, *11*, 1007–1018. [CrossRef]
10. Samach, A.; Onouchi, H.; Gold, S.E.; Ditta, G.S.; Schwarz-Sommer, Z.; Yanofsky, M.F.; Coupland, G. Distinct Roles of CONSTANS Target Genes in Reproductive Development of *Arabidopsis*. *Science* **2000**, *288*, 1613–1616. [CrossRef]
11. Borner, R.; Kampmann, G.; Chandler, J.; Gleißner, R.; Wisman, E.; Apel, K.; Melzer, S. A MADS domain gene involved in the transition to flowering in Arabidopsis. *Plant J.* **2000**, *24*, 591–599. [CrossRef]
12. De La Torre, A.R.; Piot, A.; Liu, B.; Wilhite, B.; Weiss, M.; Porth, I. Functional and morphological evolution in gymnosperms: A portrait of implicated gene families. *Evol. Appl.* **2020**, *13*, 210–227. [CrossRef]
13. Mouradov, A.; Glassick, T.; Hamdorf, B.; Teasdale, R.D. Molecular control of early cone development inPinus radiata. *Protoplasma* **1999**, *208*, 3–12. [CrossRef]
14. Mellerowicz, E.J.; Horgan, K.; Walden, A.; Coker, A.; Walter, C. PRFLL—A Pinus radiata homologue of FLORICAULA and LEAFY is expressed in buds containing vegetative shoot and undifferentiated male cone primordia. *Planta* **1998**, *206*, 619–629. [CrossRef]
15. Maizel, A.; Busch, M.A.; Tanahashi, T.; Perkovic, J.; Kato, M.; Hasebe, M.; Weigel, D. The Floral Regulator LEAFY Evolves by Substitutions in the DNA Binding Domain. *Science* **2005**, *308*, 260–263. [CrossRef]
16. Carlsbecker, A.; Tandre, K.; Johanson, U.; Englund, M.; Engström, P. The MADS-box gene DAL1 is a potential mediator of the juvenile-to-adult transition in Norway spruce (*Picea abies*). *Plant J.* **2004**, *40*, 546–557. [CrossRef]
17. Mouradov, A.; Glassick, T.; Hamdorf, B.; Murphy, L.; Fowler, B.; Marla, S.; Teasdale, R.D. *NEEDLY*, a *Pinus radiata* ortholog of *FLORICAULA/LEAFY* genes, expressed in both reproductive and vegetative meristems. *Proc. Natl. Acad. Sci. USA* **1998**, *95*, 6537–6542. [CrossRef]
18. Uddenberg, D.; Reimegård, J.; Clapham, D.; Almqvist, C.; von Arnold, S.; Emanuelsson, O.; Sundström, J.F. Early Cone Setting in Picea abies acrocona Is Associated with Increased Transcriptional Activity of a MADS Box Transcription Factor. *Plant Physiol.* **2012**, *161*, 813–823. [CrossRef]
19. Dornelas, M.C.; Rodriguez, A.P.M. A Floricaula/Leafy gene homolog is preferentially expressed in developing female cones of the tropical pine Pinus caribaea var. caribaea. *Genet. Mol. Biol.* **2005**, *28*, 299–307. [CrossRef]
20. Klintenäs, M.; Pin, P.A.; Benlloch, R.; Ingvarsson, P.K.; Nilsson, O. Analysis of conifer FLOWERING LOCUS T/TERMINAL FLOWER1-like genes provides evidence for dramatic biochemical evolution in the angiosperm FT lineage. *New Phytol.* **2012**, *196*, 1260–1273. [CrossRef]
21. Niu, S.; Li, J.; Bo, W.; Yang, W.; Zuccolo, A.; Giacomello, S.; Chen, X.; Han, F.; Yang, J.; Song, Y.; et al. The Chinese pine genome and methylome unveil key features of conifer evolution. *Cell* **2022**, *185*, 204–217.e14. [CrossRef]
22. Karlgren, A.; Gyllenstrand, N.; Källman, T.; Sundström, J.F.; Moore, D.; Lascoux, M.; Lagercrantz, U. Evolution of the PEBP Gene Family in Plants: Functional Diversification in Seed Plant Evolution. *Plant Physiol.* **2011**, *156*, 1967–1977. [CrossRef]
23. Theißen, G. Development of floral organ identity: Stories from the MADS house. *Curr. Opin. Plant Biol.* **2001**, *4*, 75–85. [CrossRef]
24. Weigel, D.; Meyerowitz, E.M. Activation of Floral Homeotic Genes in *Arabidopsis*. *Science* **1993**, *261*, 1723–1726. [CrossRef]
25. Bowman, J.L.; Smyth, D.R.; Meyerowitz, E.M. Genetic interactions among floral homeotic genes of Arabidopsis. *Development* **1991**, *112*, 1–20. [CrossRef]
26. Mouradov, A.; Hamdorf, B.; Teasdale, R.D.; Kim, J.T.; Winter, K.-U.; Theißen, G. A DEF/GLO-like MADS-box gene from a gymnosperm: Pinus radiata contains an ortholog of angiosperm B class floral homeotic genes. *Dev. Genet.* **1999**, *25*, 245–252. [CrossRef]
27. Sundström, J.; Engström, P. Conifer reproductive development involves B-type MADS-box genes with distinct and different activities in male organ primordia. *Plant J.* **2002**, *31*, 161–169. [CrossRef]

28. Mouradov, A.; Glassick, T.V.; Hamdorf, B.A.; Murphy, L.C.; Marla, S.S.; Yang, Y.; Teasdale, R.D. Family of MADS-Box Genes Expressed Early in Male and Female Reproductive Structures of Monterey Pine. *Plant Physiol.* **1998**, *117*, 55–62. [CrossRef]
29. Tandre, K.; Albert, V.A.; Sundas, A.; Engstrom, P. Conifer homologues to genes that control floral development in angiosperms. *Plant Mol. Biol.* **1995**, *27*, 69–78. [CrossRef]
30. Englund, M.; Carlsbecker, A.; Engström, P.; Vergara-Silva, F. Morphological "primary homology" and expression of AG-subfamily MADS-box genes in pines, podocarps, and yews. *Evol. Dev.* **2011**, *13*, 171–181. [CrossRef]
31. Carlsbecker, A.; Sundström, J.; Tandre, K.; Englund, M.; Kvarnheden, A.; Johanson, U.; Engström, P. The DAL10 gene from Norway spruce (Picea abies) belongs to a potentially gymnosperm-specific subclass of MADS-box genes and is specifically active in seed cones and pollen cones. *Evol. Dev.* **2003**, *5*, 551–561. [CrossRef]
32. Ne'eman, G.; Goubitz, S.; Werger, M.J.A.; Shmida, A. Relationships between tree size, crown shape, gender segregation and sex allocation in Pinus halepensis, a Mediterranean pine tree. *Ann. Bot.* **2011**, *108*, 197–206. [CrossRef]
33. Climent, J.; Prada, M.A.; Calama, R.; Chambel, M.R.; de Ron, D.S.; Alía, R. To grow or to seed: Ecotypic variation in reproductive allocation and cone production by young female Aleppo pine (Pinus halepensis, Pinaceae). *Am. J. Bot.* **2008**, *95*, 833–842. [CrossRef]
34. Weinstein, A. Geographic variation and phenology of Pinus halepensis, P. brutia and P. eldarica in Israel. *For. Ecol. Manag.* **1989**, *27*, 99–108. [CrossRef]
35. Brandizzi, F. Ruzin SE. 1999. *Plant microtechnique and microscopy*. 322 pp. Oxford, New York: Oxford University Press. £32.50 (softback). *Ann. Bot.* **2000**, *86*, 708. [CrossRef]
36. Asgari, F.; Irian, S.; Jonoubi, P.; Majd, A. Meristem Structure, Development of Cones and Microsporogenesis of Tehran Pine (Pinus Eldarica Medw.). *J. Plant Dev.* **2014**, *21*, 83–93.
37. Grabherr, M.G.; Haas, B.J.; Yassour, M.; Levin, J.Z.; Thompson, D.A.; Amit, I. Full-length transcriptome assembly from RNA-Seq data without a reference genome. *Nat. Biotechnol.* **2011**, *29*, 644–652. [CrossRef]
38. Langmead, B.; Trapnell, C.; Pop, M.; Salzberg, S.L. Ultrafast and memory-efficient alignment of short DNA sequences to the human genome. *Genome Biol.* **2009**, *10*, R25. [CrossRef]
39. Li, B.; Dewey, C.N. RSEM: Accurate transcript quantification from RNA-Seq data with or without a reference genome. *BMC Bioinform.* **2011**, *12*, 323. [CrossRef]
40. Haas, B.J.; Papanicolaou, A.; Yassour, M.; Grabherr, M.; Blood, P.D.; Bowden, J.; Couger, M.B.; Eccles, D.; Li, B.; Lieber, M.; et al. De novo transcript sequence reconstruction from RNA-seq using the Trinity platform for reference generation and analysis. *Nat. Protoc.* **2013**, *8*, 1494–1512. [CrossRef]
41. Robinson, M.D.; McCarthy, D.J.; Smyth, G.K. edgeR: A Bioconductor package for differential expression analysis of digital gene expression data. *Bioinformatics* **2010**, *26*, 139–140. [CrossRef]
42. Buchfink, B.; Xie, C.; Huson, D.H. Fast and sensitive protein alignment using DIAMOND. *Nat. Methods* **2015**, *12*, 59–60. [CrossRef]
43. Altschul, S.F.; Gish, W.; Miller, W.; Myers, E.W.; Lipman, D.J. Basic Local Alignment Search Tool. *J. Mol. Biol.* **1990**, *215*, 403–410. [CrossRef]
44. Conesa, A.; Götz, S.; García-Gómez, J.M.; Terol, J.; Talón, M.; Robles, M. Blast2GO: A universal tool for annotation, visualization and analysis in functional genomics research. *Bioinformatics* **2005**, *21*, 3674–3676. [CrossRef]
45. Upton, G.J.G. Fisher's exact test. *J. R. Stat. Soc. Ser. A (Stat. Soc.)* **1992**, *155*, 395–402. [CrossRef]
46. Benjamini, Y.; Hocberg, Y. Controlling the false discovery rate: A practical and powerful approach to multiple testing. *J. R. Stat. Soc.* **1995**, *57*, 289–300. [CrossRef]
47. Kumar, S.; Stecher, G.; Li, M.; Knyaz, C.; Tamura, K. MEGA X: Molecular Evolutionary Genetics Analysis across Computing Platforms. *Mol. Biol. Evol.* **2018**, *35*, 1547–1549. [CrossRef]
48. Edgar, R.C. MUSCLE: Multiple sequence alignment with high accuracy and high throughput. *Nucleic Acids Res.* **2004**, *32*, 1792–1797. [CrossRef]
49. Jones, D.T.; Taylor, W.R.; Thornton, J.M. The rapid generation of mutation data matrices from protein sequences. *Bioinformatics* **1992**, *8*, 275–282. [CrossRef]
50. Livak, K.J.; Schmittgen, T.D. Analysis of relative gene expression data using real-time quantitative PCR and the 2(-Delta Delta C(T)) Method. *Methods* **2001**, *25*, 402–408. [CrossRef]
51. Bramlett, D.L. *Recognizing Developmental Stages in Southern Pine Flowers: The Key to Controlled Pollination*; Southeastern Forest Experiment Station: Asheville, NC, USA, 1980.
52. Yamaoka, S.; Nishihama, R.; Yoshitake, Y.; Ishida, S.; Inoue, K.; Saito, M.; Okahashi, K.; Bao, H.; Nishida, H.; Yamaguchi, K.; et al. Generative Cell Specification Requires Transcription Factors Evolutionarily Conserved in Land Plants. *Curr. Biol.* **2018**, *28*, 479–486.e5. [CrossRef] [PubMed]
53. Ma, J.-J.; Chen, X.; Song, Y.-T.; Zhang, G.-F.; Zhou, X.-Q.; Que, S.-P.; Mao, F.; Pervaiz, T.; Lin, J.-X.; Li, Y.; et al. MADS-box transcription factors MADS11 and DAL1 interact to mediate the vegetative-to-reproductive transition in pine. *Plant Physiol.* **2021**, *187*, 247–262. [CrossRef] [PubMed]
54. Feng, K.; Hou, X.-L.; Xing, G.-M.; Liu, J.-X.; Duan, A.-Q.; Xu, Z.-S.; Li, M.-Y.; Zhuang, J.; Xiong, A.-S. Advances in AP2/ERF super-family transcription factors in plant. *Crit. Rev. Biotechnol.* **2020**, *40*, 750–776. [CrossRef]
55. Shigyo, M.; Ito, M. Analysis of gymnosperm two-AP2-domain-containing genes. *Dev. Genes Evol.* **2004**, *214*, 105–114. [CrossRef]

56. Vahala, T.; Oxelman, B.; Arnold, S.v. Two APETALA2-like genes of Picea abies are differentially expressed during development. *J. Exp. Bot.* **2001**, *52*, 1111–1115. [CrossRef]
57. Nilsson, L.; Carlsbecker, A.; Sundås-Larsson, A.; Vahala, T. APETALA2 like genes from Picea abies show functional similarities to their Arabidopsis homologues. *Planta* **2007**, *225*, 589–602. [CrossRef]
58. Carlsbecker, A.; Sundström, J.F.; Englund, M.; Uddenberg, D.; Izquierdo, L.; Kvarnheden, A.; Vergara-Silva, F.; Engström, P. Molecular control of normal and acrocona mutant seed cone development in Norway spruce (Picea abies) and the evolution of conifer ovule-bearing organs. *New Phytol.* **2013**, *200*, 261–275. [CrossRef]
59. Klocko, A.L.; Lu, H.; Magnuson, A.; Brunner, A.M.; Ma, C.; Strauss, S.H. Phenotypic Expression and Stability in a Large-Scale Field Study of Genetically Engineered Poplars Containing Sexual Containment Transgenes. *Front. Bioeng. Biotechnol.* **2018**, *6*, 100. [CrossRef]
60. Li, C.; Chen, C.; Gao, L.; Yang, S.; Nguyen, V.; Shi, X.; Siminovitch, K.; Kohalmi, S.E.; Huang, S.; Wu, K.; et al. The Arabidopsis SWI2/SNF2 Chromatin Remodeler BRAHMA Regulates Polycomb Function during Vegetative Development and Directly Activates the Flowering Repressor Gene SVP. *PLoS Genet.* **2015**, *11*, e1004944. [CrossRef]
61. Chen, F.; Zhang, X.; Liu, X.; Zhang, L. Evolutionary Analysis of MIKCc-Type MADS-Box Genes in Gymnosperms and Angiosperms. *Front. Plant Sci.* **2017**, *8*, 895. [CrossRef]
62. Niu, S.; Yuan, H.; Sun, X.; Porth, I.; Li, Y.; El-Kassaby, Y.A.; Li, W. A transcriptomics investigation into pine reproductive organ development. *New Phytol.* **2016**, *209*, 1278–1289. [CrossRef]
63. Dreni, L.; Zhang, D. Flower development: The evolutionary history and functions of the AGL6 subfamily MADS-box genes. *J. Exp. Bot.* **2016**, *67*, 1625–1638. [CrossRef]
64. Gramzow, L.; Weilandt, L.; Theißen, G. MADS goes genomic in conifers: Towards determining the ancestral set of MADS-box genes in seed plants. *Ann. Bot.* **2014**, *114*, 1407–1429. [CrossRef]

agronomy

Article

# Effects of Different Regulating Measures on the Floral and Nutritional Physiology of Lemon

Yuxia Du [1], Jinxue Li [1], Jianmei Dong [1], Chengxiao Hu [2], Danping Li [1], Qiling Tan [2,*], Jinzhi Zhang [2], Jing Li [1], Xianyan Zhou [1], Chunhua Zhu [1] and Xinpu Lai [1]

1 Institute of Tropical and Subtropical Economic Crops, Yunnan Academy of Agricultural Sciences, Changqing Road, Ruili 678600, China
2 Key Laboratory of Horticultural Plant Biology (Ministry of Education), College of Horticulture and Forestry Science, Huazhong Agricultural University, Wuhan 430070, China
* Correspondence: qltan@mail.hzau.edu.cn

**Abstract:** Lemon is an important economic crop in the world and can bloom several times each year. The Dehong prefecture of Yunnan province is one of the main lemon-producing areas in China, and is a suitable area for planting high-quality lemons in China. However, the hot and rainy summers in Yunnan are not conducive to flower bud differentiation, which leads to low lemon yield. Therefore, normal flower bud differentiation is important to guarantee lemon production and quality. In this study, we selected some lemon trees for a pot experiment, and we sprayed the lemon leaves with gibberellin and paclobutrazol to regulate lemon flower formation. We set four separate concentration gradients for each regulator: 50, 100, 150, and 200 mg/L. The PBZ concentration gradients were 200, 400, 600, and 1200 mg/L. After the experiment, we determined and analyzed the morphological index and fruit quality of the lemon trees. The results showed that under the same cultivation and management conditions, spraying paclobutrazol substantially inhibited the growth of lemon shoots, increased the fruit setting rate, and improved the fruit yield. However, gibberellin considerably reduced the number of lemon flowering branches and promoted the vegetative growth of the lemons. When the concentration of paclobutrazol was 600 mg/L, the amount of lemon fruits reached the maximum, which remarkably increased the titratable acid and soluble solids contents of the fruit and ultimately increased the fruit quality compared with the control. Altogether, selecting the appropriate concentration of regulators to control the flowering and fruit setting of fruit trees is important and has value in guiding actual production.

**Keywords:** lemon; paclobutrazol; gibberellin; flowering; fruit quality

**Citation:** Du, Y.; Li, J.; Dong, J.; Hu, C.; Li, D.; Tan, Q.; Zhang, J.; Li, J.; Zhou, X.; Zhu, C.; et al. Effects of Different Regulating Measures on the Floral and Nutritional Physiology of Lemon. *Agronomy* **2022**, *12*, 2381. https://doi.org/10.3390/agronomy12102381

Academic Editor: Youssef Rouphael

Received: 25 August 2022
Accepted: 23 September 2022
Published: 1 October 2022



## 1. Introduction

Lemons (*Citrus limon* (L.) Burm. F.), a fruit tree that can bloom several times a year, is an important economic crop in the world. The fruit ripening time of lemons is in the fruit-offseason—spring—so is suitable for the domestic market when spring fruits are scarce. The Dehong prefecture of Yunnan province is one of the main lemon-producing areas and is suitable for planting high-quality lemons in China [1] due to its early, high-quality, off-season production, and annual output [2]. However, the high temperatures and rainy summers in Yunnan hinder lemon flowering, which considerably reduce the fruit yield. Therefore, normal flower bud differentiation is crucial for the quality and production of lemons.

In recent years, plant growth regulators have been widely used in agricultural production. They have a remarkable effect on promoting flowers and controlling branches of fruit trees [3,4]. Plant growth regulators, such as gibberellin (GA), play an important role in the transportation and distribution of nutrients from the source to the reservoir in fruit trees, and inhibit the flowering effect of fruit loads [5]. This inhibiting effect of gibberellin has been confirmed in fruit trees, such as peach [6], plum [7], pear [8], and apple [9]. The application of gibberellin during the stage of citrus flower bud differentiation can reduce the flowering intensity [10], increase the fruit setting rate of non-pollinated ovaries [11], delay chlorophyll decomposition, inhibit carotenoid accumulation, and promote the vegetative growth of plants. Spraying $GA_3$ can deploy enough nutrients to the growing branches in time, which promotes vegetative growth but is not conductive to the reproductive growth (flowering) of branches [12,13]. Gibberellin treatment resulted in a decreased expression of many MADS-box genes in shoots, some of which are involved in the development of floral meristems and floral organs [14]. When performing foliar spraying with $GA_3$ (20–50 mg/L), the flowering of citrus was reduced by 25–60%. In the process of the flower bud differentiation of citrus, $GA_3$ inhibited the production of flower buds, resulting in leafy branches in the apex as well as a higher percentage of flowers, which promoted fruit development [10]. Pollination induced $GA_1$ synthesis in the ovary, and when pollination was presented by removing the male flowers, the levels of this hormone reduced in the ovary, and young fruit abscission increased. However, these effects can be counteracted by the administration of $GA_3$ [11]. Paclobutrazol (PBZ) is an inhibitor of GA biosynthesis. The use of paclobutrazol has considerable potential for controlling excessive branch growth in citrus. The effect of PBZ on the stem is reflected in its control of the longitudinal expansion of cells, thereby shortening the length of the internodes and dwarfing the plant. At the same time, the number of vascular tissues in the stem increases, the thickness of the stem wall increases, and the basal stem thickens [15]. The application of PBZ can increase the flowering and fruiting of plants [16,17]. In a previous study, the application of PBZ to mangoes changed the contents of gibberellin, abscisic acid, and cytokinin, and induced early flowering [18]. Moreover, PBZ promotes flowering in many citrus species, such as *C. aurantifolia* and *C. sinensis*. Limes planted in Nepal even had their flowering date advanced by 70 days through PBZ application [19]. The effect of PBZ on citrus flowering depends on the fruit load, and PBZ cannot promote flowering beyond its threshold. Flower bud differentiation is associated with changes in hormones and sugars in plants [20,21]. In citrus, PBZ can inhibit growth by influencing the biosynthesis of gibberellin and inducing flowering [22]. High and low levels of endogenous GA in leaves were inversely or positively proportionally correlated with the number of flowers, respectively [23]. In citrus varieties with a low parthenocarpic ability, the application of the exogenous hormone $GA_3$ can substantially reduce the ABA content in the ovary after flowering, thereby reducing the shedding of small fruits. In contrast, in parthenocarpic cultivars such as mandarin oranges, the inhibition of GAs synthesis by PBZ increases ABA levels and fruitlet abscission [24].

However, few detection methods for the positive and negative regulation of lemon flower promotion are available either at home or abroad. As such, in this study, we used different concentrations of PBZ and gibberellin to spray lemons to investigate the effects of the two regulators on flower formation and nutritional physiology, to provide some theoretical basis for high-quality, high volumes, and stable yields of lemons.

## 2. Materials and Methods

### 2.1. Plant Materials

We obtained the plant materials from three-year-old grafted Eureka lemons, which were located at the scientific research base of the Lemon Experimental Station of the Institute of Tropical and Subtropical Economic Crops of the Yunnan Academy of Agricultural Sciences, Ruili city, Dehong prefecture (97°52′12″ E, 24°1′9″ N, elevation 750.4 m).

### 2.2. Experiment Design and Methods

We conducted the experiment under potting conditions. The growth medium was a 3:1 mixture of latosolic red soil and perlite, using 65 kg of the mixture per pot. We selected lemon trees with the same growth status for exogenous hormone treatment. In the experiment, we used gibberellin ($GA_3$) and paclobutrazol (PBZ) as plant growth regulators, and we set four concentration gradients to each regulator. The $GA_3$ concentration gradient was 50 (GA 50), 100 (GA 100), 150 (GA 150), and 200 mg/L (GA 200). The PBZ concentration gradient was 200 (PBZ 200), 400 (PBZ 400), 600 (PBZ 600), and 1200 mg/L (PBZ 1200). We sprayed the plant growth regulators on the leaves once every 10 days for 30 days, for a total of 3 times. According to the experimental design, we set 9 treatments (1 control) and 9 replicates, with a total of 81 lemon trees. The experiment duration was 72 days.

### 2.3. *Measurement Indexes and Methods*

#### 2.3.1. Investigation of Plant Growth Index

At the end of the experiment, we observed the morphology of the plants and leaves that we had treated with different hormones. We measured the length of the topmost branches, number of leaves, leaf longitudinal and transverse diameter, petiole length, length of the internode, and diameter of the stem were measured with Vernier calipers and rulers. We manually counted the total, flowering, and new branches on the 52nd and 72nd days of the experiment (when the lemons began to blossom). We divided the flowering branches into four types: the top flowering branch with leaves, the top flowering branch without leaves, the branch with inflorescences and leaves, and the non-leaf branch with inflorescences. We separately recorded the number of these four types of flowering branches on the 52nd and 72nd days.

#### 2.3.2. Fruit Load and Fruit Quality of Lemons Regulated by Different Hormones

We counted the number of fruits per tree for each treatment, then, we fit the data of the hormone concentration-fruit number curve. For each treatment, we randomly selected nine fruits on the basis of health, size consistency, color uniformity, and features of commercial maturity. We brought these fruits back to the laboratory to determine the fruit quality. We measured the longitudinal and transverse diameters and skin thickness of the lemons with a Vernier caliper. The fruit shape index is the ratio of the transverse and longitudinal diameters of the fruit. We also weighed the single fruits. We used a citrus juice squeezer, extract the juice. Juice yield was calculated by using the weight of juice divided by the weight of lemons. We determined the soluble solids (TSS) and titratable acid (TA) of the lemon fruits with a hand-held refractometer (PAL-BX | ACID1, Atago Co., Ltd., Tokyo, Japan). We titrated vitamin C (VC) with 2, 6-dichlorindophenol [25].

#### 2.3.3. Dynamic Changes C/N Ratio in Leaves

We sampled the leaves every seven days from the beginning of the experiment. We collected 15 leaves from each treatment, and we sampled 3 replicates to obtain the total soluble sugar, starch, and nitrogen contents. We determined the total soluble sugar contents in the leaves with the modified anthrone method [26], and the starch content in the leaves using the anthrone-sulfuric acid method [27]. We determined the nitrogen content using the Kjeldahl method [28]. Our calculations of the leaf carbon-to-nitrogen (C:N) ratio were based on the total content of both carbon (the total soluble sugar and starch) and nitrogen.

### 2.4. *Data Processing and Analysis*

We used SPSS 17.0 statistical software (SPSS Inc., Chicago, IL, USA) for variance analysis and the least significant difference (LSD) test. We used Origin 8.5 software (Origin lab cooperation, Hampton, MA, USA) for drawing.

## 3. Results

### 3.1. *Effects of Hormone Treatment on Lemons*

Different hormone types and levels had different effects on the lemon shoot characteristics (Figure 1). Compared with the control, we found a substantial difference in the length of the topmost shoots according to the PBZ spraying concentration. The length of the topmost shoots increased under PBZ600 and PBZ1200, but decreased under PBZ200 and PBZ400 treatments (Figure 1A). The length of the topmost shoots varied from 27.34 (GA50) to 35.06 (GA200) cm after spraying $GA_3$. The length of the topmost shoots increased with the increase in the gibberellin concentration. For the number of leaves, PBZ600 (22.60) and PBZ1200 (33.06) plants had more leaves than the control plants (Figure 1B). GA100 treatment decreased the number of leaves, but other gibberellin treatments increased the number of leaves. The leaf longitudinal diameter of the lemons was no different in any PBZ or GA treatments (Figure 1C). Moreover, spraying PBZ notably increased the leaf transverse diameter (Figure 1D), and shortened the petiole and internode lengths (Figure 1E,F), but we noted no significant association with the spraying concentration. The $GA_3$ treatment increased the leaf longitudinal diameter of the lemon compared with that in the control (Figure 1C), and we found no significant difference among the spraying concentration treatments. With the increase in the $GA_3$ concentration, $GA_3$ substantially shortened the transverse diameter of the leaves (Figure 1D), but had little effect on the petiole and internode lengthes or stem diameter (Figure 1E–G).

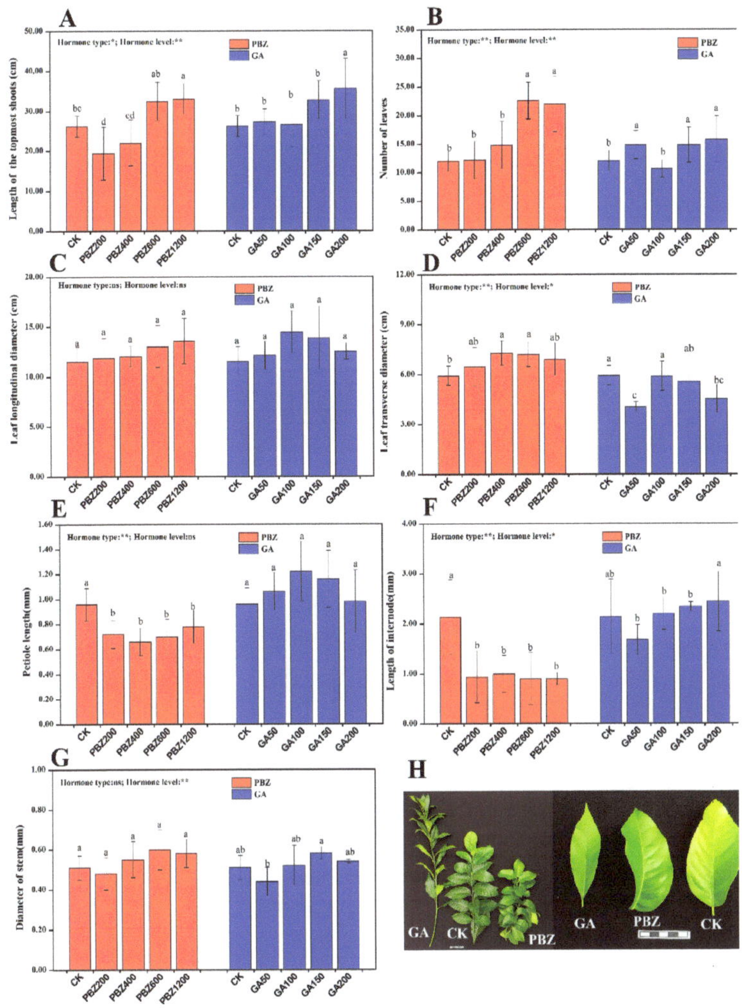


**Figure 1.** Effects of hormone regulation on the length of the topmost shoots (**A**), number of leaves (**B**), leaf longitudinal diameter (**C**), leaf transverse diameter (**D**), petiole length (**E**), length of internode (**F**), diameter of stem (**G**). (**H**) Plant and leaf morphology after treatment with different hormones. The

red and blue bars represent PBZ and GA, respectively. Different letters (a, b, c and d) denote significant differences based on a $p$ value < 0.05. * denotes significant difference ($p < 0.05$), and ** denotes extremely significant difference ($p < 0.01$).

## 3.2. Types of Lemon Branches Regulated by Different Hormones

We found no significant difference in the total branches of the trees treated with different hormones (Figure 2A), but the effects on the number of flowers and new branches were different (Figure 2). On the 52nd day of the experiment, the application of PBZ, which we applied in the same way as $GA_3$, considerably reduced the number of flowering branches. At the end of the experiment (72nd day), the application of PBZ increased the number of branches for all types of lemons. However, the number of new branches increased, and the number of flowering branches was reduced under the GA treatment (Figure 2B,C). When the concentration of PBZ was 600 mg/L, the number of flowers and inflorescence branches reached the maximum 211.00 (Figure 2B).

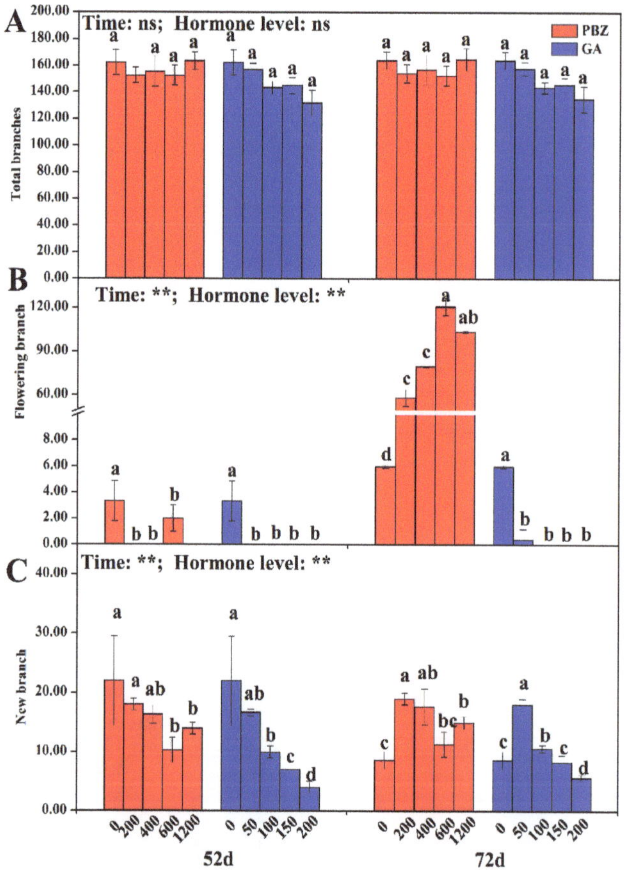


**Figure 2.** Effects of hormone regulation on types of lemon total (**A**), flowering (**B**), and new branches (**C**). Different letters (a, b, c and d) denote significant differences based on a $p$ value < 0.05. ** denotes extremely significant difference ($p < 0.01$).

We mainly observed the effects of hormone regulation on the number of lemon flower branches on the 72nd day of the treatment (Figure 3). The number of top flowering branches with leaves first increased and then decreased with the increase in the PBZ concentration on the 72nd day (Figure 3A). The number of top flowering branches without the leaves of the lemons increased with the increase in the PBZ concentration on the 72nd day (Figure 3B). The number of top flowering branches without leaves was 13.15- to 43.45-fold higher than that of the control, indicating the promoting effect of PBZ on flowering branches. The numbers of branches with inflorescences and leaves were also higher than in the control, especially for plants in the PBZ600 treatment (Figure 3C). The branches with inflorescences and leaves were the largest in number of the other three types of flowering branches. The PBZ application also increased the number of non-leaf branches with inflorescences, especially for plants under the PBZ200 and PBZ1200 treatments (Figure 3D). The application of gibberellin substantially reduced the number of flowering branches of all types (Figure 3), which all decreased with the increase in the gibberellin concentration. Thus, GA treatments would not be favorable for the four types of flower branches.

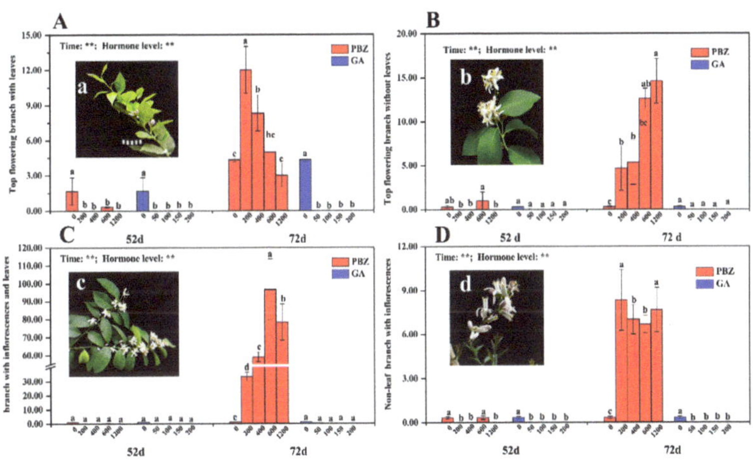


**Figure 3.** Effects of hormone regulation on the types and number of lemon flower branches: top flowering branch with leaves (**A**), top flowering branch without leaves (**B**), branch with inflorescences and leaves (**C**), and non-leaf branch with inflorescences (**D**).Different letters (a, b, c and d) denote significant differences based on a $p$ value < 0.05. ** denotes extremely significant difference ($p$ < 0.01).

### 3.3. Fruit Load, Fruit Morphology, and Fruit Quality of Lemons Regulated by Different Hormones

The effects of different types of hormones and hormone levels on the fruit load of lemon were remarkably different (Figure 4A). PBZ substantially increased the fruit load of the lemons, and the fruit load of the lemons first increased and then decreased with the increase in the PBZ concentration, which reached the maximum when the PBZ concentration was 600 mg/L. However, gibberellin application was not conducive to increasing the fruit load. We observed only a small amount of fruit in the GA50 treatment. The curve fitting between PBZ concentration and fruit load showed that when the PBZ spraying concentration was 850 mg/L (Figure 4B), the fruit load reached a maximum of 48 per plant, which was 5.2 times higher than that in the control. Different types of hormone treatments affected the lemon fruit quality and morphology (Figure 4D and Table 1). The PBZ treatment considerably reduced the fruit weight, fruit longitudinal diameter, fruit transverse diameter, and skin thickness of the lemons, but notably increased the juice yield. The GA50 treatment plants had higher values than the control plants in terms of fruit weight, fruit thickness, and juice yield. The fruit shape index values ranged from 0.70 (PBZ200) to 0.92 (PBZ600). A high concentration of PBZ (600 and 1200 mg/L) improved the fruit shape index, but GA had no noticeable effect. With PBZ1200 concentration, the juice yield of the lemon fruits was 1.4 larger than that of the control. The application of gibberellin decreased the VC content and increased the TA and TSS contents (Figure 4C).

The VC content also decreased in the PBZ treatment group, and the TA and TSS varied with the PBZ concentration.

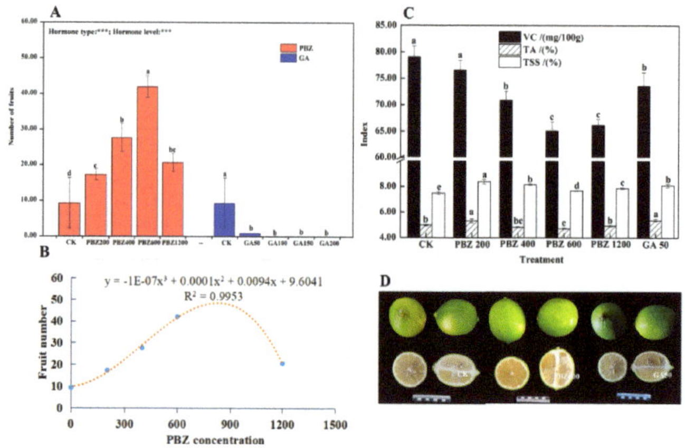

**Figure 4.** Effects of hormone regulation on fruit load and quality of lemons. (**A**) Effects of hormone regulation on number of fruits; (**B**) PBZ concentration–fruit number curves fit to pooled data calculated by an equation; (**C**) effects of hormone regulation on TSS, TA, and VC contents; (**D**) fruit morphologies under different treatments. Different letters (a, b, c and d) denote significant differences based on a $p$ value < 0.05. *** denotes extremely significant difference ($p < 0.001$).

**Table 1.** Effects of hormone regulation on quality of lemon fruit. Different letters (a, b, c and d) denote significant differences based on a $p$ value < 0.05.

| Treatment | Fruit Weight (g) | Longitudinal Diameter (cm) | Transverse Diameter (cm) | Fruit Shape Index | Fruit Thick (cm) | Juice Yeild (%) |
|---|---|---|---|---|---|---|
| CK | 133.87 ± 3.40 a | 7.91 ± 0.23 a | 5.69 ± 0.13 a | 0.72 ± 0.00 c | 0.61 ± 0.03 ab | 39.83 ± 1.23 e |
| PBZ200 | 132.90 ± 2.02 a | 8.18 ± 0.03 a | 5.71 ± 0.07 a | 0.70 ± 0.01 c | 0.57 ± 0.01 cd | 45.11 ± 1.22 d |
| PBZ400 | 128.90 ± 1.37 b | 7.48 ± 0.23 b | 5.47 ± 0.12 b | 0.73 ± 0.04 c | 0.59 ± 0.02 bc | 48.02 ± 0.32 c |
| PBZ600 | 123.10 ± 1.47 c | 6.23 ± 0.16 d | 5.75 ± 0.09 a | 0.92 ± 0.04 a | 0.64 ± 0.02 a | 52.35 ± 1.02 b |
| PBZ1200 | 127.83 ± 1.65 b | 7.05 ± 0.14 c | 5.45 ± 0.07 b | 0.77 ± 0.01 b | 0.56 ± 0.02 d | 55.81 ± 1.48 a |
| GA50 | 134.6 ± 4.21 a | 7.84 ± 0.42 a | 5.56 ± 0.11 a | 0.71 ± 0.02 c | 0.64 ± 0.04 ab | 42.70 ± 3.11 d |

### 3.4. Contents of Soluble Sugar, Starch, and C/N Ratio Regulated by Different Hormones

With the growth of lemons from July to August in 2017, the soluble sugar contents in the lemon leaves increased under the PBZ treatment, but decreased under the $GA_3$ treatment (Figure 5A,B). When the concentration of PBZ was 600 mg/L, the soluble sugar content of the lemon leaves reached the maximum during the prophase of sampling. Most of the plants in the treatments (except for PBZ1200) had a lower TSS content than those in the control on 12 August 2017. The starch content in the lemon leaves decreased with the increase in both the PBZ and $GA_3$ hormone levels (Figure 5C,D). The results also revealed that the starch content of the plants in the PBZ and $GA_3$ treatments decreased as the experiment continued. Carbohydrates are the main product of plant photosynthesis and are an important substance involved in plant life metabolism, the C/N ratio of the lemon leaves showed a trend of first increasing and then decreasing with the increase in PBZ concentrations (Figure 5E). Among the different PBZ concentrations treatments, when the concentration of PBZ was 1200 mg/L, the C/N of the leaves was the highest. For the $GA_3$ treatment, the C/N ratio of the lemon leaves was lower compared to those in the control as the experiment continued (Figure 5F).

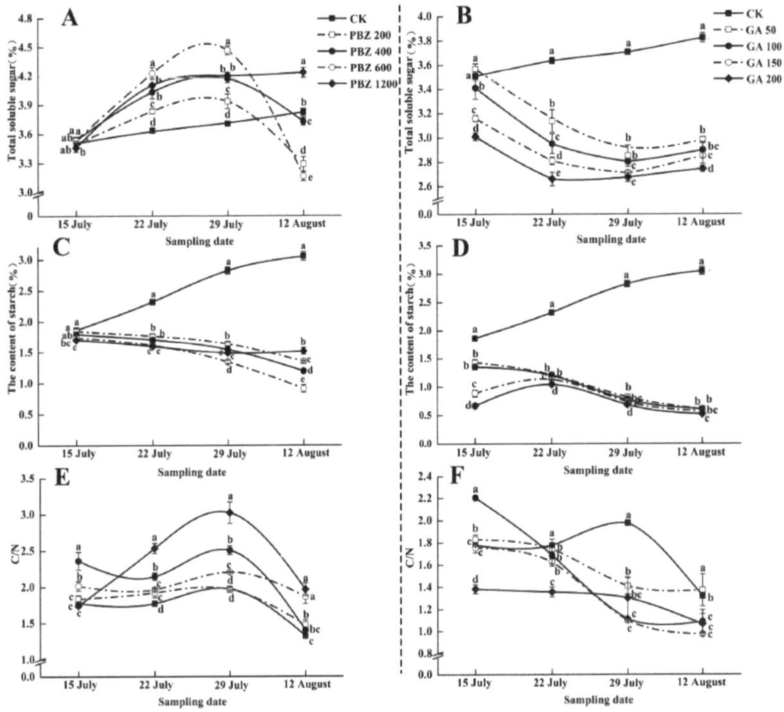

**Figure 5.** Effects of hormone regulation on total soluble sugar (**A**,**B**), and the starch contents (**C**,**D**), and C/N ratio (**E**,**F**) of lemon leaves. The left-hand plants were treated with paclobutrazol, and the right-hand plants were treated with gibberellin. Different letters (a, b, c and d) denote significant differences based on a $p$ value < 0.05.

## 4. Discussion

### *4.1. Lemon Growth and Development under the Regulation of Different Hormones*

Spraying PBZ on fruit trees can inhibit the growth of new shoots, promote fruit setting, and increase the fruit yield [29]. The application of PBZ can not only increase the number and ratio of fruit flower buds, but can also improve the quantity and quality of flowers, and promote the flowering process and early flowering [30]. In this experiment, the lengths of the petiole and interleaf of the lemons were shortened by spraying PBZ. Liu [31] also reached the same conclusion when studying the regulation effect of different flower promotion measures on lemons. Figure 2B shows that the application of PBZ increased the number of all flowering branches of lemons. Our results showed that the application of gibberellin reduced the numbers of lemon flower, terminal flower, and inflorescence branches: all decreased with the increase in gibberellin concentration, which is consistent with the finding reported by Du [32], who found that the application of gibberellin inhibited the flowering of lemons and reduced the number of lemon flower buds. Sun [33] also reached the same conclusion in *Vernicia fordii.*

The physiological effect of PBZ is mainly directly or indirectly regulating the distribution of photosynthetic products by blocking the synthesis of gibberellin in plants. Thus, the flowering, fruit setting rate, and fruit yield of fruit trees increase [34,35]. In this study, the application of PBZ increased the fruit load of the lemons, and with the increase in the PBZ concentration, the fruit load of the lemons first increased and then decreased, which reached a maximum when the PBZ concentration was 600 mg/L. Gibberellin reduced the fruit load of the lemons, which was consistent with the findings of Du [32]. PBZ application increased the titratable acid and soluble solids contents, but reduced the VC content of the lemon fruits, which is consistent with the results reported by Yan [36] and Yang [37].

### 4.2. *Effects of Hormone Regulation on the Nutritional Physiology of Lemon Leaves*

Spraying PBZ can regulate the nutritional growth of fruit trees, improve the soluble sugar and starch contents in plants, and increase the C/N ratio, which promotes flower bud differentiation, improves the flowering rate, and increases the yield [38,39]. The application of gibberellin inhibits the formation of flower buds by altering the distribution of carbohydrates in plants [40]. In this study, the soluble sugar content in the lemon leaves treated with PBZ first increased and then decreased with the increase in hormone levels, and the soluble sugar content in the lemon leaves reached a maximum when the PBZ concentration was 600 mg/L, but the starch content in the lemon leaves was reduced for different PBZ concentrations. This is consistent with the results of Yang [37]. Gibberellin treatment reduced the soluble sugar and starch contents in lemon leaves, and both of them decreased with the increase in hormone level. Zhang [41] also reached the same conclusion using litchi.

The C/N ratio of lemon leaves was increased by the application of PBZ. Yang [42] studied the effects of PBZ application, and drought stress on seedling vigor, photosynthetic capacity, and non-structural carbohydrates of *Phyllostachys edulis*, and concluded that drought stress increases the carbon–nitrogen ratio in the leaves *of Phyllostachys edulis*. However, spraying PBZ at the same time of drought stress may reduce the carbon–nitrogen ratio in the leaves. Gibberellin also reduced the C/N ratio of the lemon leaves.

### 4.3. *Effects of Hormone Regulation on the Nutritional Physiology of Lemon Leaves*

Spraying PBZ reduced the single fruit weight, fruit longitudinal diameter, fruit transverse diameter, skin thickness, and VC content of the lemons, but increased the juice yield and TSS content of the lemons, which is similar to the research results obtained by Fu [43] and Cao [44]. After spraying GA, the content of VC decreased, but the TSS of the fruit increased. However, Wu [45] found that spraying GA resulted in a substantial increase in the acid content of the pulp and a notable decrease in the sugar content of the pulp, which may have related to the spraying time and concentration. From the field experiment in this study, we found that the regulation technical measures used for autumn-flowering and fruiting can considerably improve the yield of lemon autumn flowers and fruits, but has little effect on the quality of autumn lemon flowers and fruits.

## 5. Conclusions

By regulating the distribution of photosynthetic products in plants, PBZ applications can promote the flowering process of fruit trees and make them bloom earlier. In this study, we found that the optimum PBZ concentration to promote lemon flowering was 600 mg/L, and the fruit quality increased by spraying PBZ on fruit trees. The physiological effect of gibberellin was the opposite to that of PBZ. Therefore, hormones should be reasonably applied to control the flowering and setting of fruit trees, and the concentration of hormones should be controlled.

**Author Contributions:** Conceptualization, J.L. (Jinxue Li) and Y.D.; methodology, Y.D.; formal analysis, J.D.; investigation, D.L. and J.L. (Jing Li); writing—original draft preparation, J.L. (Jinxue Li); writing—review and editing, C.H., Q.T., J.Z., J.L. (Jing Li), X.Z. and C.Z.; supervision, X.L. All authors have read and agreed to the published version of the manuscript.

**Funding:** This research was funded by National Natural Science Foundation of China (J.L.; 32060646), the Major Special Science and Technology Project of the Yunnan Province (Y.D.; 202102AE090020 and 202102AE090054), and the Special Projects for Foreign Cooperation of the Yunnan Province (Y.D.; 202003AD150014).

**Institutional Review Board Statement:** Not applicable.

**Informed Consent Statement:** Not applicable.

**Data Availability Statement:** Not applicable.

**Conflicts of Interest:** The authors declare no conflict of interest.

## References


1. Gao, J.Y.; Zhou, D.G.; Yue, J.Q.; Peng, M.X.; Sun, Y.F.; Li, J.X.; Zhang, P. Initial research on the dynamic regular of physiological deflowering and fruit drop of planting lemon in Dehong state. *Southwest China J. Agric. Sci.* **2008**, *2*, 328–331.
2. Li, J.X.; Peng, M.X.; Zhou, D.G.; Duo, J.Z.; Zhao, J.; Yue, J.Q. Effects of different regulating technologies on fruit of autumn flowering lemon. *J. Hunan Agric. Univ. (Nat. Sci.)* **2012**, *38*, 271–275.
3. Siqueira, D.L.D.; Cecon, P.R.; Carlos, L. Chamhum Salomão. Desenvolvimento do limoeiro 'Volkameriano' (*Citrus volkameriana* Pasq.) submetido a doses de paclobutrazol e ácido giberélico. *Rev. Bras. De Frutic.* **2008**, *30*, 764–768. [CrossRef]
4. Valero, D.; Martinez-romero, D.; Serrano, M.; Riquelme, F. Postharvest gibberellin and heat treatment effects on polyamines, abscisic acid and firmness in lemons. *J. Food Sci.* **1998**, *64*, 5. [CrossRef]
5. Martínez-Fuentes, C.; Mesejo, N.; Muñoz-Fambuena, C.; Reig, M.C.; González-Mas, D.J.; Iglesias, E.; Primo-Millo, M.A. Fruit load restricts the flowering promotion effect of paclobutrazol in alternate bearing *Citrus* spp. *Sci. Hortic.* **2013**, *151*, 122–127. [CrossRef]
6. Jr, J.H.; Lewis, L.N. Response of one-year-old cherry and mature bearing cherry, peach and apple trees to gibberellin. *Proc. Am. Soc. Hortic. Sci.* **1959**, *74*, 93–100.
7. Bradley, M.V.; Crane, J.C. Gibberellin-Induced Inhibition of Bud Development in Some Species of Prunus. *Science* **1960**, *131*, 825–826. [CrossRef]
8. Griggs, W.H.; Iwakiri, B.T. Effects of gibberellin and 2,4,trichlorophenoxypropionic acid sprays on 'Bartlett' pear trees. *Proc. Am. Soc. Hortic. Sci.* **1961**, *77*, 73–89.
9. Marcelle, R.; Sironval, C. Effect of Gibberellic Acid on Flowering of Apple Trees. *Nature* **1963**, *197*, 405. [CrossRef]
10. Bons, H.K.; Kaur, N.; Rattanpal, H.S. Quality and quantity improvement of citrus: Role of plant growth regulators. *Int. J. Agric.* **2015**, *8*, 433–447. [CrossRef]
11. Manuel, T.; Marco, C.; Fred, G.J. *The Genus Citrus*, 1st ed.; Elsevier Inc.: London, UK, 2020.
12. Koshita, Y.; Takahara, T.; Ogata, T.; Goto, A. Involvement of endogenous plant hormones (IAA, ABA, GAs) in leaves and flower bud formation of satsuma mandarin (*Citrus unshiu* Marc.). *Sci. Hortic.* **1999**, *79*, 185–194. [CrossRef]
13. Ashraf, M.; Karim, F.; Rasul, E. Interactive effects of gibberellic acid ($GA_3$) and salt stress on growth, ion accumulation and photosynthetic capacity of two spring wheat (*Triticum aestivum* L.) cultivars differing in salt tolerance. *Plant Growth Regul.* **2002**, *36*, 49–59. [CrossRef]
14. Goldberg-Moeller, R.; Shalom, L.; Shlizerman, L.; Samuels, S.; Zur, N.; Ophir, R.; Blumwald, E.; Sadka, A. Effects of gibberellin treatment during flowering induction period on global gene expression and the transcription of flowering-control genes in Citrus buds. *Plant Sci. Int. J. Exp. Plant Biol.* **2013**, *198*, 46–57. [CrossRef]
15. Blanchard, M.G.; Runkle, E.S. Dipping bedding plant liners in paclobutrazol or uniconazole inhibits subsequent stem extension. *Horttechnology* **2007**, *17*, 178–182. [CrossRef]
16. Kasraoui, M.F.; Duquesnoy, I.; Winterton, P.; Lamaze, T. Soluble and cell wall bound peroxidase activities are markers of flower bud development stages in lemon. *J. Appl. Bot. Food Qual.* **2014**, *87*, 1–8.
17. Okuda, H.; Kihara, T.; Iwagaki, I. Effects of paclobutrazol application to soil at the beginning of maturation on sprouting, shoot growth, flowering and carbohydrate contents in roots and leaves of satsuma mandarin. *J. Hortic. Sci.* **1996**, *71*, 785–789. [CrossRef]
18. Reddy, Y.T.N.; Prasad, S.R.S.; Bindu, G. Hormonal changes in response to paclobutrazol induced early flowering in mango cv. Totapuri. *Sci. Hortic.* **2012**, *150*, 414–418.
19. Bazurto, F.P.; Celi, A.; Corozo, L.; Solís, L. Importance of paclobutrazol in out-of-season citrus production. *Manglar* **2022**, *19*, 117–127. [CrossRef]
20. Chen, X.; Tao, Z.L.; Wu, Z.X.; Wang, L.X.; Fu, H.Z.; Zhou, Z.D.; Fan, W.B. Effect of paclobutrazol plus ethephon treatment on endogenous hormones and carbon and nitrogen nutrients in litchi variety 'Feizixiao'. *Agric. Sci. Technol.* **2013**, *14*, 1125–1131.
21. Liu, T.; Hu, Y.Q.; Li, X.X. Comparison of dynamic changes in endogenous hormones and sugars between abnormal and normal *Castanea mollissima*. *Prog. Nat. Sci.* **2008**, *18*, 685–690. [CrossRef]
22. Cruz, M.; Siqueira, D.; Salomão, L.; Cecon, P. Influence of paclobutrazol and of the environment temperature on flowering and fruitification of acid lime Tahiti. *Ciência E Agrotecnologia* **2008**, *32*, 1148–1153. [CrossRef]
23. Liu, H.M.; Li, J.; Zhu, C.H.; Peng, M.X.; Li, J.X.; Du, X.Y.; Gao, J.Y.; Yue, J.Q. Effects of paclobutrazol on vegetative and reproductive growth of lemon. *J. South. Agric.* **2013**, *44*, 1694–1699.
24. Zacarias, L.; Talon, M.; Ben-Cheikh, W.; Lafuente, M.T.; Primo-Millo, E. Abscisic acid increases in non-growing and paclobutrazol-treated fruits of seedless mandarins. *Physiol. Plant.* **1995**, *95*, 613–619. [CrossRef]
25. AOAC. *Vitamin C (Ascorbic Acid) in Vitamin Preparations and Juices*; AOAC International: Arlington, VA, USA, 1990.
26. Yemm, E.W.; Willis, A.J. The estimation of carbohydrates in plant extracts by anthrone. *Biochem. J.* **1954**, *57*, 508–514. [CrossRef] [PubMed]
27. Mccready, R.M.; Guggolz, J.; Silviera, V.; Owens, H.S. Determination of Starch and Amylose in Vegetables. *Anal. Chem.* **1950**, *22*, 1156–1158. [CrossRef]

28. Huwei, S.; Fan, F.; Juan, L.; Quanzhi, Z. Nitric oxide affects rice root growth by regulating auxin transport under nitrate supply. *Front. Plant Sci.* **2018**, *9*, 659.
29. Kamran, M.; Wennan, S.; Ahmad, I.; Xiangping, M.; Wenwen, C.; Xudong, Z.; Siwei, M.; Khan, A.; Qingfang, H.; Tiening, L. Application of paclobutrazol affect maize grain yield by regulating root morphological and physiological characteristics under a semi-arid region. *Sci. Rep.* **2018**, *8*, 4818. [CrossRef]
30. Sharaf-Eldien, M.N.; El-Bably, S.Z.; Magouz, M.R. *Effect of Pinching and Spraying of Paclobutrazol on Vegetative Growth, Flowering and Chemical Composition of Zinnia elegans, Jacq*; Faculty of Agriculture, Mansoura University: Mansoura, Egypt, 2017.
31. Liu, H.M.; Du, Y.X.; Li, J.X.; Zhang, J.Z.; Li, J.; Yue, J.Q. Regulation Effect of Spring-flowering Lemon Treated with Flower Bud-inducing Methods in Dry-hot Valleys in Yunnan. *Southwest China J. Agric. Sci.* **2016**, *29*, 1437–1442.
32. Du, Y.X.; Li, J.; Zhang, J.Z.; Zhu, C.H.; Fu, X.M.; Huang, H.B.; Li, J.X.; Yue, J.Q. Positive and negative regulation effects of various plant growth regulators on yield and quality of lemon. *Acta Agric. Jiangxi* **2018**, *30*, 33–37.
33. Sun, Y.; Chen, X.; Liu, R.; Li, J.A. Effects of different concentrations of gibberellin on flower bud differentiation in tung oil tree. *Non Wood For. Res.* **2014**, *32*, 97–100.
34. Bausher, M.G.; Yelenosky, G. Morphological changes in citrus associated with relatively high concentrations of paclobutrazol. *J. Plant Growth Regul.* **1987**, *5*, 139–147. [CrossRef]
35. Cruz, M.D.C.M.D.; Siqueira, D.L.D.; Salomão, L.C.C.; Cecon, P.R.; Santos, D.D. Levels of carbohydrates in acid lime tree Tahiti treated with paclobutrazol. *Rev. Bras. Frutic.* **2007**, *29*, 222–227. [CrossRef]
36. Yan, Z.; Nie, J.Y.; Xu, G.F.; Li, J.; Li, Z.X.; Wang, X.D.; Kuang, L.X.; Li, H.F.; Li, M.Q. Effect of spraying Paclobutrazol on fruit quality during the period of fruit expansion in spring snow peach. *China Fruits* **2015**, *6*, 40–42.
37. Yang, D. *The Effect of Paclobutrazol on Growth and Development of Pingguoli*; Yanbian University: Yanji, China, 2008.
38. Chen, X.G.; Li, H.M.; Zhang, A.J.; Shi, X.M.; Tang, Z.H.; Wei, M.; Shi, C.Y. Effect of paclobutrazol under different N-application rates on photosynthesis and starch accumulation in edible sweet potato. *Acta Agron. Sin.* **2012**, *38*, 1728–1733.
39. Le, R.S.; Barry, G.H. Vegetative growth responses of citrus nursery trees to various growth retardants. *Horttechnology* **2007**, *20*, 197–201.
40. Nagel, O.W.; Henk, K.; Hans, L. Growth rate and biomass partitioning of wildtype and low-gibberellin tomato (*Solanum lycopersicum*) plants growing at a high and low nitrogen supply. *Physiol. Plant.* **2001**, *111*, 33–39. [CrossRef]
41. Zhang, H.N.; Su, Z.X.; Chen, H.B. Effects of $GA_3$ and $PP_{333}$ on flower formation, carbon and nitrogen accumulation in leaves of litchi Feizixiao. *J. South. Agric.* **2016**, *47*, 2098–2102.
42. Yang, L.Z.; Pan, C.X.; Shao, S.L.; Tao, C.Y.; Wang, W.; Ying, Y.Q. Effects of $PP_{333}$ and drought stress on the activity, photosynthetic characteristics, and non-structural carbohydrates of *Phyllostachys edulis* seedlings. *Acta Ecol. Sin.* **2018**, *38*, 2082–2091.
43. Fu, X.F.; Chen, W.H. The effect of $PP_{333}$ on the flower promoting and tip-controling of Xuegan orange. *South China Fruits* **2000**, *4*, 10.
44. Cao, S.Y.; Tang, Y.Z.; Zhang, J.C. Effects of $GA_{33}$ and $PP_{333}$ on the apple flower bud differentiation course and contents of endogenous hormone. *J. Fruit Sci.* **2001**, *6*, 313–316.
45. Wu, J.H.; Chen, Q.X.; Tang, Z.Q.; Wang, Z.Q.; Ren, S.X. Effects of gibberellin and benzylaminopurine on Yield and quality of Citrus. *J. Zhejiang Agric. Sci.* **2009**, *4*, 671–672.

*agronomy*

*Article*

# Transcriptomic Analysis of the Differences in Leaf Color Formation during Stage Transitions in *Populus* × *euramericana* 'Zhonghuahongye'

Mengjiao Chen [1,2,†], Hui Li [1,3,†], Wan Zhang [1,3], Lin Huang [1] and Jingle Zhu [1,*]

1 Research Institute of Non-Timber Forestry, CAF, Zhengzhou 450003, China
2 Research Institute of Tropical Forestry, Chinese Academy of Forestry, Guangzhou 510520, China
3 College of Forestry, Henan Agricultural University, Zhengzhou 450002, China
* Correspondence: zhujingle@caf.ac.cn
† These authors contributed equally to this work.

**Abstract:** To explore the mechanism underlying the leaf color variation of *Populus* × *euramericana* 'Zhonghuahongye' ('Zhonghong' poplar) leaves at different maturation stages, we used 'Zhonghong' poplar leaves and analyzed the $L^*$ (lightness), $a^*$ (redness), and $b^*$ (yellowness) color difference values and the pigment contents in the three maturation stages. The Illumina HiSeqTM 2000 high-throughput sequencing platform was used for transcriptome sequencing analysis, and leaf color changes during stage transitions were preliminarily explored. Overall, with the increase in $L^*$ and $b^*$ during leaf development, the $a^*$, $C^*$ (colour saturation), and chromatic values decreased, the total anthocyanin content decreased, and the chlorophyll and carotenoid content increased. A total of 11,868 differentially expressed genes were identified by transcriptome sequencing. Comparing the expression differences of structural genes involved in anthocyanin synthesis in the leaves at different stages, we identified 5612 genes in the R1 vs. R2 comparison, 10,083 in the R1 vs. R3 comparison, and 6068 in the R2 vs. R3 comparison (R1, R2, R3 refer to samples obtained on 1 April, 6 April, and 11 April, respectively). Key genes such as DFR•(Dihydroflavanol 4-reductase), ANS (anthocyanidin synthase), FLS (flavonol synthase), CHS (chalcone synthase), BZ1 (Bronze 1), bHLH35, and bHLH63 were identified. These structural genes and those that encode transcription factors may be related to the regulation of anthocyanin synthesis. Here, the key genes related to leaf color change in 'Zhonghong' poplar were discovered, providing an important genetic basis for the subsequent genetic improvement of 'Zhonghong' poplar.

**Keywords:** *Populus* × *euramericana* 'Zhonghuahongye'; anthocyanins; transcriptome sequencing; differential analysis

**Citation:** Chen, M.; Li, H.; Zhang, W.; Huang, L.; Zhu, J. Transcriptomic Analysis of the Differences in Leaf Color Formation during Stage Transitions in *Populus* × *euramericana* 'Zhonghuahongye'. *Agronomy* **2022**, *12*, 2396. https://doi.org/10.3390/agronomy12102396

Academic Editor: Víctor Manuel Rodríguez

Received: 12 August 2022
Accepted: 26 September 2022
Published: 3 October 2022



## 1. Introduction

'Zhonghong' poplar, full name *Populus* × *euramericana* 'Zhonghuahongye', a deciduous tree species with brightly colored leaves, is an attractive tree species used both as an ornamental and for independent breeding in China [1,2]. 'Zhonghong' poplar, which was officially released by the Chinese Academy of Forestry (Beijing, China), is a bud mutation variety of L2025 poplar (variety right number: 20,060,007). New leaves of 'Zhonghong' poplar trees are an attractive rose, then they gradually change from bright purplish red to brownish green as the seasons change, and they do not produce feathery seeds that are spread in the air, making this species very suitable for urban, garden, and road beautification [3].

Leaf color is the result of a combination of genetic and environmental factors and is related to the type, mass fraction, and distribution of pigments in leaf cells [4]. At present, research on the color of plant leaves has mainly focused on the mechanism underlying leaf color and the regulation of chemicals that drive plant leaf color; moreover, research on the physiological and morphological characteristics of leaves, their functional natural products, and leaf pigment composition has greatly advanced [5]. Many studies on the mechanism of leaf color variation have explored the relationship between leaf color changes and pigment content in leaves, and candidate genes related to anthocyanin synthesis have been identified [5]. Poplar, as a woody model plant, has been the focus of many studies on the mechanism of phloem regulation. MYB6, which is mainly expressed in young leaves, was isolated from Populus tremula, and overexpression of MYB6 promoted the expression of flavonoid synthesis genes [6]. Among the reported MYB transcription factors in poplar, not only positive regulatory factors but also a class of transcriptional repressive effect factors were found. For example, the activation of the MYB182 transcription factor suppresses the expression of structural and regulatory genes, thus reducing the accumulation of poplar anthocyanins [7].

The amount of anthocyanin accumulated is the end result of biosynthesis and degradation processes. The degradation of anthocyanins occurs in different plant organs due to various environmental and developmental conditions. For example, anthocyanins usually accumulate in young/juvenile leaves and are degraded as the leaves mature [8]. In cases where there is no apparent benefit to the plant, anthocyanin degradation may occur due to changes in the vesicles that contain this pigment. Also the vesicle pH can have an effect on the pigment stability, and an increase in pH in the vesicles of senescent tissues may reduce the stability of anthocyanins and cause their chemical degradation [9]. The degradation of anthocyanins in Cabernet Sauvignon grape skins at high temperatures may be the result of chemical degradation but may also be due to heat stress-induced activity of degradative enzymes (β-glucosidase, peroxidase) in the vesicles [10]. In other cases, such as in flowers or photosynthetic tissues of mandarin jasmine, the drastic color changes caused by phloem degradation are likely to be a signal for pollinators with a clear cause, and this process may be regulated by specific genes and proteins [11].

In order to investigate the causes of leaves color change, this study was conducted to identify genes related to anthocyanin synthesis by RNA sequencing (RNA-seq) of three different developmental stages of 'Zhonghong' poplar. The results provided basic data for elucidating the molecular mechanism of leaf color development at the color-changing stage of the leaves of 'Zhonghong' poplar.

## 2. Materials and Methods

### 2.1. Plant Materials and Sampling

In the state-owned Mengzhou forest farm on the south side of Xiguo town, Mengzhou city, Henan Province, three healthy, growing 'Zhonghong' poplar trees were selected as test material, and their leaves were collected on April 1 (R1), April 6 (R2), and April 11 (R3). Each leaf was divided into 2 parts along the main leaf vein (excluding the leaf secondary veins). Half of the fresh samples were used for pigment content and color difference determinations, and half was immediately flash frozen in liquid nitrogen and stored at −80 °C for subsequent transcriptome-related tests. Each sample was biologically replicated 3 times.

### 2.2. Leaf Colour Analysis

A CR2500 color difference meter (Minolta, Japan) was used to measure the $L^*$ value (representing lightness; positive means white, and negative means black), $a^*$ value (representing the red/green phase; positive means red, and negative means green), and $b^*$ value (representing the yellow/blue phase; positive mean yellows, and negative means blue) of fresh leaves. For each leaf, 10 randomly selected opaque points (avoiding the leaf veins) were measured, and the average of these 10 values was taken as the final value of the indicators [12]. Through the measured redness (hue $a^*$), yellowness (hue $b^*$), and lightness ($L^*$) values, the chroma ($C^*$) value and colored light value were calculated. $C^*$ is used to describe the vividness of a color. The higher the chroma is, the purer the color, and the brighter the leaf color; the lower the chroma is, the more astringent the color, and the more turbid the leaf color [13].

$$C^* - \left(a^{*2} + b^{*2}\right)^{\frac{1}{2}} \quad (1)$$

$$\text{coloured light value} = \frac{2000 \cdot a^*}{L^* \cdot (a^{*2} + b^{*2})^{\frac{1}{2}}} \quad (2)$$

*2.3. Determination of Leaf Pigment Contents*

The contents of photosynthetic pigments in the leaves were determined by the direct extraction method [14,15]. The leaves were washed, ground (after removing the veins) and weighed to 0.1 g (to a precision of 0.01 g). Ten milliliters of 80% acetone was added to the samples, after which the samples were incubated in the dark for 72 h. Afterwards, the supernatant was removed, and chlorophyll was extracted. For this, the samples were ground in 10 mL of a 10% HCl methanol solution, incubated at room temperature for 4 h, and then centrifuged at 3500× $g$ for 15 min. The absorbance values of the chlorophyll extracts at 445 nm, 644 nm, and 662 nm and the absorbance values of anthocyanin extracts at 530 nm and 657 nm were measured by UV spectrophotometry (Jinghua Instruments 752, Shanghai, China); each measurement involved 3 biological replicates and 3 technical replicates. The anthocyanin content ($C_A$), chlorophyll a content ($C_a$), chlorophyll b content ($C_b$), total chlorophyll content ($C_T$), and carotenoid content ($C_{car}$) were calculated according to Formulas (3)–(7):

$$\text{Chlorophyll a content } \left(\text{mg·g}^{-1}\right): C_a = (9.78 \times \text{A}_{662} - 0.99 \times \text{A}_{644})\text{V}/(\text{W} \times 1000) \quad (3)$$

$$\text{Chlorophyll b content } \left(\text{mg·g}^{-1}\right): C_b = (21.40 \times \text{A}_{644} - 4.65 \times \text{A}_{662})\text{V}/(\text{W} \times 1000) \quad (4)$$

$$\text{Total chlorophyll content } \left(\text{mg·g}^{-1}\right): C_T = C_a + C_b \quad (5)$$

$$\text{Carotenoid content } \left(\text{mg·g}^{-1}\right): C_{car} = (4.69 \times \text{A}_{445})\text{V}/(\text{W} \times 1000) - 0.27 \times (C_a + C_b) \quad (6)$$

$$\text{Anthocyanin content } \left(\text{mg·g}^{-1}\right): C_A = (\text{A}_{530} - 0.25 \times \text{A}_{657})\text{V}/(\text{W} \times 1000) \quad (7)$$

Notes: A, absorbance value; V, extract volume; W, sample mass.

*2.4. Total RNA Extraction and Detection*

Extraction and transcriptome sequencing of the RNA samples were performed by staff of Wuhan Metware Biotechnology Co., Ltd. (Wuhan, China). The RNA integrity and DNA contamination were analyzed by agarose gel electrophoresis. A NanoPhotometer spectrophotometer and Qubit 2.0 fluorometer (Thermo Fisher Scientific, Waltham, MA, USA) were used to determine the concentration and purity of the RNA samples, and an Agilent 2100 Bioanalyzer (Agilent, Santa Clara, CA, USA) was used to accurately determine the RNA integrity and total RNA amount.

*2.5. Library Construction and Sequencing*

After verifying the quality of the RNA samples, mRNA was enriched using Oligo (dT) magnetic beads. A sequencing library was subsequently constructed, and an Illumina HiSeq$^{TM}$ 2000 sequencer (San Diego, CA, USA) was used for high-throughput sequencing. The off-machine data were filtered to obtain clean data, and sequence alignments with the specified reference genome were performed using Hisat2 software (version 2.1.0, The Kim Lab, Dallas, TX, USA) to obtain mapped data [16]. Differential expression analysis, functional annotation, and functional enrichment analysis of differentially expressed genes (DEGs) were performed according to the expression levels of the genes in different samples or different sample groups.

*2.6. Screening and Functional Annotation of DEGs*

DESeq2 (Bioconductor, Coronado, CA, USA) [17,18] was used to analyze the differential expression between sample groups, and the DEG set between groups was obtained according to the conditions of $|\log2(\text{fold-change})| \geq 1$ and a false discovery rate (FDR) < 0.05. Functional annotation and enrichment analysis of the DEGs were performed through the Kyoto Encyclopedia of Genes and Genomes (KEGG) [19], Gene Ontology (GO) and EuKaryotic Orthologous Groups (KOG) [20,21] databases. The number of DEGs and significant enrichment results for each KEGG pathway, GO term, and KOG functional classification were determined.

*2.7. Construction of a Coexpression Network*

The DEGs were then used to perform a weighted gene coexpression network analysis using WGCNA v1.69. The resulting coexpression network was visualized using the free software Cytoscape 3.9.1.

*2.8. Transcription Factor Analysis*

Plant transcription factor prediction was performed using the iTAK [22] software (Shanghai, China); iTAK integrates two databases, PlnTFDB [23] and PlantTFDB [24].

*2.9. Quantitative Real-Time PCR (RT-qPCR)*

The synthesized cDNA was used as a template for amplification in a real-time fluorescence quantitative PCR machine (Bio-Rad CFX96, Hercules, CA, USA) according to the instructions of the Taq Pro Universal SYBR qPCR Master Mix kit (Q712 of Vazyme, Nanjing Vazyme Biotech Co., Nanjing, China), and each sample was analyzed three times. Gene-specific primers for RT-qPCR were designed by Oligo 7 software and synthesized by Henan Shangya Biological Co., Ltd. (Zhengzhou, China) Using β-actin as the internal reference gene, we tested the amplification reaction background of the target gene, and the relative expression of the genes in the list was calculated by the $2^{-\Delta\Delta Ct}$ method [25]. The RT-qPCR specific primers and internal reference primers used are listed in Table S4.

## 3. Results

*3.1. Analysis of Leaf Colour Phenotypes at Different Stages of Leaf Maturation*

As the leaves matured, they gradually changed from a bright purplish red to a purplish green (Figure 1A), and the changes in the leaf color parameters at different stages of leaf maturation were analyzed (Figure 1B). We found that the lightness ($L^*$) increased, and the hue ($b^*$) value was positive, which means that the yellow/blue color was yellowish, and the yellowing degree increased. The hue ($a^*$) of the R3 leaves was negative, indicating that the red/green color was greenish, and the hue ($a^*$) and saturation ($C^*$) tended to decrease gradually. The saturation value comprehensively reflects the color and lightness of a leaf surface. The saturation value gradually decreased, indicating that the comprehensive characteristics of leaf color and lightness showed a decreasing trend with increasing leaf development (the red color became lighter, and the lightness intensified). The greater the colored light value is, the darker the leaf redness, and the less intense the lightness; these parameters can be used as representative features to reflect the changes in the color and lightness of 'Zhonghong' poplar leaves.

The main leaf pigments of 'Zhonghong' poplar are anthocyanins, chlorophyll, and carotenoids; chlorophyll includes both chlorophyll a and chlorophyll b. The change trend of leaf pigment contents at different maturation stages is shown in Figure 1C: the content of chlorophyll a was much higher than those of carotenoids and anthocyanins. Since both chlorophyll and carotenoids are photosynthetic pigments, the change in carotenoid content in the leaves was essentially the same as that of total chlorophyll. Although the content of total anthocyanins decreased gradually, the contents of the other pigments showed a tendency to increase. Through the results of the above comparative analysis, it could be preliminarily inferred that with the decrease in anthocyanin content in the leaves of 'Zhonghong' poplar, the content of chlorophyll increased, and the shift from a predominance of anthocyanins (61.2%) to that of chlorophyll (67.7%) was the main factor driving the gradual transformation of the leaves from bright purplish-red to purplish-green (Figure 1D).

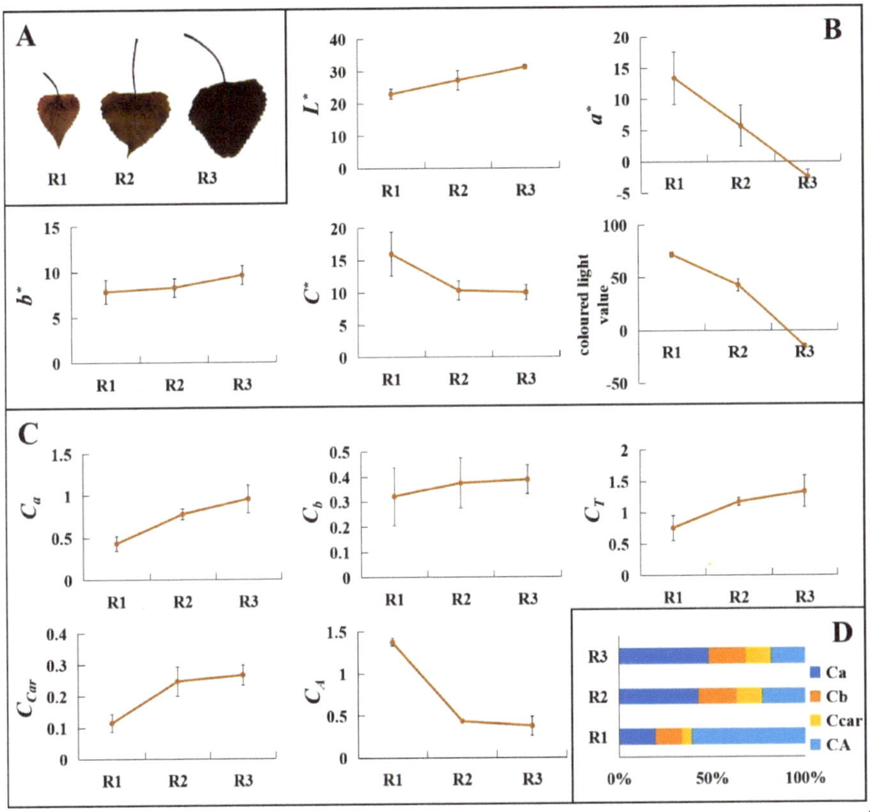

**Figure 1.** Leaves' physiological characteristics. (**A**) 'Zhonghong' poplar leaves; (**B**) Leaf color difference values of 'Zhonghong' poplar leaves; (**C**) content of photosynthetic pigments in 'Zhonghong' poplar leaves; (**D**) percentage content of each pigment in the leaves.

### 3.2. Sequencing Data Quality Assessment

To further explore the molecular mechanisms of anthocyanin biosynthesis during leaf color change, we constructed nine cDNA libraries to analyze the transcriptome at three stages of leaf development (three biological replicates at each stage). A total of 65.86 Gb of clean data was obtained by sequencing the cDNA libraries. There were 6 Gb of clean data per sample, and the Q30 base percentage was 93.97% or more (Table S1). HISAT2 was used to compare the clean read sequences with the sequence information in the Mullein reference genome database. Most of the gene sequences matched the reference genome sequence, and the comparison percentage of each sample ranged from 82.98 to 85.53% (Table S2). Based on the location information of the reads in the genome, a total of 1273 new genes were identified (Table S3).

The expression level of the genes was measured according to the fragments per kilobase of transcript per million mapped reads (FPKM), and the gene expression of all samples was restricted to the range $2 \geq \log10(\text{FPKM}) \geq -2$. (Figure S1). The samples were subsequently subjected to principal component analysis (Figure S2) and correlation analysis

(Figure S3). The Pearson correlation coefficient (R2) between samples was higher than 0.8, and the clustering was obvious, indicating that the transcriptome data were reliable for subsequent analysis. The clean data are available in the NCBI database.

### 3.3. DEG Analysis

DeSeq2 was used to analyze the genes differentially expressed in the leaves of 'Zhonghong' poplar at different times; a total of 11,868 DEGs were identified. The statistics and distribution of the DEGs in each group are shown in Figure 2A,B. With respect to the three periods of leaf color change, we identified fewer DEGs in the R1 vs. R2 comparison group (5612) than in the R2 vs. R3 comparison group (6068), indicating that with the development of the leaves, the number of DEGs involved in color regulation increased, and the related process of gene regulation became more complex. The expression level of most DEGs gradually decreased, which suggests that the biological response involved in the leaf color transformation process was mainly driven by the downregulation of genes. In addition, a heatmap was constructed to visualize the overall picture of all DEG expression patterns, to better understand the overall variation in DEG expression (Figure 2C). The expression patterns of some DEGs were consistent with the transition period, and most DEGs showed great differences in terms of their expression profiles between R1 and R3. Further analysis showed that 1633 DEGs were commonly identified in all three sets of comparative analyses (Figure 2D). To study the expression patterns of genes differentially expressed in different periods, the FPKM of the genes was centralized and standardized, and a total of 10 expression patterns were obtained by k-means cluster analysis. The results suggested that a group of genes with similar changes in their expression could have similar functions (Figure 2E).

### 3.4. Coexpression Network Analysis

We performed a weighted gene coexpression network analysis (WGCNA) to search for genes related to anthocyanin synthesis in 'Zhonghong' poplar from an integrated network perspective and constructed a coexpression network involving all DEGs in the three different developmental stages. Genes with similar expression patterns were clustered into the same modules, and different modules were distinguished by color. Finally, 11 different merged modules were identified, and the subsequent analysis was performed according to the merged modules (Figure 3A). The results of a correlation analysis revealed that these modules corresponded to specific distribution patterns in different periods. The characteristic gene expression profiles of the 11 modules were analyzed, the results of which are shown in Figure 3B. The genes of the blue-green module had higher expression levels in stage 1 and lower expression levels in stage 3, consistent with the change in anthocyanin production (Figure 3C). The genes of the blue module had higher expression levels in stage3 and lower expression levels in stage 1, indicating an opposite trend with respect to the change in anthocyanin production (Figure 3D).

To understand the biological processes associated with the genes of the different modules and identify the categories of enriched pathways linked to the genes in each module, the genes in the blue-green and blue modules were further annotated by GO, KEGG, and KOG enrichment analysis (Figure S4). The DEGs of the blue and blue-green modules were extensively annotated in pathways such as RNA degradation, RNA transport, biosynthesis of secondary metabolites, carotenoid biosynthesis, glutathione metabolism, metabolic pathways, phenylpropanoid biosynthesis, and porphyrin and chlorophyll metabolism; these DEGs were also significantly enriched in the biosynthesis of various secondary in metabolites, terpenoid backbone biosynthesis, metabolic pathways, carotenoid biosynthesis, and porphyrin and chlorophyll metabolism. According to the results of the KOG analysis of the DEGs, the blue modules were mainly enriched in pathways related to posttranslational modification, protein turnover, chaperones, and signal transduction mechanisms. Similarly, the blue-green modules were mainly enriched in posttranslational modification, protein turnover, chaperones, translation, ribosomal structure and biogenesis, and protein function.

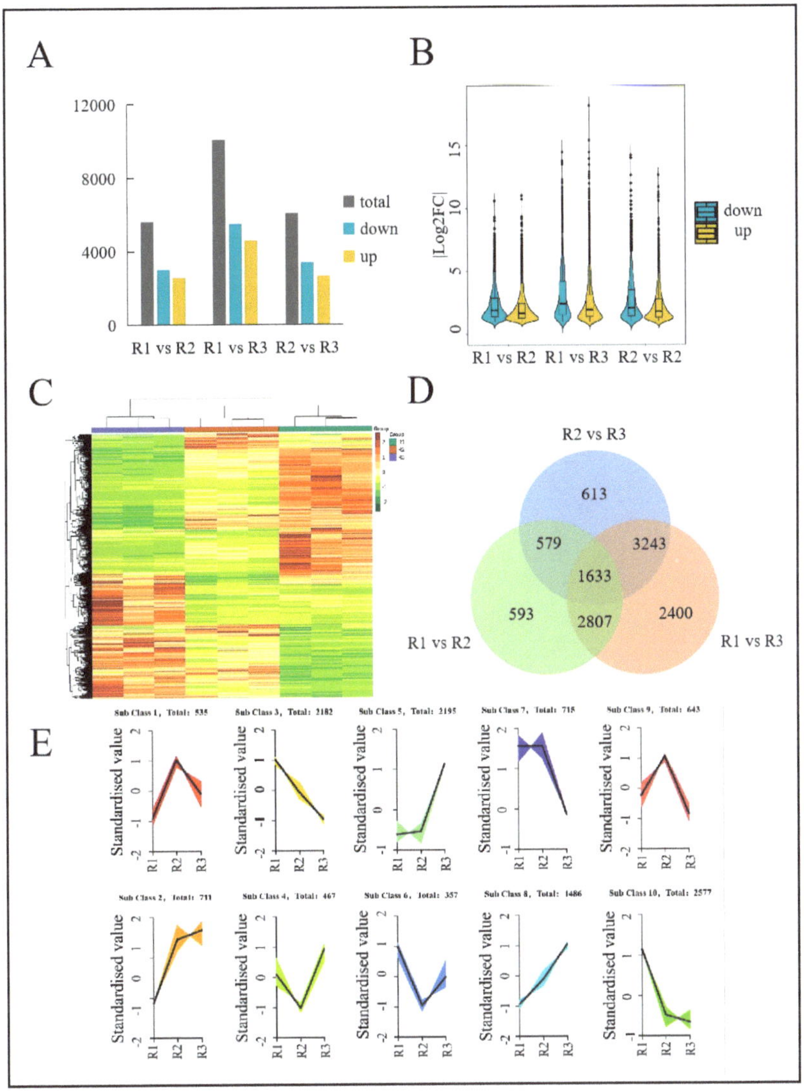


**Figure 2.** Differential gene expression. (**A**) Comparison of the genes differentially expressed in leaf maturation stages; (**B**) number and distribution of the differentially expressed genes; (**C**) cluster heat map of all differentially expressed genes; (**D**) Venn diagram of annotated transcripts in multiple samples; (**E**) K-means cluster plot of all differentially expressed genes.

### 3.5. Functional Annotation and Enrichment Analysis of DEGs

The DEGs were classified into three categories of GO functional annotations, namely, biological processes, cellular components, and molecular functions, and the types and quantities of annotations for each category were different. The numbers of DEGs in the three developmental stage comparison groups were different, but the main enriched entries were very similar. In the biological process category, the DEGs were mainly associated with subcategories including metabolic processes, cellular processes, stress responses, biological regulation, and biological process regulation. In the category of cellular components, the DEGs were mainly associated with subcategories such as cells, cell parts, organelles, membranes, and membrane parts. In the molecular function category, the DEGs were mainly associated with subcategories such as catalytic activity, linkage, transcriptional regulator activity, transport activity, and molecular function regulation.

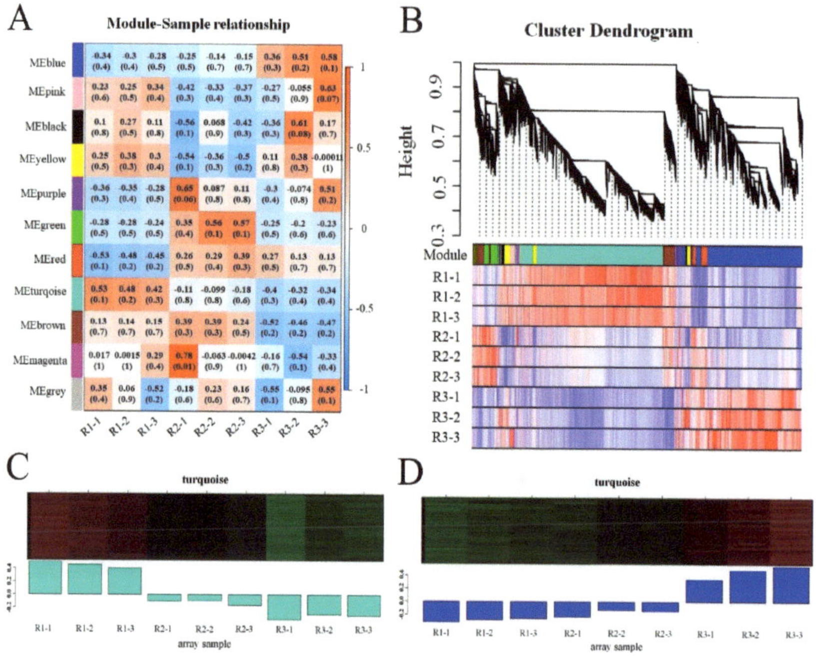


**Figure 3.** Weighted gene co-expression network analysis (WGCNA) of DEGs identified in 'Zhonghong' poplar leaves. (**A**) Relationships between modules and different samples, including petals. Each row in the table corresponds to a module, and each column corresponds to a sample; (**B**) gene dendrogram obtained by hierarchical clustering with the module color indicated by the color of the row underneath. A total of 11 distinct modules were identified; (**C**) Eigengene expression profile for the turquoise module in different tissues. The *Y*-axis indicates the module eigengene value; the *X*-axis indicates the samples; (**D**) Eigengene expression profile for the blue module in different tissues.

The DEGs associated with the accumulation of flavonoids and anthocyanins were subjected to KEGG functional enrichment analysis, and the metabolic pathways of the DEGs were determined. In the R1 vs. R2 comparison, metabolic pathways were enriched in the most DEGs, photosynthesis-antenna proteins were highly enriched in DEGs, and monoterpene biosynthesis was the pathway most significantly enriched in DEGs. In the R1 vs. R3 comparison, biosynthesis of secondary metabolites was enriched in most DEGs, the enrichment of caffeine metabolism was the greatest, and the biosynthesis of secondary metabolites was the most significantly enriched in DEGs. In the R2 vs. R3 comparison, ribosomes were enriched in most DEGs, and ABC transport was the pathway most significantly enriched in DEGs. Taken together, these findings indicated that some DEGs in leaves at different stages were related to the synthesis pathways of monoterpenes and secondary metabolites. Based on the results of the functional annotation and enrichment analysis, the following metabolic pathways related to anthocyanin synthesis were identified: the phenylpropane biosynthesis pathway (ko00940), the flavonoid biosynthesis pathway (ko00941), the anthocyanin biosynthesis pathway (ko00942), the flavonoid biosynthesis pathway (map00943), and the flavonoid and flavonol biosynthesis pathway (map00944).

### 3.6. Analysis of DEGs Related to Anthocyanin Synthesis

Factors that affect leaf color formation mainly include structural genes and regulatory genes. Based on the functional classification and enrichment analysis results of the DEGs via the GO, KEGG, and KOG databases, 24 differentially expressed structural genes associated with flavonoid metabolic pathways, anthocyanin biosynthesis, and phenylpropane biosynthesis were identified. Each gene had multiple transcripts, and those that were differentially expressed in the R1 vs. R2, R1 vs. R3, R2 vs. R3 comparisons as well as those that were downregulated included dihydroflavonol 4-reductase (DFR), BZ1, HCT, and FLS. In addition, the gene POPTR_008G138600v3 (DFR) was upregulated in all three comparisons, and the expression of multiple transcripts of the POPTR_009G133300v3 (BZ1) and ANS, CHS, and 3AT genes decreased during the three periods, which is consistent with the leaf color changes (Figure 4).

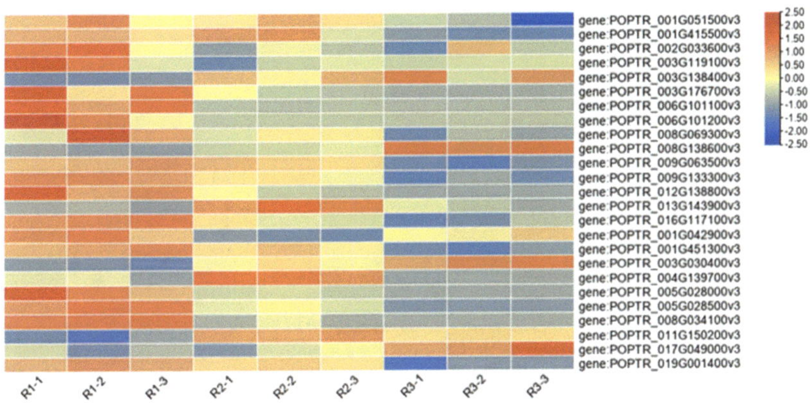


**Figure 4.** Heat map of gene expression of anthocyanin synthesis-related enzymes.

The correlation network of DEGs involved in flavonoid biosynthesis (ko00941), anthocyanin biosynthesis (ko00942), isoflavone biosynthesis (ko00943), and biosynthesis of flavonoids and flavonols (ko00944) is shown in Figure 5. There was a positive correlation between anthocyanin 3-O-glycoside 6″-O-POPTR_009G063500v3 and three anthocyanin 3-O-glucosyltransferases (BZ1:POPTR_009G133300v3, POPTR_013G118700v3, POPTR_013G143900v3) in the anthocyanin biosynthesis regulatory network. In the flavonoid biosynthesis pathway, a complex regulatory relationship involving CYP73A, caffeoyl-coenzyme A O-methyltransferase, CHS, CHI, ANS, FLS, CYP75B1, ANR, HCT, LAR, DFR, CYP75A, codeine 3-O-demethylase, and salicylic acid 3-hydroxylase appeared.

### 3.7. Transcription Factor (TF) Analysis

TFs are the main proteins that control biological metabolic processes. During the leaf color change period in 'Zhonghong' poplar, we identified a total of 149 DEGs encoding TFs in the three R1 vs. R2, R1 vs. R3, and R2 vs. R3 comparisons. Of the 121 DEGs encoding TFs, 90 were consistently downregulated in terms of their expression in all three periods, and 31 were upregulated: 13 AP2, 7 B3, 6 basic helix–loop–helix (bHLH), 3 bZIP, 6 C2C2, 5 C2H2, 4 GRF, 9 HB, 2 HSF, 2 HMG, 11 MYB and MYB-related, 8 NAC, 4 WRKY, and 6 zf-HD genes encoding TFs, as well as others. These 121 TFs are thought to be important factors regulating differences in anthocyanin synthesis at different developmental stages of 'Zhonghong' poplar leaves (Figure 6).

### 3.8. qRT-PCR-Based Verification of DEGs

To confirm the expression patterns of candidate genes identified in the transcriptome data, 10 key DEGs (Table S4, Figure S5) involved in the main pathways of anthocyanin accumulation were successfully validated via qRT-PCR analysis. Although the level of expression was not exactly the same, the expression trends were consistent with the RNA-seq expression data, confirming the reliability of the current RNA-seq data and the computational analysis results, enabling subsequent DEG analysis.

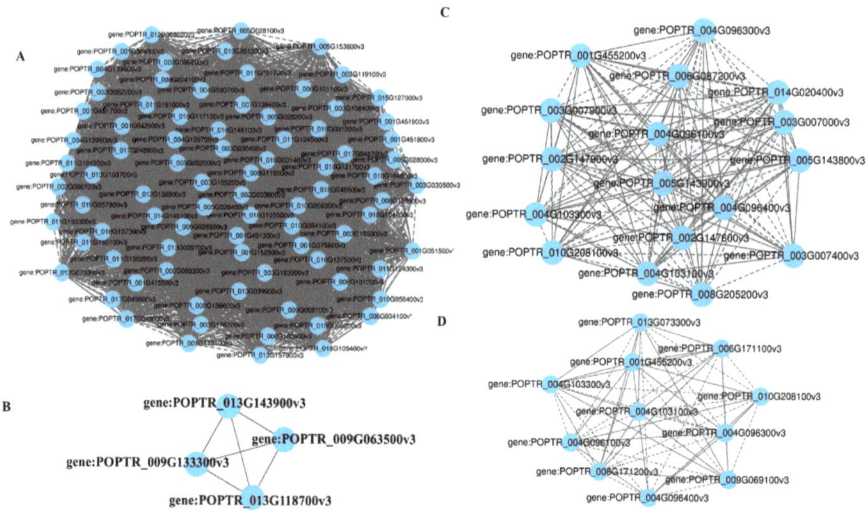


**Figure 5.** Correlation network diagram. (**A**) Flavonoid biosynthesis (ko00941); (**B**) anthocyanin biosynthesis (ko00942); (**C**) isoflavone biosynthesis (ko00943); (**D**) biosynthesis of flavonoids and flavonols (ko00944).

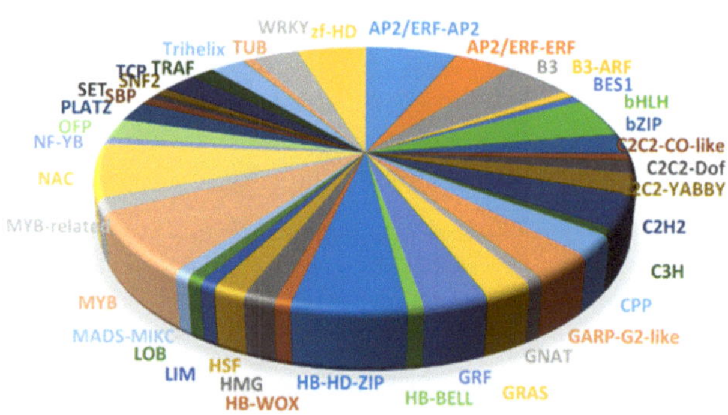


**Figure 6.** Frequency distribution of transcription factors in the transcriptome data.

## 4. Discussion

As 'Zhonghong' poplar leaves mature, their brightness ($L^*$) and yellowing ($b^*$) increase, while their redness ($a^*$), saturation ($C^*$), and chromatic value decrease; the color of new leaves is red and bright, and the chromaticity varies in a wide range. The values of these parameters are significantly different and can be used as representative to describe the color change of 'Zhonghong' poplar leaves. As the leaves matured, the content of total anthocyanin in the leaves of 'Zhonghong' poplar decreased sharply, while the content of chlorophyll and carotenoids first increased and then decreased. A high content and a high proportion of anthocyanins play important roles in the appearance of leaf traits [26]. The different colors of plant tissues are determined by the ratio and distribution of three pigments (chlorophyll, carotenoids, flavonoids/anthocyanins) [27]. In green-leaf plants, the proportion of chlorophyll is high; in yellow-leaf plants, the proportion of carotenoids is high; in purple- and blue-leaf plants, anthocyanins account for more than the other two pigments [28]. The flavonoids involved in color formation are mainly anthocyanins, which give flowers, fruits, and other plant tissues their characteristic red, purple, and blue colors. Many transgenic experiments have shown that the red color of plant leaves is related to the overexpression of genes promoting anthocyanin synthesis [29–31]. However, in this study, it was found that the leaves of 'Zhonghong' poplar also had a high chlorophyll content, and the final leaf color was determined by the ratio of the leaf chlorophyll content to the anthocyanin content. Therefore, the high ratio of total anthocyanin content to chlorophyll content in the leaves of 'Zhonghong' poplar may be the physiological mechanism underlying the bright red color of the leaves of 'Zhonghong' poplar.

The leaves of many plants in nature gradually become less red as they develop. Hughes found that the anthocyanin content of newborn young leaves was high, and as the fenestra tissue differentiated significantly, the accumulation of anthocyanin decreased, but the chlorophyll content tended to increase at the same time, resulting in the phenomenon of reddening [32]. Since anthocyanins absorb light in both the visible and the UV regions, their accumulation in developing young leaves can act as a kind of "sunscreen", protecting the leaves, especially their photosynthetic machinery, from the damaging UV rays as well as the high-intensity photoinhibition caused by visible light. As the leaves mature and form protective waxes that reflect the sunlight, providing photoprotection, they typically change from red to green. This loss of red pigmentation may be due to the increased chlorophyll accumulation during leaf expansion and growth, the termination of anthocyanin biosynthesis, and the slowing of growth, or anthocyanin degradation that is actively induced when it is no longer needed as a photoprotectant. In heathy plants, anthocyanin concentration is directly related to the age of the leaf; anthocyanin concentration is higher in young leaves, and as the leaf develops, it gradually decreases due to dilution and degradation, resulting in green leaf traits. The decrease in anthocyanin concentration as the leaves matured in this study was due partly to the dilution of the pigment in the growing tissue and partly to the degradation process that actually occurred. In some plants, the change from red young leaves to mature green leaves is not due to the degradation of anthocyanins but simply to increased chlorophyll synthesis, termination of anthocyanin synthesis, and accumulation of anthocyanins caused by the dilution effect in the leaves.

The transcriptome revealed a large-scale upregulation of genes involved in the flavonoid and anthocyanin biosynthetic pathways. Our study found that some structural genes, i.e., DFR, BZ1, ANS, CHS, 3AT, HCT, and FLS, were differentially expressed. Among these DEGs, some were involved in flavonoid biosynthesis and anthocyanin biosynthesis pathways. According to some studies, HCT genes are mainly involved in chlorogenic acid biosynthesis and lignin biosynthesis pathways [33]. ANR catalyzes the reduction of anthocyanins to synthesize epicatechin and epigallocatechin. UFGT glycosylation makes anthocyanins more stable [34,35]. Studies have revealed that there is a competitive relationship between ANR and UFGT. The upregulation of ANR expression increases the procyanidin catechin content and decreases the anthocyanin content [36,37]. Genes related to the flavonoid pathway can be divided into structural genes directly involved in flavonoid production and regulatory genes that control the expression of structural genes. In a study of blue grape hyacinth, it was found that substrate competition among metabolites may occur. Flavonol synthase (FLS) competes with flavonoid substrates for DFR, resulting in the elimination of blue pigments and a lower cyanidin content in white flower mutants than in blue flowers [38]. Through the analysis of flavonoid pathway gene expression, studies have shown that the red or purple leaf color traits of many transgenic plants are related to the overexpression of genes related to anthocyanin synthesis [29–31]. Compared with wild-type plants, transgenic tomato plants overexpressing AtPAP2 accumulated more anthocyanins [39], and poplar plants overexpressing PtrMYB119 accumulated more anthocyanins [31]. We found that the expression of BZ1, DFR, ANS, CHS, HCT, FLS, and 3AT biosynthetic genes decreased with increasing leaf growth and development. Their downregulated expression with increasing leaf growth and development is consistent with a role in anthocyanin and flavonoid biosynthesis.

In addition to substrate competition, plant leaf color differences may also be due to TFs. Flavonoid biosynthesis is driven by three types of TFs, namely, R2R3-MYBs, bHLH proteins, and WD40 repeat proteins, which form a MYB–bHLH–WD40 TF complex [38,40]. Among the components of this complex, MYBs are the main factors determining anthocyanin production [6]. Our results identified transcription factors that may be associated with leaf aging and anthocyanin changes during aging, such as MYB102, MYB73, MYB8, MYB306, MYB108, MYB90, and MYB113. Previous studies in rice showed a signaling response of MYB102 through the downregulation of ABA accumulation during leaf senescence [41]. In apple, MYB306 was shown to regulate anthocyanin synthesis by suppressing DFR gene expression [42] Our study also found that MYB102 and MYB306 expression was upregulated as the leaves matured, which was consistent with the leaf phenotype. However, at the same time, we found that MYB108 was highly expressed in senescing petals by ethylene signaling in moonflower petals, while in our results, MYB108 expression gradually decreased with leaf senescence, which may be due to the difference between flowering plants and trees [43]. Some transcription factors that promote anthocyanin synthesis, i.e., MYB8, MYB90, and MYB113, showed a tendency to be downregulated as the leaves matured, but MYB113 still retained a high expression, probably because MYB113 is mainly expressed in mature leaves [44,45]. Jiao' results suggested that MYB73 is involved in SA and JA signaling pathways [46], and a study showed that ethylene and jasmonic acid affect anthocyanin synthesis [47]. A previous study showed that, when the JMJ25 gene is overexpressed, its product can bind to the TF MYB182 locus and demethylate its chromatin, thereby inhibiting the expression of structural genes involved in the anthocyanin biosynthesis pathway in poplar, resulting in a decrease in anthocyanin content; gene expression levels negatively regulate the synthesis and accumulation of anthocyanins [48]. By analyzing our transcriptomic data, we identified several key genes encoding TFs of these three families that may regulate poplar flavonoid and anthocyanin production. Among them, the transcript abundance of three R2R3-MYB TFs was altered in all the leaves, suggesting that these TFs may play a key role in leaf color determination. These results suggest that these MYBs have important roles, but their functions in flavonoid or anthocyanin biosynthesis need to be further investigated via functional analysis.

## 5. Conclusions

In this study, the transcriptomes of 'Zhonghong' poplar red leaves in three developmental periods were sequenced and analyzed. A total of 11,868 DEGs were identified by pairwise comparison. GO and KEGG annotations and enrichment analysis found that the differentially expressed genes primarily regulate biological processes such as metabolic processes, cellular processes, stress responses, bioregulation, and bioprocess regulation. During the leaf color transchromation period, a total of 24 genes related to anthocyanins were screened, and by comparing gene expression in leaves at different developmental stages, structural genes such as DFR, BZ1, ANS, CHS, 3AT, HCT, and FLS appeared to be expressed differently in different samples; we speculate that these genes mediate the regulation of leaf color in 'Zhonghong' poplar during the transient phase through mechanisms such as flavonoid biosynthesis, anthocyanin biosynthesis, isoflavone biosynthesis, and flavonoid and flavonol biology. Transcriptome data provide valuable information and gene sequences for future studies on the variation of related genes during the transcolorization of 'Zhonghong' poplar leaves and lay a foundation for further research to confirm the precise mechanisms by which transcription factors regulate anthocyanin biosynthesis and degradation.

**Supplementary Materials:** The following are available online at https://www.mdpi.com/article/10.3390/agronomy12102396/s1, Figure S1: The expression violin plot shows the distribution status and probability density of the sample data, Figure S2: Pearson correlation between samples, Figure S3: 2D PCA Plot, Figure S4: Enrichment maps of genes in the turquoise and blue modules for GO, KEGG, and KOG, Figure S5: Consistency of genes in the qRT-PCR and transcriptome results. Table S1: Sequencing output statistics table, Table S2: Comparison statistics table, Table S3: Novel genes annotation table, Table S4: Primer sequences used for qRT-PCR.

**Author Contributions:** J.Z., L.H. and M.C. were the executors of the experimental design and experimental research of this study; M.C. completed the data analysis and wrote the first draft of the paper; H.L. and W.Z. participated in the experimental design and the analysis of the experimental results; J.Z. and L.H. were the creators and leaders of the project, directing experimental design, data analysis, and thesis writing and revision. All authors have read and agreed to the published version of the manuscript.

**Funding:** This research was funded by the Opening Project of State Key Laboratory of Tree Genetics and Breeding (K2022201); Key R & D and Promotion Project of Henan Province (Science and Technology Research) (202102110229).

**Data Availability Statement:** Transcriptome datasets are available in the National Center for Biotechnology Information under BioProject PRJNA881405 (https://dataview.ncbi.nlm.nih.gov/?search=SUB12061013&archive=bioprojet (accessed on 31 October 2021)).

**Conflicts of Interest:** The authors declare no conflict of interest.

## References

1. Yang, S.; Zhu, Y. *Selection and Breeding of Populus Hongye*; Northwestern Polytechnical University Press: Xian, China, 2018.
2. Zhu, Y.; Cheng, X. A new poplar variety 'Zhonghongyang'. *For. Sci.* **2008**, *10*, 173–174.
3. Chen, X. *Research on the Attrition Phenomenon of Red Leaf Poplar*; Henan Agricultural University: Zhengzhou, China, 2012.
4. Yang, S.; Zhu, Y.; Ma, Y.; Yan, Z.; Ding, X. Correlation between leaf color and pigment composition of Poplar in the growing season. *J. Northeast For. Univ.* **2013**, *41*, 63–68.
5. Gentile, D.; Patamia, V.; Fuochi, V.; Furneri, P.M.; Rescifina, A. Analysis of differences in anthocyanin metabolism by transcriptome sequencing of two species of Toona sinensis at different stages. *Mol. Plant Breed.* **2022**, *20*, 422–434.
6. Wang, L.; Lu, W.; Ran, L.; Dou, L.; Yao, S.; Hu, J.; Fan, D.; Li, C.; Luo, K. R2R3-MYB transcription factor MYB6 promotes anthocyanin and proanthocyanidin biosynthesis but inhibits secondary cell wall formation in Populus tomentosa. *Plant J.* **2019**, *99*, 733–751. [CrossRef] [PubMed]
7. Yoshida, K.; Ma, D.; Constabel, C.P. The MYB182 Protein Down-Regulates Proanthocyanidin and Anthocyanin Biosynthesis in Poplar by Repressing Both Structural and Regulatory Flavonoid Genes. *Plant Physiol.* **2015**, *167*, 693–710. [CrossRef]
8. Kiferle, C.; Fantini, E.; Bassolino, L.; Povero, G.; Spelt, C.; Buti, S.; Giuliano, G.; Quattrocchio, F.M.; Koes, R.; Perata, P.; et al. Tomato R2R3-MYB Proteins SlANT1 and SlAN2: Same Protein Activity, Different Roles. *PLoS ONE* **2015**, *10*, e0136365. [CrossRef]
9. Mazza, G.; Miniati, E. *Anthocyanins in Fruits, Vegetables, and Grains*, 1st ed.; CRC Press: Boca Raton, FL, USA, 1993; p. 384.
10. Mori, K.; Goto-Yamamoto, N.; Kitayama, M.; Hashizume, K. Loss of anthocyanins in red-wine grape under high temperature. *J. Exp. Bot.* **2007**, *58*, 1935–1945. [CrossRef]
11. Vaknin, H.; Bar-Akiva, A.; Ovadia, R.; Nissim-Levi, A.; Forer, I.; Weiss, D.; Oren-Shamir, M. Active anthocyanin degradation in Brunfelsia calycina (yesterday–today–tomorrow) flowers. *Planta* **2005**, *222*, 19–26. [CrossRef] [PubMed]
12. Gonnet, J.F. Colour effects of co-pigmentation of anthocyanins revisited-1. A colorimetric definition using the CIELAB scale. *Food Chem* **1998**, *63*, 409–415. [CrossRef]
13. Zhao, D.Q.; Tao, J. Recent advances on the development and regulation of flower color in ornamental plants. *Front. Plant Sci.* **2015**, *6*, 261. [CrossRef]
14. Zhu, Z.; Lu, Y. Plant color mutants and the anthocyanin pathway. *Chin. Bull. Bot.* **2016**, *51*, 107–119.
15. Liu, X.Y.; Han, H.Q.; Ge, H.Y.; Jiang, M.M.; Chen, H.Y. Cloning, expression and interaction of anthocyaninrelated transcription factors SmTTG1, SmGL3 and SmTT8 in eggplant. *Acta Hortic. Sin.* **2014**, *41*, 2241–2249.
16. Daehwan, K.; Ben, L.; Salzberg, S.L. HISAT: A fast spliced aligner with low memory requirements. *Nat. Methods* **2015**, *12*, 357–360.
17. Love, M.I.; Huber, W.; Anders, S. Moderated estimation of fold change and dispersion for RNA-seq data with DESeq2. *Genome Biol.* **2014**, *15*, 550. [CrossRef]
18. Varet, H.; Brillet-Guéguen, L.; Coppée, J.Y.; Dillies, M.A. SARTools: A DESeq2- and edgeR-based R pipeline for comprehensive differential analysis of RNA-seq data. *PLoS ONE* **2016**, *11*, e157022. [CrossRef]
19. Kanehisa, M.; Araki, M.; Goto, S.; Hattori, M.; Hirakawa, M.; Itoh, M.; Katayama, T.; Kawashima, S.; Okuda, S.; Tokimatsu, T.; et al. KEGG for linking genomes to life and the environment. *Nucleic Acids Res.* **2008**, *36*, 480–484. [CrossRef] [PubMed]
20. Tatusov, R.L.; Koonin, E.V.; Lipman, D.J. A genomic perspective on protein families. *Science* **1997**, *278*, 631–637. [CrossRef] [PubMed]
21. Tatusov, R.L.; Galperin, M.Y.; Natale, D.A.; Koonin, E.V. The COG database: A tool for genome-scale analysis of protein functions and evolution. *Nucleic Acids Res.* **2000**, *28*, 33–36. [CrossRef]

22. Zheng, Y.; Jiao, C.; Sun, H.; Rosli, H.G.; Pombo, M.A.; Zhang, P.; Banf, M.; Dai, X.; Martin, G.B.; Giovannoni, J.J.; et al. iTAK: A program for genome-wide prediction and classification of plant transcription factors, transcriptional regulators, and protein kinases. *Mol. Plant* **2016**, *9*, 1667–1670. [CrossRef] [PubMed]
23. Pérez-Rodríguez, P.; Riaño-Pachón, D.M.; Corrêa, L.G.G.; Rensing, S.A.; Kersten, B.; Mueller-Roeber, B. PlnTFDB: Updated content and new features of the plant transcription factor database. *Nucleic Acids Res.* **2010**, *38*, D822–D827. [CrossRef] [PubMed]
24. Jin, J.; Zhang, H.; Kong, L.; Gao, G.; Luo, J. PlantTFDB 3.0: A portal for the functional and evolutionary study of plant transcription factors. *Nucleic Acids Res.* **2014**, *42*, D1182–D1187. [CrossRef] [PubMed]
25. Livak, K.J.; Schmittgen, T.D. Analysis of relative gene expression data using real-time quantitative PCR and the 2(-Delta Delta C(T)) *Method. Methods* **2001**, *25*, 402–408. [CrossRef] [PubMed]
26. Zhang, F.; Zhao, J.; Wan, X.; Luo, X.; Li, W.; Sun, L.; Chen, Q. From green to red: Large-scale transcriptome comparison of a bud sport in poplar (*Populus deltoides*). *Acta Physiol. Plant.* **2016**, *38*, 244. [CrossRef]
27. Yang, S.; Liang, K.; Wang, A.; Zhang, M.; Zhang, L. Physiological Characterization and Transcriptome Analysis of Camellia oleifera Abel. During Leaf Senescence. *Forests* **2020**, *11*, 812. [CrossRef]
28. Wang, Z.X.; Yu, Y.F.; Chen, L.; Qin, H.Y.; Liu, Y.X.; Ai, J.; Fan, S.T. Advances in leaf pigment composition, structure and photosynthetic characteristics of colored-leaf plants. *Plant Physiol. J.* **2016**, *52*, 1–7.
29. He, X.; Zhao, X.; Gao, L.; Shi, X.; Dai, X.; Liu, Y.; Xia, T.; Wang, Y. Isolation and characterization of key genes that promote favonoid accumulation in purple-leaf tea (*Camellia sinensis* L.). *Sci. Rep.* **2018**, *8*, 130. [CrossRef]
30. An, C.H.; Lee, K.W.; Lee, S.H.; Jeong, Y.J.; Woo, S.G.; Chun, H.; Park, Y.I.; Kwak, S.S.; Kim, C.Y. Heterologous expression of IbMYB1a by diferent promoters exhibits diferent patterns of anthocyanin accumulation in tobacco. *Plant Physiol. Bioch.* **2015**, *89*, 1–10. [CrossRef]
31. Cho, J.S.; Nguyen, V.P.; Jeon, H.W.; Kim, M.H.; Eom, S.H.; Lim, Y.J.; Kim, W.C.; Park, E.J.; Choi, Y.I.; Ko, J.H. Overexpression of PtrMYB119, a R2R3-MYB transcription factor from Populus trichocarpa, promotes anthocyanin production in hybrid poplar. *Tree Physiol.* **2016**, *36*, 1162–1176. [CrossRef]
32. Hughes, N.M.; Morley, C.B.; Smith, W.K. Coordination of anthocyanin decline and photosynthetic maturation in juvenile leaves of three deciduous tree species. *New Phytologist.* **2007**, *175*, 675–685. [CrossRef]
33. Wei, L. Cloning and functional study of quinic acid/shikimate-hydroxycinnamoyl transferase gene in tea plants. *Henan Agric. Sci.* **2016**, *45*, 35–39.
34. Huang, X.; Hu, L.; Kong, W.; Yang, C.; Xi, W. Red light-transmittance bagging promotes carotenoid accumulation of grapefruit during ripening. *Commun. Biol.* **2022**, *5*, 303. [CrossRef] [PubMed]
35. Castellarin, S.D.; Di Gaspero, G.; Marconi, R.; Nonis, A.; Peterlunger, E.; Paillard, S.; Adam-Blondon, A.F.; Testolin, R. Colour variation in red grapevines (*Vitis vinifera* L.): Genomic organisation, expression of flavonoid 3′-hydroxylase, flavonoid 3′,5′-hydroxylase genes and related metabolite profiling of red cyanidin-/blue delphinidin-based anthocyanins in berry skin. *BMC Genom.* **2006**, *7*, 12. [CrossRef] [PubMed]
36. Liu, Y.; Shi, Z.; Maximova, S.; Payne, M.J.; Guiltinan, M.J. Proanthocyanidin synthesis in Theobroma cacao: Genes encoding anthocyanidin synthase, anthocyanidin reductase, and leucoanthocyanidin reductase. *BMC Plant Biol.* **2013**, *13*, 202. [CrossRef] [PubMed]
37. Quattrocchio, F.R.; Baudry, A.N.; Lepiniec, L.O.; Grotewold, E.R. The Regulation of Flavonoid Biosynthesis. In *The Science of Flavonoids*; Springer: New York, NY, USA, 2006; pp. 97–122.
38. Lou, Q.; Liu, Y.; Qi, Y.; Jiao, S.; Tian, F.; Jiang, L.; Wang, Y. Transcriptome sequencing and metabolite analysis reveals the role of delphinidin metabolism in flower colour in grape hyacinth. *J. Exp. Bot.* **2014**, *65*, 3157–3164. [CrossRef] [PubMed]
39. Li, N.; Wu, H.; Ding, Q.; Li, H.; Li, Z.; Ding, J.; Li, Y. The heterologous expression of Arabidopsis PAP2 induces anthocyanin accumulation and inhibits plant growth in tomato. *Funct. Integr. Genom.* **2018**, *18*, 341–353. [CrossRef] [PubMed]
40. Wang, H.; Wang, X.; Song, W.; Bao, Y.; Jin, Y.; Jiang, C.; Wang, C.; Li, B.; Zhang, H. PdMYB118, isolated from a red leaf mutant of Populus deltoids, is a new transcription factor regulating anthocyanin biosynthesis in poplar. *Plant Cell Rep.* **2019**, *38*, 927–936. [CrossRef]
41. Piao, W.L.; Kim, S.H.; Lee, B.D.; An, G.; Sakuraba, Y.; Paek, N.C. Rice transcription factor OsMYB102 delays leaf senescence by down-regulating abscisic acid accumulation and signaling. *J. Exp. Bot.* **2019**, *70*, 2699–2715. [CrossRef]
42. Wang, S.; Zhang, Z.; Li, L.X.; Wang, H.B.; Zhou, H.; Chen, X.S.; Feng, S.Q. Apple MdMYB306-like inhibits anthocyanin synthesis by directly interacting with MdMYB17 and MdbHLH33. *Plant J.* **2022**, *110*, 1021–1034. [CrossRef]
43. Zhang, S.; Zhao, Q.; Zeng, D.; Xu, J.; Zhou, H.; Wang, F.; Ma, N.; Li, Y. RhMYB108, an R2R3-MYB transcription factor, is involved in ethylene- and JA-induced petal senescence in rose plants. *Hortic Res.* **2019**, *1*, 131. [CrossRef]
44. Sun, C.; Wang, C.; Zhang, W.; Liu, S.; Wang, W.; Yu, X.; Song, T.; Yu, M.; Yu, W.; Qu, S. The R2R3-type MYB transcription factor MdMYB90-like is responsible for the enhanced skin color of an apple bud sport mutant. *Hortic Res.* **2021**, *8*, 156. [CrossRef]
45. Koo, Y.; Poethig, R.S. Expression pattern analysis of three R2R3-MYB transcription factors for the production of anthocyanin in different vegetative stages of Arabidopsis leaves. *Appl. Biol. Chem.* **2021**, *64*, 5. [CrossRef]
46. Jia, J.; Xing, J.H.; Dong, J.G.; Han, J.M.; Liu, J.S. Functional Analysis of MYB73 of Arabidopsis thaliana Against Bipolaris oryzae. *Agric. Sci. China* **2011**, *10*, 721–727. [CrossRef]

47. Ni, J.; Zhao, Y.; Tao, R.; Yin, L.; Gao, L.; Strid, A.; Qian, M.; Li, J.; Li, Y.; Shen, J.; et al. Ethylene mediates the branching of the jasmonate-induced flavonoid biosynthesis pathway by suppressing anthocyanin biosynthesis in red Chinese pear fruits. *Plant Biotechnol. J.* **2020**, *18*, 1223–1240. [CrossRef]
48. Fan, D.; Wang, X.; Tang, X.; Ye, X.; Ren, S.; Wang, D.; Luo, K. Histone H3K9 demethylase JMJ25 epigenetically modulates anthocyanin biosynthesis in poplar. *Plant J.* **2018**, *96*, 1121–1136. [CrossRef]

agronomy

MDPI

Article

# Overexpression of Two MADS-Box Genes from *Lagerstroemia speciosa* Causes Early Flowering and Affects Floral Organ Development in *Arabidopsis*

**Lichen Yang [1,2,3,4,5,6,†], Zhuojiao Li [1,2,3,4,5,6,†], Tangchun Zheng [1,2,3,4,5,6,*], Jia Wang [1,2,3,4,5,6], Tangren Cheng [1,2,3,4,5,6] and Qixiang Zhang [1,2,3,4,5,6,*]**

1 Beijing Key Laboratory of Ornamental Plants Germplasm Innovation and Molecular Breeding, Beijing Forestry University, Beijing 100083, China
2 National Engineering Research Center for Floriculture, Beijing Forestry University, Beijing 100083, China
3 Beijing Laboratory of Urban and Rural Ecological Environment, Beijing Forestry University, Beijing 100083, China
4 Engineering Research Center of Landscape Environment of Ministry of Education, Beijing Forestry University, Beijing 100083, China
5 Key Laboratory of Genetics and Breeding in Forest Trees and Ornamental Plants of Ministry of Education, Beijing Forestry University, Beijing 100083, China
6 School of Landscape Architecture, Beijing Forestry University, Beijing 100083, China
* Correspondence: zhengtangchun@bjfu.edu.cn (T.Z.); zqx@bjfu.edu.cn (Q.Z.); Tel.: +86-10-62336321 (T.Z.); +86-10-62338005 (Q.Z.)
† These authors contributed equally to this work.

**Abstract:** *Lagerstroemia speciosa* is an important ornamental plant, and there is only one double-petaled variety, 'Yunshang', produced by natural mutation, in the whole genus of *Lagerstroemia*. The members of the MADS-box family play important roles in floral organ development. However, little is known about the biological function of the MADS-box gene in *L. speciosa*. In our study, two MADS-box genes (*LsAG2* and *LsDEF1*) were isolated from *L. speciosa*, and their expression levels in different tissues and developmental stages were analyzed by RT-qPCR. Subcellular localization showed that LsAG2 and LsDEF1 are localized in the nucleus. The overexpression of *LsAG2* and *LsDEF1* in *Arabidopsis thaliana* caused transgenic plants to exhibit different phenotypes, such as floral organ aberrations, early flowering, and dwarf plants, and resulted in the up-regulation of endogenous genes related to flowering (i.e., *AP1*, *PI*, *FLC*, *FUL*, *LFY*, and *FT*). Our results provide a theoretical basis for the flowering time, flower development, and genetic improvement of double-petaled flowers in the genus of *Lagerstroemia*.

**Keywords:** *Lagerstroemia speciosa*; MADS-box gene; *LsAG2*; *LsDEF1*; flowering time; flower development

**Citation:** Yang, L.; Li, Z.; Zheng, T.; Wang, J.; Cheng, T.; Zhang, Q. Overexpression of Two MADS-Box Genes from *Lagerstroemia speciosa* Causes Early Flowering and Affects Floral Organ Development in *Arabidopsis*. *Agronomy* **2023**, *13*, 976. https://doi.org/10.3390/agronomy13040976

Academic Editors: Jinzhi Zhang and Avi Sadka

Received: 2 February 2023
Revised: 20 March 2023
Accepted: 23 March 2023
Published: 25 March 2023



## 1. Introduction

A flower is the reproductive organ of angiosperms. Studying the flower developmental process of ornamental plants is of great significance for measures aiming to adjust the flowering period and flower shape, and to improve the ornamental and economic value of the plants. Floral meristem determination is a complex developmental event that is regulated by a variety of genetic factors [1]. The flower of *A. thaliana* is typically representative of angiosperm flowers. The flower development process of *A. thaliana* was initially summarized as the "ABC model" [2] and then gradually expanded to the "ABCDE model" [3]. The "ABC model" divides the genes involved in flower organ differentiation into three categories: A, B, and C, which, respectively, regulate the development of the calyx, petal, stamen, and pistil, either separately or coordinately. In addition, *Arabidopsis* mutations in class A and class C genes lead these genes to exhibit ectopic expressions of each other, thereby altering the flower morphology and suggesting that the class A and C genes are antagonistic to each other [4,5]. In the study of *Petunia hybrida* flower development, the newly proposed class D genes were proven to be involved in the regulation of *P. hybrida* ovule formation [6], and the *A. thaliana* homologous genes of the class D gene in *P. hybrida* have also had their involvement in the regulation of the development and formation of *A. thaliana* ovules [7] confirmed. Moreover, researchers have found that if the functions of *SEP1* (*SEPALLATA1*), *SEP2*, and *SEP3* in *A. thaliana* are simultaneously inhibited, all the floral organs of *Arabidopsis* will be completely changed to sepals [8]. There is speculation that these genes are independent of ABC genes and have regulatory effects on the expression level of ABC genes. These genes were named as E-type genes, which are considered to be involved in floral organ regulation during flower development and floral transition [9].

The different flower forms of angiosperms, especially the double petal trait, can greatly improve the ornamental value of plants. There are many origins of existing double-petaled ornamental plants, and the existing studies on a variety of plants that naturally form double flowers and single flowers have shown that the formation of double flowers is mostly related to changes in the plant MADS-box gene sequence or expression. For example, the *Arabidopsis* mutant resulting from *AG* (*AGAMOUS*) gene deletion shows a double-petal phenotype [10], and the formation of the double-petal variety 'Double White' in *Thalictrum thalictroides* is due to the deletion of the K domain of the *ThtAG1* gene transcription protein, which affects interactions with the E-type protein ThtSEP3 [11].

Studies on the MADS-box family genes have shown that members of this family are also widely involved in the transformation of plants into flowers and the formation of the flower organ morphology. For example, the overexpression of two *SUPPRESSOR OF OVEREXPRESSION OF CONSTANS 1* (*SOC1*) homologs in *Pyrus bretschneideri* leads to early flowering in *Arabidopsis* [12]. Most of the other genes affect the development of plant flowers by participating in the regulation of MADS-box genes. For example, the *LEAFY* (*LFY*) gene in *A. thaliana* interacts with the *APETALA1* (*AP1*) gene to promote the transformation of inflorescences into floral meristems and participates in the determination of floral organ meristems through the regulation of *APETALA3* (*AP3*) and *PISTILLATA* (*PI*) [13]. The *WUSCHEL* (*WUS*) gene regulates floral organ determination by inducing the expression of the *AG* gene, while *AG* counteracts *WUS* and inhibits its role in floral meristem differentiation [14]. Among the ABCE-type genes, there are some *APETALA2* (*AP2*) family genes, while the other genes are MIKC-type MADS-box family genes [15,16]. With the advancement of whole-genome sequencing technology, the MADS-box family genes of many ornamental plants have been identified and analyzed, including carnations (*Dianthus caryophyllus*) [17], wild chrysanthemum (*Chrysanthemum nankingense*) [18], mei (*Prunus mume*) [19], and so on.

Crape myrtle is an ornamental woody plant used worldwide. It blooms in midsummer, and the flowering period is as long as three months. *Lagerstroemia speciosa* is a special species of the *Lagerstroemia* genus, among which the flowers of *L. speciosa* are very large (5–7 cm in diameter) and colorful, with high garden application and ornamental value. There are approximately 55 species in the whole *Lagerstroemia* genus and approximately 200 cultivated varieties of crape myrtle, but there is only one double-petaled variety, 'Yunshang', produced by a natural bud mutation of *L. speciosa*, with defined double petals resulting from stamen petalization. Differential transcriptome sequencing of single-petaled *L. speciosa* and double-petaled *L. speciosa* at different stages of flower organ development was carried out, and four flower-development-related genes were identified [20]. However, our knowledge of the biological function of the flower-development-related genes in *Lagerstroemia* is still limited. In our study, we cloned two MADS-box family genes related to the regulation of flower organ development in *L. speciosa*. The temporal and spatial expression patterns of the two genes were analyzed, and the biological functions were further verified in *Arabidopsis*, which provided a theoretical basis for the flower time, flower development, and genetic improvement of double-petal flowers in *Lagerstroemia*.

## 2. Materials and Methods

### 2.1. Plant Material

Plant materials were obtained from the resource garden of Guangxi Academy of Forestry (Nanning City, Guangxi Province). The samples were quick-frozen in liquid nitrogen and then stored at −80 °C in a freezer. The double-petaled *L. speciosa* (DPL) is derived from the single-petaled *L. speciosa* (SPL). Flower buds of different developmental stages (S1: buds 4–6 mm in diameter, S2: buds 6–10 mm in diameter, S3: buds 10–14 mm in diameter, and S4: full bloom but not powered) and different floral organ tissues (calyx, petals, stamens/petalized stamens, and pistils) at S4 stage of the SPL and DPL were collected, as in our previous study [20]. In total, 16 samples were used to analyze the gene expression patterns of MADS-box genes related to flower development in *L. speciosa*.

*Arabidopsis thaliana* (Col-0) plants were grown in an artificial climate chamber with an average temperature of 22 ± 2 °C, with 16 h of light and 8 h of darkness, 100 $\mu mol \cdot m^{-2} \cdot s^{-1}$ light intensity, and 65–75% relative humidity.

### 2.2. Sequence Analysis and Phylogenetic Tree Construction

Physicochemical properties such as the molecular weight (MW) and theoretical isoelectric point (*pI*) were predicted in Prot Param (https://web.expasy.org/protparam/, accessed on 7 May 2020). Protein hydrophobicity was predicted with an online tool in Prot Scale (https://web.expasy.org/protscale/, accessed on 7 May 2020). The SOPMA online software (https://npsa-prabi.ibcp.fr/cgi-bin/npsa_automat.pl?page=npsa_sopma.html, accessed on 7 May 2020) was used to predict the secondary structures of the proteins. The conserved protein domains of LsAG2 and LsDEF1 were predicted using an online tool in NCBI (https://www.ncbi.nlm.nih.gov/cdd, accessed on 7 May 2020). Phosphorylation sites were predicted in NetPhos (https://services.healthtech.dtu.dk/service.php?NetPhos-3.1, accessed on 10 May 2020). Plant-mPLoc in Cell-PLoc 2.0 (https://services.healthtech.dtu.dk/service.php?SignalP-5.0, accessed on 10 May 2020) was used for subcellular localization analysis. TMHMM Server 2.0 was employed to predict transmembrane helices in the proteins (https://services.healthtech.dtu.dk/service.php?TMHMM-2.0, accessed on 10 May 2020). The results of the gene structure were visualized in GSDS 2.0 (http://gsds.gao-lab.org/, accessed on 10 May 2020). We used the NCBI Blastx website (https://blast.ncbi.nlm.nih.gov/Blast.cgi, accessed on 10 May 2020) to align the gene-encoded protein sequences with other plant protein sequences in the database and the NJ method to construct a phylogenetic tree between the two gene-encoded proteins and the aligned homologous proteins.

### 2.3. RNA Extraction and RT-qPCR Analysis

A total of 16 samples representing 4 different developmental stages and 4 different floral organ tissues of the DPL and SPL were collected. Total RNA was extracted from the 16 samples using an RNAprep Pure Plant Kit (TianGen, Beijing, China). First-strand cDNAs were synthesized using a PrimeScript™ RT Reagent Kit (TaKaRa, Dalian, China), according to the manufacturer's instructions.

RT-qPCR was performed using TB Green® Premix Ex Taq™ II (TaKaRa) and a CFX Connect Real-Time System (Bio-Rad, Hercules, CA, USA). The primers for RT-qPCR (Table S1) were designed using Integrated DNA Technologies tools (https://sg.idtdna.com/scitools/Applications/RealTimePCR/, accessed on 15 July 2020) and synthesized by Sangon Biotech (Shanghai, China) Co., Ltd. RT-qPCR was performed in 20 μL volumes containing 10 μL of TB Green Premix, 0.8 μL of forward primer (10 μM), 0.8 μL of reverse primer (10 μM), 2 μL of cDNA template (50 ng/μL), and 6.4 μL $ddH_2O$. Each test included three technical repetitions and three biological replicates. The *Elongation factor-1-alpha* (*EF-1α*) gene of *L. speciosa* was employed as a reference gene [21]. For the same type of tissue or developmental stage, the sample with the lowest expression was selected as the control, and the gene expression levels were calculated using the $2^{-\Delta\Delta Ct}$ method [22].

The total RNA was isolated from whole flowers of transgenic and wild-type plants using a MiniBEST Plant RNA Extraction Kit (TaKaRa), and we detected the expression levels of the *LsAG2* and *LsDEF1* genes in the transgenic plants and the endogenous flower-development-related genes in *A. thaliana*. Specific primers (Table S1) were designed for *AP1*, *FRUITFULL* (*FUL*), *LFY*, *FLOWERING LOCUS T* (*FT*), *FLOWERING LOCUS C* (*FLC*), and *PI*. Then, RT-qPCR analysis was performed, with the *Actin* gene employed as a reference gene [23].

### 2.4. Cloning of the Two MADS-Box Genes

The specific primers (Table S1) of *LsAG2* and *LsDEF1* were designed using Primer premier 5.0 software, according to the transcriptome data. Then, the sequences of *LsAG2* and *LsDEF1* were amplified from the cDNA of the SPL and DPL flower buds with specific primers (Table S1). The purified PCR products were constructed into a pCloneEZ Blunt Cloning vector (Taihe Biotechnology, Beijing, China), and the final LsAG2 and LsDEF1 sequences were further verified by sequencing (Sangon Biotech Co., Ltd., Shanghai, China).

### 2.5. Subcellular Localization Analysis

To analyze the subcellular localization of *LsAG2* and *LsDEF1*, the CDSs (without a stop codon) of *LsAG2* and *LsDEF1* were cloned into the vector pSuper1300-GFP to generate the *LsAG2-GFP* and *LsDEF1-GFP* fusion genes driven by *CaMV35S*, respectively. The recombinant plasmids pSuper1300-*LsAG2-GFP* and pSuper1300-*LsDEF1-GFP* were transformed into *Agrobacterium tumefaciens* competent cells, GV3101, and the recombinant *Agrobacterium* and empty vector control were injected into healthy *Nicotiana benthamiana* leaves, respectively. The infected tobacco plants were cultured in the dark for 24 h and then placed in an incubator for normal long-day culture for 2–3 d. Finally, the green fluorescent signals were observed using a TCS SP8 confocal microscope (Leica, Wetzlar, Germany). 4′,6-diamidino-2-phenylindole (DAPI) was used to dye the nuclear DNA as a positive control for nuclear localization.

### 2.6. Vector Construction and Plant Transformation

The plant expression vector was constructed using the method of seamless cloning, and the PCR primers (Table S1) of the target fragments were synthesized according to the instructions for the CV12-Seamless Assembly and Cloning kit (Aidlab, Beijing, China). With pCAMBIA1304 linearized by double digestion using restriction enzymes, *Nco* I and *Bst* E II, the PCR fragments of *LsAG2* and *LsDEF1* were inserted into the linearized vector to form the recombinant vectors (pCAMBIA1304-*LsAG2*, pCAMBIA1304-*LsDEF1*). Finally, the recombinant vectors were transferred into GV3101 using the freeze–thaw transformation method and transformed into *A. thaliana* via the floral dip method.

Transgenic *Arabidopsis* seeds were selected and placed on 1/2 MS medium containing 50 mg/L Kanamycin. After these positive seedlings developed four true leaves, they were then transferred into pots containing a mixture of turf peat, vermiculite, and pearlite (3:1:1 *v*/*v*) in an artificial climate chamber.

### 2.7. Phenotypic Observation

The number of days that a single flower lasted from the beginning to the end of the experiment was regarded as the flowering time. The phenotypes of transgenic lines were observed every 2 d from the opening of the first flower, such as flower and leaf deformity variation, etc., and we recorded indicators such as the flower volume and flowering period. The flowering period refers to the time from the blooming of the first flower to the fading of the last flower.

### 2.8. Statistical Analyses

Statistical analyses were performed using the SPSS 19.0 (SPSS, Chicago, IL, USA). ANOVA was used for multiple-group comparisons, and statistical significance ($p < 0.05$) was determined by Student's $t$-test.

## 3. Results

### 3.1. Cloning of LsAG2 and LsDEF1 and Bioinformatics Analysis

On the basis of the differentially expressed genes (DEGs) derived from previous transcriptome data, two genes related to flower development in the MADS-box family of *L. speciosa* were cloned. Comparing the CDS and DNA sequences of these two genes cloned from the SPL and DPL, it was found that there was no difference in the amino acid sequences, indicating that the formation of stamen petalization in *L. speciosa* was not caused by the gene structure differences between *LsAG2* and *LsDEF1*. Prediction of the upstream promoter sequences of these genes showed that the basic cis-elements, such as TATA-box and CAAT-box, were found in *LsAG2* and *LsDEF1*, and there are multiple cis-acting elements involved in the hormone response and light response (Table S2). In terms of gene structure, *LsAG2* and *LsDEF1* contain 6 and 0 intron regions, respectively, and encode proteins of 237 and 252 amino acids in length, with relative molecular weights of 27.2 kDa and 29.07 kDa, respectively. These two proteins are hydrophilic, basic, and unstable proteins. They do not contain signal peptide sites or transmembrane domains and are not secreted proteins. LsAG2 and LsDEF1 contain 19 and 25 phosphorylation sites, and both contain a conserved MADS-MEF2-like domain and a K-box domain, but there are some differences between the two protein sequences, which we observed in the start and end positions of the two domains (Table S3, Figure S1). The secondary structures of LsAG2 and LsDEF1 were predicted, demonstrating that they both have four types of coils: α-helix, β-sheet, extended chain, and random coil, of which the main components are α-helix (46.41%–51.98%) and random coil (29.11%–31.75%) (Table S4). A phylogenetic tree was constructed by comparing the homologous proteins of LsAG2 and LsDEF1 with other homologous protein sequences, and we found that both LsAG2 and LsDEF1 were closest to the homologous protein in pomegranate in terms of genetic distance (Figure 1). That is, LsAG2 and LsDEF1 have the typical characteristics of MADS-box family transcription factors, and both belong to the MADS-box family.

### 3.2. Spatiotemporal Specificity Expression Analysis of LsAG2 and LsDEF1

In order to study the spatiotemporal specificity expression of *LsAG2* and *LsDEF1*, the expression levels of *LsAG2* and *LsDEF1* in different flower development stages and different floral organ tissues of the SPL and DPL were determined using RT-qPCR. The results showed that *LsAG2* was not expressed in the calyx of SPL, and the expression level was extremely low in the single-petaled pistils and petals; however, its expression was significantly up-regulated in the pistils and petals of DPL. In the S3 stage, the expression level of *LsAG2* in the SPL was significantly higher than in the other three periods, while in the DPL, the expression level of *LsAG2* in the S4 stage was slightly higher than in the other three stages (Figure 2a,b; Table S5). *LsDEF1* was very low or not expressed in the calyces and pistils of the DPL and SPL and was relatively high in the petals. The expression level of *LsDEF1* in the stamens of the DPL was higher than that in the stamens of SPL but slightly lower than that in the petals of SPL. *LsDEF1* was expressed in each stage of flower development in the DPL and showed a gradual upward trend, reaching the highest value in the S4 stage. In the SPL, the expression level of *LsDEF1* first increased and then decreased (Figure 2c,d; Table S5).

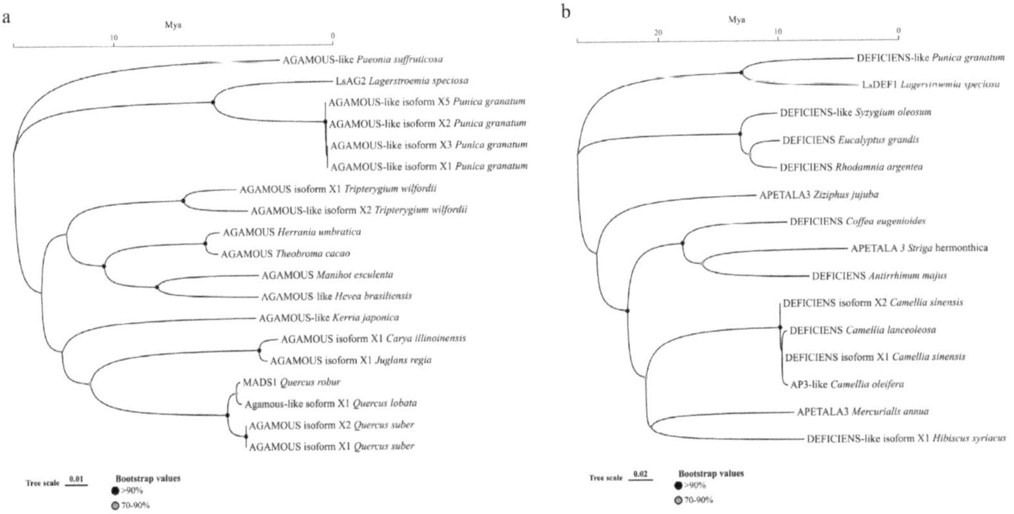


**Figure 1.** Phylogenetic analysis of LsAG2, LsDEF1, and other MADS-box proteins from different species. (**a**) Phylogenetic analysis of LsAG2 and MADS-box proteins of *Paeonia suffruticosa*, *Punica granatum*, *Tripterygium wilfordii*, *Herrania umbratical*, *Theobroma cacao*, *Manihot esculenta*, *Hevea brasiliensis*, *Kerria japonica*, *Carya illinoinensis*, *Juglans regia*, *Quercus robur*, *Quercus lobata*, and *Quercus suber*. (**b**) Phylogenetic analysis of LsDEF1 and MADS-box proteins of *Punica granatum*, *Syzygium oleosum*, *Eucalyptus grandis*, *Rhodamnia argentea*, *Ziziphus jujuba*, *Coffea eugenioides*, *Striga hermonthica*, *Antirrhinum majus*, *Camellia lanceoleosa*, *Camellia sinensis*, *Camellia oleifera*, *Mercurialis annua*, and *Hibiscus syriacus*.

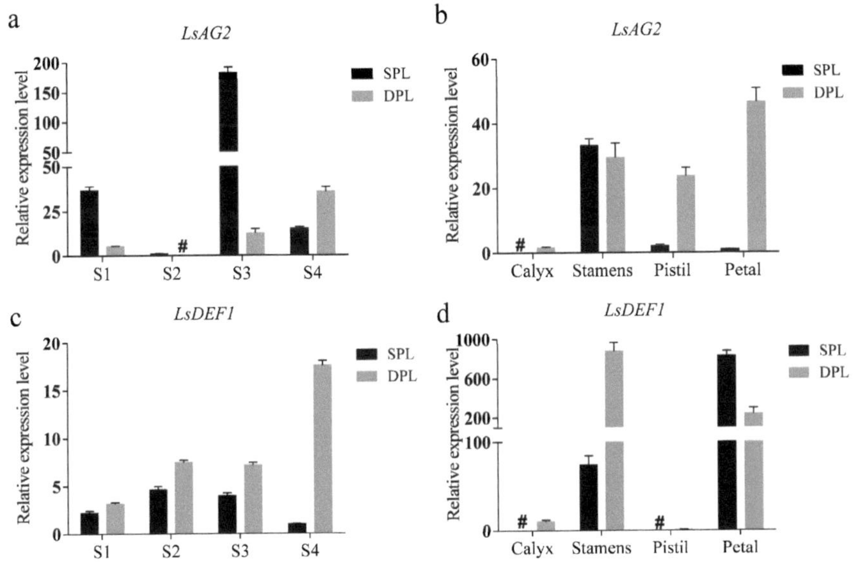


**Figure 2.** The expression patterns of *LsAG2* and *LsDEF1*. (**a**) Expression level of *LsAG2* during different developmental stages. (**b**) Expression level of *LsAG2* in different flower tissues. (**c**) Expression level of *LsDEF1* during different developmental stages. (**d**) Expression level of *LsDEF1* in different flower tissues. S1 to S4, four different development stages of flower budding. SPL, single-petaled *L. speciosa*;

DPL, double-petaled *L. speciosa*. In the same type of tissue or developmental stage, the sample with the lowest expression was selected as the control, and the gene expression levels were calculated using the $2^{-\Delta\Delta Ct}$ method. The symbol '#' indicates that the gene cannot be detected.

### 3.3. Subcellular Localization of LsAG2 and LsDEF1

The *LsAG2-GFP* and *LsDEF1-GFP* fusion genes (pSuper-*35S::LsAG2-GFP*, pSuper-*35S::LsDEF1-GFP*) were used to examine the subcellular localization of LsAG2 and LsDEF1 transiently expressed in *N. benthamiana* leaves. The results showed that the GFP signals of LsAG2-GFP and LsDEF1-GFP were detected only in the nucleus of the tobacco cells, while the GFP signal of the empty plasmid control was distributed throughout the whole cell (Figure 3).

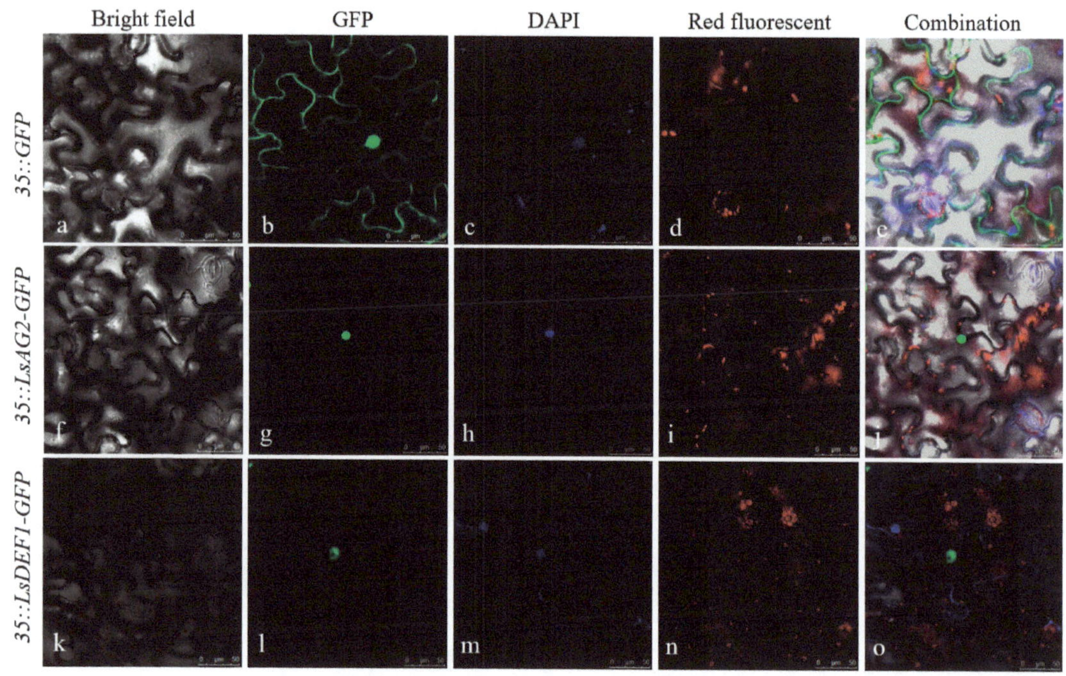


**Figure 3.** Subcellular localization of LsAG2 and LsDEF1 proteins in tobacco leaf cells. (**a–e**) Subcellular localization of GFP. (**f–j**) Subcellular localization of LsAG2-GFP. (**k–o**) Subcellular localization of LsDEF1-GFP. Bright field: white light. GFP: green fluorescence signal. DAPI: 4′,6-diamidino-2-phenylindole (DAPI) nuclear DNA fluorescent stain. Red fluorescent: chloroplast auto-fluorescence. Combination: combined signals of different fluorescence.

### 3.4. Phenotype of Overexpression of LsAG2 and LsDEF1 in Arabidopsis

*LsAG2* and *LsDEF1* were transferred into *A. thaliana* using *A. tumefaciens*-mediated floral organ dipping. Finally, 15 and 16 transgenic lines overexpressing the *LsAG2* and *LsDEF1* genes were obtained (Figure S2), respectively. Under long-day light conditions, the wild-type *A. thaliana* took approximately 45 d from sowing to flowering. However, many *35S::LsAG2* and *35S::LsDEF1* $T_2$ transgenic lines emerged in the flowering period, approximately 30 d after sowing (Table S6), and very few lines appeared before the seedlings were transplanted (just about 10 d). Compared to the wild-type, some *35S::LsDEF1* $T_2$ transgenic lines had more flowers and a longer flowering time.

In addition to the significant changes in the flowering period, the leaves and flowers of the transgenic lines also changed. The *35S::LsAG2* transgenic plant had smaller flower diameters than the wild-type *A. thaliana* (Figure 4a,e,f), and similar abnormal flower phenotypes, such as smaller petals, a sepal-like shape, hypoplasia, or the absence of petals, enlarged stamens, and less loose powder, were detected in the *35S::LsAG2* transgenic plants (Figure 4b–d). These transgenic plants were still fertile. On the other hand, abnormal development of the flower organs may be the reason of for *35S::LsDEF1*'s infertility, as a few of the plants lacked pistils, and the number of stamens was 7–10 (Figure 4j,k). A few were missing petals or showed malformed development of the entire flower (Figure 4l,m), and in some lines, the number of pistils increased to four, and the number of stamens increased to eight. Both the *35S::LsAG2* and *35S::LsDEF1* transgenic plants showed significant early flowering (Figure 4g,h,n), and the $T_2$ generation of the *35S::LsDEF1* transgenic plants showed a lack of fruiting, with a larger flower volume and longer flowering period than the wild-type *A. thaliana* (Figure 4i).

**Figure 4.** Phenotypes of overexpression of *LsAG2* and *LsDEF1* in *Arabidopsis*. (**a**) Flower of wild-type *A. thaliana*. (**b–d**) Flowers of *35S::LsAG2* transgenic *A. thaliana*. (**e**) Inflorescence of wild-type *Arabidopsis*. (**f**) Inflorescence of *35S::LsAG2* transgenic *A. thaliana*. (**g**) A 30-day-old *35S::LsAG2* transgenic *A. thaliana* starts flowering. (**h**) A 30-day-old wild-type *A. thaliana*. (**i**) The *35S::LsDEF1* transgenic *A. thaliana* has a longer flowering time and more flowers than the wild-type. (**j–m**) Flowers of *35S::LsDEF1* transgenic *A. thaliana*. (**n**) A 35-day-old *35S::LsDEF1* transgenic *A. thaliana* starts flowering earlier than the wild-type.

### 3.5. Detection of Endogenous Genes Related to the Flower Time and Flower Development

In order to identify whether the transcription levels of endogenous genes related to the flower time and flower development in *A. thaliana* are affected by the overexpression of the *LsAG2* and *LsDEF1*, RT-qPCR was used to detect the transcription levels of endogenous genes (*AtAP1, AtLFY, AtFLC, AtFT, AtPI,* and *AtFUL)* in transgenic *A. thaliana.* The results showed that in the overexpression of the *LsAG2* gene in *A. thaliana*, the expression levels of endogenous genes significantly decreased, except for *AtPI* and *AtFUL*. Upon overexpressing *LsDEF1* in *A. thaliana* lines, the expression level of *AtLFY* was slightly down-regulated, while the expression levels of *AtAP1, AtPI, AtFLC,* and *AtFUL* were up-regulated (Figure 5; Table S7).

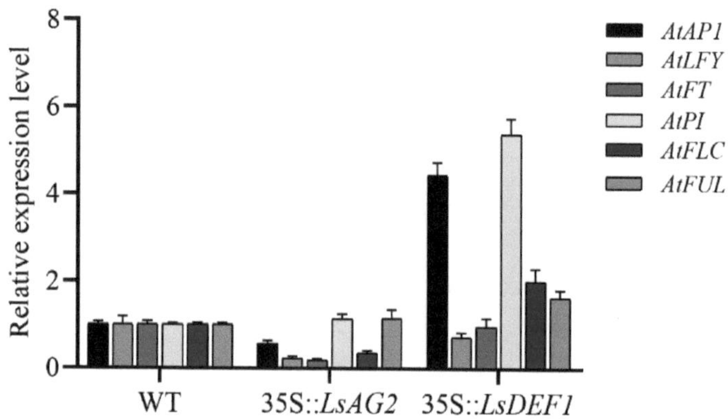


**Figure 5.** Expression levels of endogenous genes related to the flower time and flower development. The wild-type was selected as the control, and the gene expression levels were calculated using the $2^{-\Delta\Delta Ct}$ method.

## 4. Discussion

*Lagerstroemia speciosa* is an ornamental tree in summer, originating from Asia. The analysis of its flower development process is of great significance for the flowering period and flower type in breeding. Using the DPL and SPL as materials, the CDS, full-length DNA, and the promoter sequence approximately 1800 bp upstream of the *LsAG2* and *LsDEF1* genes were cloned by PCR. Previous studies have found that the formation of double-petal phenomena in many plants is closely related to structural differences in MADS-box or AP2 family genes. For example, the formation of the double rose is related to the deletion of the target site of *miR172* on *AP2* [24]. In sakura (*Prunus lannesiana*), there is a 190 bp skip reading in the exon of *AGAMOUS* (*PrseAG*), and the protein encoded by it also shows several amino acid differences from the protein encoded by the same gene in single-petaled *P. lannesiana* [25]. Comparing the CDS and DNA sequences cloned from the DPL and SPL, we found that there was no difference between the sequences, indicating that the formation of the stamen petalization of *L. speciosa* may not be caused by the differences in the gene structures of *LsAG2* and *LsDEF1*. Meanwhile, there was no difference in the upstream promoter region sequence, indicating that the differences in the gene expression levels of *LsAG2* and *LsDEF1* may be caused by other unknown regulation-related genes.

The *LsAG2* is the homologous gene of C-type gene *AG* in *A. thaliana.* Its expression pattern in different flower organs of the SPL showed more resemblance to the characteristics of the C-type gene in the ABCDE model, and the expression pattern in different floral organs of the DPL was very different from that in the SPL, indicating that the ectopic expression of *LsAG2* may be closely related to the formation of the DPL. *LsDEF1* is a B-type gene in the flower development model, and its expression pattern in different floral organs of *L. speciosa* was completely consistent with the characteristics of B-type genes. The *DEFICIENS* (*DEF*) homologous genes in other species also show the same trend [26], suggesting that the function of *DEF* may be conserved in angiosperms. In addition, the results of subcellular localization showed that the fusion proteins formed by the LsAG2-GFP and LsDEF1-GFP proteins were all localized in the nucleus, indicating that the proteins should have the function of transcription factors and may participate in the related processes of *L. speciosa* flower development.

*AG* has important and conserved functions in floral organ development in a variety of plants and is closely related to the double petal formation of angiosperm. For example, the overexpression of *JcAG* from Jatropha (*Jatropha curcas*) in *A. thaliana* also causes early flowering [27]. The overexpression of *TeAG1* and *TeAGL11-1* of marigold (*Tagetes erecta*) leads to leaf curling and early flowering in *A. thaliana* [28]. Here, the *35S::LsAG2* transgenic plants exhibited multiple phenotypes, such as flower aberration (shortened petal length, sepalization, or even its absence), early flowering, leaf curling, dwarf plants, and a reduced seed yield, indicating that *LsAG2* is widely involved in a variety of lifecycle activities, including flower development, in *A. thaliana*. The overexpression the *LsDEF1* in *A. thaliana* resulted in the phenotypes of degenerated pistils and increased stamens in the flowers. The overexpression of *BcAP3* (*APTEROUS3*) of turnip (*Brassica rapa* ssp.) in *A. thaliana* can lead to abnormal anther development and low pollen viability, resulting in male sterility [29]. The ectopic expression of *FaesAP3* of buckwheat (*Fagopyrum esculentum*) can rescue the stamen development and lack of petal development in *A. thaliana ap3* mutants [30], indicating that *DEF1/AP3* may be conserved in different plants.

In order to explore the reason for the double petalization of DPL, we measured the expression levels of *LsAG2* and *LsDEF1* in different tissues and performed functional verification on *A. thaliana*. RT-qPCR showed that the expression level of *LsAG2* in the petals of the DPL was significantly higher than that in the petals of the SPL, and the main phenotype of *35S::LsAG2* transgenic *Arabidopsis* is the distortion of petals. The expression level of *LsDEF1* in the stamens of the DPL was significantly higher than that in the stamens of the SPL, and the main phenotype of *35S::LsDEF1* transgenic *Arabidopsis* is the increase in the number of stamens, indicating that these two genes are, indeed, involved in the process of petal morphological transformation, but there is still a lack of key steps that result in petaloid stamens or increase the number of petals. A large number of studies have shown that changes in the expression patterns of *AG* homologous genes can lead to the formation of double flowers. This change in expression patterns includes the loss of gene expression [31,32] or the loss of key regions of the protein during transcription [25,33]. Therefore, future research should focus on exploring the effects of changes in the expression patterns of *AG* homologous genes on *L. speciosa* flower development and the screening of interacting proteins using yeast two-hybrid technology to analyze the mechanism of double flower formation in *L. speciosa*.

## 5. Conclusions

The *LsAG2* and *LsDEF1* genes of *L. speciosa* were cloned, and their expression patterns were verified by RT-qPCR. The results of subcellular localization showed that both genes were localized in the nucleus, and when *LsAG2* and *LsDEF1* were overexpressed in *A. thaliana*, the floral organs of the transgenic plants were changed, and the expression levels of endogenous genes related to the flower time and flower development also changed. In brief, *LsAG2* and *LsDEF1* have important roles in *L. speciosa* flower development, and the results of this research provide a theoretical basis for the regulation of flowering time and flower development in *L. speciosa*.

**Supplementary Materials:** The following supporting information can be downloaded at: https://www.mdpi.com/article/10.3390/agronomy13040976/s1, Figure S1: Bioinformatics analysis of *LsAG2* and *LsDEF1*; Figure S2: PCR detection of transgenic *A. thaliana*; Table S1: The primer sequences for cloning, RT-qPCR and vector construction; Table S2: Analysis of promoter cis-regulatory elements; Table S3: Physicochemical properties of MADS-box proteins of *L. speciosa*; Table S4: Secondary structures of MADS-box proteins of *L. speciosa*; Table S5: Relative expression data of *LsAG2* and *LsDEF1* in different samples; Table S6: Flowering time of *LsAG2* and *LsDEF1* transgenic *A. thaliana* plants under long-day conditions; Table S7: Relative expression data of *LsAG2* and *LsDEF1* in wild-type and transgenic *Arabidopsis*.

**Author Contributions:** Funding acquisition, T.Z. and Q.Z.; investigation, Z.L.; methodology, L.Y. and Z.L.; project administration, Z.L., T.Z. and Q.Z.; resources, T.Z., J.W. and T.C.; supervision, T.Z., J.W. and T.C.; validation, L.Y.; visualization, L.Y. and Z.L.; writing—original draft, L.Y. and Z.L.; writing—review and editing, T.Z. and Q.Z. All authors have read and agreed to the published version of the manuscript.

**Funding:** The research was supported by the Beijing Municipal Science and Technology Project (grant No. Z181100002418006), Beijing High-Precision Discipline Project, Discipline of Ecological Environment of Urban and Rural Human Settlements and Special Fund for Beijing Common Construction Project.

**Data Availability Statement:** Not applicable.

**Conflicts of Interest:** The authors declare no conflict of interest.

## References

1. Andrés, F.; Coupland, G. The genetic basis of flowering responses to seasonal cues. *Nat. Rev. Genet.* **2012**, *13*, 627–639. [CrossRef] [PubMed]
2. Coen, E.S.; Meyerowitz, E.M. The war of the whorls: Genetic interactions controlling flower development. *Nature* **1991**, *353*, 31–37. [CrossRef]
3. Rijpkema, A.S.; Vandenbussche, M.; Koes, R.; Heijmans, K.; Gerats, T. Variations on a theme: Changes in the floral ABCs in angiosperms. *Semin. Cell Dev. Biol.* **2009**, *21*, 100–107. [CrossRef]
4. Drews, G.N.; Bowman, J.L.; Meyerowitz, E.M. Negative regulation of the Arabidopsis homeotic gene AGAMOUS by the APETALA2 product. *Cell* **1991**, *65*, 991–1002. [CrossRef] [PubMed]
5. Bowman, J.L.; Smyth, D.R.; Meyerowitz, E.M. Genetic interactions among floral homeotic genes of Arabidopsis. *Development* **1991**, *112*, 1–20. [CrossRef] [PubMed]
6. Callens, C.; Tucker, M.R.; Zhang, D.; Wilson, Z.A. Dissecting the role of MADS-box genes in monocot floral development and diversity. *J. Exp. Bot.* **2018**, *69*, 2435–2459. [CrossRef]
7. Liljegren, S.J.; Ditta, G.S.; Eshed, Y.; Savidge, B.; Bowman, J.L.; Yanofsky, M.F. *SHATTERPROOF* MADS-box genes control seed dispersal in Arabidopsis. *Nature* **2000**, *404*, 766–770. [CrossRef] [PubMed]
8. Pelaz, S.; Ditta, G.S.; Baumann, E.; Wisman, E.; Yanofsky, M.F. B and C floral organ identity functions require *SEPALLATA* MADS-box genes. *Nature* **2000**, *405*, 200–203. [CrossRef] [PubMed]
9. Li, G.; Meng, Z.; Kong, H.; Chen, Z.; Lu, A. ABC model and floral evolution. *Chin. Sci. Bull.* **2003**, *48*, 2651–2657. [CrossRef]
10. Yanofsky, M.F.; Ma, H.; Bowman, J.L.; Drews, G.N.; Feldmann, K.A.; Meyerowitz, E.M. The protein encoded by the Arabidopsis homeotic gene agamous resembles transcription factors. *Nature* **1990**, *346*, 35–39. [CrossRef]
11. Galimba, K.D.; Tolkin, T.R.; Sullivan, A.M.; Melzer, R.; Theißen, G.; Di Stilio, V.S. Loss of deeply conserved C-class floral homeotic gene function and C- and E-class protein interaction in a double-flowered ranunculid mutant. *Proc. Natl. Acad. Sci. USA* **2012**, *109*, E2267–E2275. [CrossRef] [PubMed]
12. Liu, Z.; Wu, X.; Cheng, M.; Xie, Z.; Xiong, C.; Zhang, S.; Wu, J.; Wang, P. Identification and functional characterization of SOC1-like genes in *Pyrus bretschneideri*. *Genomics* **2020**, *112*, 1622–1632. [CrossRef] [PubMed]
13. Winter, C.M.; Austin, R.S.; Blanvillain-Baufumé, S.; Reback, M.A.; Monniaux, M.; Wu, M.F.; Sang, Y.; Yamaguchi, A.; Yamaguchi, N.; Parker, J.E.; et al. LEAFY target genes reveal floral regulatory logic, cis motifs, and a link to biotic stimulus response. *Dev. Cell* **2011**, *20*, 430–443. [CrossRef]
14. Lenhard, M.; Bohnert, A.; Jürgens, G.; Laux, T. Termination of stem cell maintenance in Arabidopsis floral meristems by interactions between WUSCHEL and AGAMOUS. *Cell* **2001**, *105*, 805–814. [CrossRef] [PubMed]
15. Irish, V.F. The flowering of Arabidopsis flower development. *Plant J.* **2010**, *61*, 1014–1028. [CrossRef]
16. Causier, B.; Schwarz-Sommer, Z.; Davies, B. Floral organ identity: 20 years of ABCs. *Semin. Cell Dev. Biol.* **2010**, *21*, 73–79. [CrossRef]
17. Zhang, X.; Wang, Q.; Yang, S.; Lin, S.; Bao, M.; Bendahmane, M.; Wu, Q.; Wang, C.; Fu, X. Identification and characterization of the MADS-box genes and their contribution to flower organ in carnation (*Dianthus caryophyllus* L.). *Genes* **2018**, *9*, 193. [CrossRef]
18. Won, S.Y.; Jung, J.A.; Kim, J.S. Genome-wide analysis of the MADS-box gene family in Chrysanthemum. *Comput. Biol. Chem.* **2021**, *90*, 107424. [CrossRef]
19. Xu, Z.; Zhang, Q.; Sun, L.; Du, D.; Cheng, T.; Pan, H.; Yang, W.; Wang, J. Genome-wide identification, characterisation and expression analysis of the MADS-box gene family in *Prunus mume*. *Mol. Genet. Genom.* **2014**, *289*, 903–920. [CrossRef]
20. Hu, L.; Zheng, T.; Cai, M.; Pan, H.; Wang, J.; Zhang, Q. Transcriptome analysis during floral organ development provides insights into stamen petaloidy in *Lagerstroemia speciosa*. *Plant Physiol. Biochem.* **2019**, *142*, 510–518. [CrossRef]
21. Zheng, T.; Chen, Z.; Ju, Y.; Zhang, H.; Cai, M.; Pan, H.; Zhang, Q. Reference gene selection for qRT-PCR analysis of flower development in *Lagerstroemia indica* and *L. speciosa*. *PLoS ONE* **2018**, *13*, e0195004. [CrossRef] [PubMed]
22. Livak, K.J.; Schmittgen, T.D. Analysis of relative gene expression data using real-time quantitative PCR and the 2(-Delta Delta C(T)) Method. *Methods* **2001**, *25*, 402–408. [CrossRef]

23. Ma, Y.-Q.; Pu, Z.-Q.; Meng, Q.; Tan, X.-M.; Yang, L.; Zhang, K.-L.; Ma, Y.-Y.; Huang, X.; Xu, Z.-Q. Dissecting SEPALLATA3 splicing variant functions during Arabidopsis vegetative growth by amiRNA technology. *J. Plant Growth Regul.* **2022**, *115*, 1–14. [CrossRef]
24. François, L.; Verdenaud, M.; Fu, X.; Ruleman, D.; Dubois, A.; Vandenbussche, M.; Bendahmane, A.; Raymond, O.; Just, J.; Bendahmane, M. A miR172 target-deficient AP2-like gene correlates with the double flower phenotype in roses. *Sci. Rep.* **2018**, *8*, 12912. [CrossRef]
25. Liu, Z.; Zhang, D.; Liu, D.; Li, F.; Lu, H. Exon skipping of *AGAMOUS* homolog *PrseAG* in developing double flowers of *Prunus lannesiana* (Rosaceae). *Plant Cell Rep.* **2013**, *32*, 227–237. [CrossRef]
26. Schwarz-Sommer, Z.; Hue, I.; Huijser, P.; Flor, P.; Hansen, R.; Tetens, F.; Lönnig, W.; Saedler, H.; Sommer, H. Characterization of the *Antirrhinum floral* homeotic MADS-box gene deficiens: Evidence for DNA binding and autoregulation of its persistent expression throughout flower development. *EMBO J.* **1992**, *11*, 251–263. [CrossRef]
27. Hui, W.K.; Liu, M.Q.; Wu, G.J.; Wang, J.Y.; Zhong, Y.; Li, H.Y.; Tang, H.L.; Zeng, W.; Ma, L.X.; Zhang, Y.; et al. Ectopic expression of an *AGAMOUS* homologue gene in *Jatropha curcas* causes early flowering and heterostylous phenotypes. *Gene* **2021**, *766*, 145141. [CrossRef] [PubMed]
28. Zhang, C.; Wei, L.; Wang, W.; Qi, W.; Cao, Z.; Li, H.; Bao, M.; He, Y. Identification, characterization and functional analysis of *AGAMOUS* subfamily genes associated with floral organs and seed development in Marigold (*Tagetes erecta*). *BMC Plant Biol.* **2020**, *20*, 439. [CrossRef] [PubMed]
29. Huang, F.; Zhang, Y.; Hou, X. *BcAP3*, a MADS box gene, controls stamen development and male sterility in Pak-choi (*Brassica rapa* ssp. chinensis). *Gene* **2020**, *747*, 144698. [CrossRef]
30. Fang, Z.W.; Qi, R.; Li, X.F.; Liu, Z.X. Ectopic expression of *FaesAP3*, a *Fagopyrum esculentum* (Polygonaceae) AP3 orthologous gene rescues stamen development in an Arabidopsis *ap3* mutant. *Gene* **2014**, *550*, 200–206. [CrossRef]
31. Dubois, A.; Raymond, O.; Maene, M.; Baudino, S.; Langlade, N.B.; Boltz, V.; Vergne, P.; Bendahmane, M. Tinkering with the C-function: A molecular frame for the selection of double flowers in cultivated roses. *PLoS ONE* **2010**, *5*, e9288. [CrossRef] [PubMed]
32. Tanaka, Y.; Oshima, Y.; Yamamura, T.; Sugiyama, M.; Mitsuda, N.; Ohtsubo, N.; Ohme-Takagi, M.; Terakawa, T. Multi-petal cyclamen flowers produced by *AGAMOUS* chimeric repressor expression. *Sci. Rep.* **2013**, *3*, 2641. [CrossRef] [PubMed]
33. Zhang, B.; Liu, Z.-x.; Ma, J.; Song, Y.; Chen, F.-J. Alternative splicing of the *AGAMOUS* orthologous gene in double flower of *Magnolia stellata* (Magnoliaceae). *Plant Sci.* **2015**, *241*, 277–285. [CrossRef] [PubMed]

*agronomy*

MDPI

*Article*

# Transcriptomic and Metabolomic Profiling Provides Insights into Flavonoid Biosynthesis and Flower Coloring in *Loropetalum chinense* and *Loropetalum chinense* var. *rubrum*

**Xia Zhang [1,2,3,†], Li Zhang [1,2,3,4,†], Damao Zhang [1,2,3], Yang Liu [1,2,3], Ling Lin [5], Xingyao Xiong [1,6,7], Donglin Zhang [1,8], Ming Sun [6,9], Ming Cai [6,9], Xiaoying Yu [1] and Yanlin Li [1,4,7,*]**

1 College of Horticulture, Hunan Agricultural University, Changsha 410128, China; xiazhang@stu.hunau.edu.cn (X.Z.); zhli911@hunaas.cn (L.Z.); zdm1558@stu.hungu.edu.cn (D.Z.); liuyang1203@stu.hunau.edu.cn (Y.L.); xiongxingyao@caas.cn (X.X.); dongling@uga.edu (D.Z.); yuxiaoying@hunau.edu.cn (X.Y.)
2 Engineering Research Center for Horticultural Crop Germplasm Creation and New Variety Breeding, Ministry of Education, Changsha 410128, China
3 Ministry of Education, Hunan Mid-subtropical Quality Plant Breeding and Utilization Engineering Technology Research Center, Changsha 410128, China
4 Hunan Horticulture Research Institute, Hunan Academy of Agricultural Sciences, Changsha 410125, China
5 School of Economics, Hunan Agricultural University, Changsha 410128, China; lljjxy@hunau.edu.cn
6 Agricultural Genomics Institute at Shenzhen, Chinese Academy of Agricultural Sciences, Shenzhen 518120, China; sunming@bjfu.edu.cn (M.S.); mingcai82@bjfu.edu.cn (M.C.)
7 Kunpeng Institute of Modern Agriculture, Foshan 528226, China
8 Department of Horticulture, University of Georgia, Athens, GA 30602, USA
9 School of Landscape and Architecture, Beijing Forest University, Beijing 100083, China
* Correspondence: liyanlin@hunau.edu.cn; Tel.: +86-015802687311
† These authors contributed equally to this work.

**Abstract:** The *Loropetalum chinense* and *Loropetalum chinense* var. *rubrum* are typical as well as traditional ornamental and Chinese herbal medicines in Asia; however, more information is needed on the mechanisms underlying their flower coloring. Here, we profiled the flavonoid metabolome and carried out full-length sequencing in addition to transcriptome analyses to investigate the flavonoid biosynthesis and global transcriptome changes among different petal coloring cultivars of *L. chinense* and *L. chinense* var. *rubrum*. The total anthocyanins in addition to the RHSCC values and CIE 1976 $L^*a^*b^*$ values of petals were highly consistent with petal color. Moreover, a total of 207 flavonoid components were identified. Of these, 13 flavonoid compounds were considered significantly different expression compounds highly consistent with color information in the 4 samples. Meanwhile, the first reference full-length transcriptome of *L. chinense* var. *rubrum* was built, which had 171,783 high-quality nonredundant transcripts with correcting with next-generation sequencing (NGS). Among them, 52,851 transcripts were annotated in the seven databases of NR, KOG, GO, NT, Pfam, Swiss-Port, and KEGG. Combined with NGS analyses, the DETs involved in flavonoids and anthocyanins contributed greatest to the flower coloring. Additionally, the different expressions of eight *LcDFRs* and four *LcANS* genes were positively correlated with flavonoid biosynthesis, and the four *LcBZ1* as well as one *Lc3Mat1* were positively correlated with the content of seven anthocyanins revealed by coupling with metabolomics and transcriptomics analyses. Together, these results were used to mine candidate genes by analyzing flower coloring changes at comprehensive metabolic and transcriptomic levels in *L. chinense* and *L. chinense* var. *rubrum*.

**Keywords:** *Loropetalum chinense* var. *rubrum*; flavonoid biosynthesis; anthocyanins; flower coloring; metabolomic; transcriptomic

**Citation:** Zhang, X.; Zhang, L.; Zhang, D.; Liu, Y.; Lin, L.; Xiong, X.; Zhang, D.; Sun, M.; Cai, M.; Yu, X.; et al. Transcriptomic and Metabolomic Profiling Provides Insights into Flavonoid Biosynthesis and Flower Coloring in *Loropetalum chinense* and *Loropetalum chinense* var. *rubrum*. *Agronomy* **2023**, *13*, 1296. https://doi.org/10.3390/agronomy13051296

Academic Editor: Carmine Guarino

Received: 12 April 2023
Revised: 30 April 2023
Accepted: 2 May 2023
Published: 4 May 2023

## 1. Introduction

*Loropetalum chinense* var. *rubrum* belongs to the Hamamelidaceae family and is commonly used in landscaping plazas, neighborhoods, gardens, roads, green areas, etc. [1,2]. It was first identified in Changsha, Hunan Province, and its wild resources are distributed in the Luoxiao Mountains among Liuyang, Pingjiang, and Liling [3]. Moreover, *L. chinense* var. *rubrum* has gorgeous flower and leaf coloring, in addition to flowering two–three times per year [4]. The petal coloring of *L. chinense* and *L. chinense* var. *rubrum* could be divided into five groups: poly-chromatic, yellowish-white, green-white, purplish-pink, and purplish red [5]. The different distributions of anthocyanins in the petals of parenchymal cells lead to the petal coloring differences in *L. chinense* var. *rubrum* [5]. Furthermore, the different components and contents of flavonoids in leaves lead to the different leaf colorings in *L. chinense* var. *rubrum* [6], while little is known about the flavonoid biosynthesis in petals of *L. chinense* var. *rubrum* [5–7].

Flavonoids are the primary pigments that contribute to the millions of flower colors in the petals of plants [8]. The water-soluble flavonoids are responsible for the range of colors from yellow to red to violet to blue, including aurones, chalcones, flavones, flavonols, flavanones, flavanones, isoflavones, and anthocyanins [8,9]. Flavones and flavonols are the co-pigments that contribute to flower colors and bluing [10]. The O-glycosides of anthocyanins are the colored flavonoids, which are the main determinants of flower color, found in the vacuoles of petal cells [11,12]. They have a basic structure of 3,5,7-trihydroxy benzene-2-phenyl benzofuran; further modification by glycosylation, acylation, and methylation depends on the plant species and varieties [9], which acquire diversified anthocyanidin structures and variable flower colors [13]. Hundreds of anthocyanins have been reported, and they are divided into six classes, including Pg (pelargonidin), Cy (cyanidin), Dp (delphinidin), Pn (peonidin), Pt (petunidin), and Mv (malvidin) [13,14]. Cyanidin derives peonidin, and delphinidin derives petunidin and malvidin formed by methylation, hydroxylation, glycosylation, and acylation. Additionally, then, this deriving of anthocyanins further creates orange-yellow, red, purple, and blue petals [15–18]. Thus, the flavonoid component and relative content are significant factors in determining the flower color, and the synthetic genes, regulatory genes, and modification genes create these.

The flavonoid biosynthetic pathway is well understood in plants [19]. Flavonoids (including anthocyanins) are generated through the phenylpropanoid pathway, which is divided into three stages [20,21]: The first stage is the general phenylpropanoid pathway catalyzed via phenylalanine ammonia-lyase (*PAL*), cinnamic acid 4-hydroxylase (*C4H*), and 4-coumarate-CoA ligase (*4CL*) [22]. The second stage involved is the production of colorless dihydroflavonols successively catalyzed via chalcone synthase (*CHS*) [23,24], chalcone isomerase (*CHI*) [25], flavanone 3-hydroxylase (*F3H*) [26], flavonoid-3′5′-hydroxylase (*F3′5′H*) [27,28], and flavonoid-3′-hydroxylase (*F3′H*) [29]. The third stage is the anthocyanins derivates catalyzed via dihydroflavonol 4-reductase (*DFR*) [28,30] and anthocyanidin (*ANS*) [31,32]. Additionally, then, the anthocyanidins and anthocyanins are catalyzed via glycosyltransferase (*GT*) [33] and methyltransferase (*MT*) [34], as well as acyltransferase (*AT*) [15] into stable anthocyanins and their further modifications. Of note, some of the flavonoid biosynthetic genes of *LcvrCHS1* [35], *LcCHS1*, *LcCHS2* [36], *LcCHI* [37], *LcDFR1*, and *LcDFR2* [38] were reported in *L. chinense* var. *rubrum*.

Transcriptional control also plays a vital role in the modulation of flavonoid biosynthesis, and several transcription factors involved in the pathway have been elucidated [39–41]. The *MBW* (MYB-bHLH-WD40) complex is well conserved and the central transcriptional regulator of activating structural genes to flavonoid pathway enzymes [42,43], which regulate the biosynthesis of flavan-3-ols [44]. *WD40* is a 'master regulator' in activating the flavonoid pathway independently [45]. The *MYB* transcription factor is the core component of the *MBW* complex, and its subgroup of *R2R3-MYB* is mainly involved in regulating flavonoid metabolism [46]. The overexpression of *AN4* (an R2R3-MYB gene) promotes the expression of the anthocyanin biosynthesis gene of *CHI*, *F3H*, and *DFR* [47]. *LhMYB12* activated the lily's *CHS* as well as *DFR* gene promoter directly and obstructed anthocyanin biosynthesis under high temperatures [48]. Although *MYB* can stimulate its companion of *bHLH*, its expression depends on different pigmented tissues [44]. *TT8* (a *bHLH* gene) accumulates anthocyanin and proanthocyanin biosynthesis by expressing *DFR* in *Arabidopsis thaliana* [49]. The *MADS*-box genes of *ScAGL11* and *ScAG* inhibit anthocyanin accumulation in the cineraria capitulum by downregulating the expression of *ScCHS2*, *ScDFR3*, and *ScF3H1* [50]. To date, the molecular mechanisms of flavonoid biosynthesis are more demonstrated in plants, while there are no reports on *L. chinense* var. *rubrum*.

Here, one *L. chinense*, 'Xiangnong Xiangyun', and three *L. chinense* var. *rubrum*, 'Huaye Jimu 2′, 'Xiangnong Fenjiao', and 'Xiangnong Nichang', were studied (Figure 1A). We identified the components of the flavonoid biosynthesis pathway in the three *L. chinense* var. *rubrum* and one *L. chinense* cultivar with MRM (multiple reaction monitoring). Then, the full-length transcriptome data and the next-generation sequencing data were obtained by combining NGS and SMRT sequencing. Moreover, this approach, combined with the flavonoid metabolic data, was applied to explain the molecular mechanisms of the flower coloring of *Loropetalum* cultivars, in which flavonoids are produced and accumulated. Accordingly, this study provides a valuable resource for further investigating flavonoid biosynthesis and ornamental flower coloring modification as well as breeding.

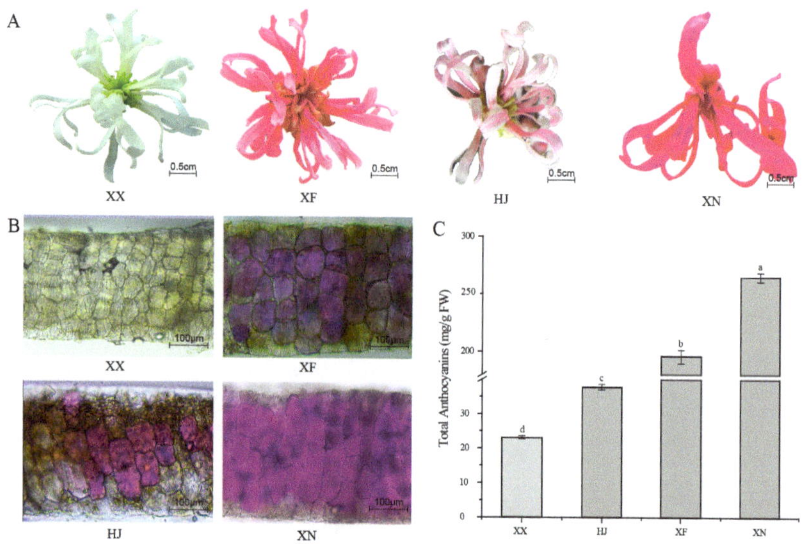


**Figure 1.** Phenotypes of *Loropetalum* cultivars and total anthocyanin content. (**A**). Phenotypes of 4 *Loropetalum* cultivars, namely 'Xiangnong Xiangyun' (XX), 'Huye Jimu 2′ (HJ), 'Xiangnong Fenjiao', (XF) and 'Xiangnong Nichang' (XN), 2 days after the flowering stage, scale bars = 0.5 cm. (**B**). Transverse section of petals of four *Loropetalum* cultivars, namely 'Xiangnong Xiangyun' (XX), 'Huye Jimu 2′ (HJ), 'Xiangnong Fenjiao' (XF), and 'Xiangnong Nichang' (XN), scale bars = 100 μm. (**C**). Total anthocyanin content in the petals of four *Loropetalum* cultivars, namely 'Xiangnong Xiangyun' (XX), 'Huye Jimu 2′ (HJ), 'Xiangnong Fenjiao' (XF), and 'Xiangnong Nichang' (XN). Error bars indicate the standard error (+SE) of the mean. Different letters indicated significant differences at a $p \leq 0.05$ level based on Duncan's test.

## 2. Materials and Methods

### 2.1. Plant Materials

One *Loropetalum chinense* ('Xiangnong Xiangyun', XX) and three *L. chinense* var. *rubrum* ('Huaye Jimu 2′, HJ; 'Xiangnong Fenjiao', XF; and 'Xiangnong Nichang', XN) were planted in the germplasm garden of *Loropetalum* spp. at Hunan Agricultural University, Changsha, Hunan Province, China (E 113.08, N 28.17). XX, XN, and XF were the naturally pollinated offspring of HJ. The flowers (flowers in full bloom for one day), stems, leaves, and roots of HJ were harvested for Iso-Seq library construction. The flowers of XX, HJ, XF, and XN were also collected for total anthocyanin detection and sequencing library construction. We considered 50 flowers comprising 1 biological sample for each *Loropetalum* cultivar, and each sample was segregated as 3 independent biological replicates. All of these samples were frozen in liquid nitrogen immediately and stored at −80 °C.

### 2.2. Petal Color Comparison

The RHS color chart (sixth edition 2015, Royal Horticultural Society, 80 Vincent Square, London SW1P 2 PE, UK) was used to describe the color differences among the four *Loropetalum* spps. A spectrophotometer (YS3020, Technology Co. Ltd., Shenzhen, China) was used to measure the color-related values of fresh petals' $L^*$, $a^*$, and $b^*$. The parameters of $L^*$, $a^*$, and $b^*$ were described in CIELAB (CIE 1976). In this system, the value of $L^*$ represented lightness (those that had higher $L^*$ values represented a white or near-white color, while those with lower ones had a black or near-black color), $a^*$ was a positive or negative coordinate representing a purplish-red-bluish-green color, and $b^*$ was a positive or negative coordinate representing a yellow-blue color.

### 2.3. Transverse Section Observation

For transverse section observations, the different *Loropetalum* spp. petals were cut into thin slices by using a razor blade. Additionally, the thin slice was then fully extended flat on a temporary slide with a clean filter covering. Next, the temporary glazing of petals was observed and photos were taken with a microscope (DMIL LED, Leica, Wetzlar, Germany) and its imaging system (LEICA DFC295 and Leica Microsystems CMS GmbH, Leica, Wetzlar, Germany).

### 2.4. Total Anthocyanins Analysis

The total anthocyanins content was determined as described by Zhang et al. [51] and Dong et al. [52]. Briefly, 0.5 g of *Loropetalum* spp. petals was ground to a powder in liquid nitrogen. These powders were then extracted with 5 mL of a mixture solution of 0.05% HCl in methanol at 4 °C for 24 h in darkness. The supernatant was transferred into a clean tube after centrifugation at 1000× *g* at 4 °C for 15 min. Additionally, 1 mL of supernatant was then transferred into a clean tube with 4 mL of buffer A (0.4 M KCl, pH 1.0) and buffer B (1.2 N citric acid, pH 4.5), respectively. Absorbance measurements of the mixture were taken at 510 and 700 nm for mixtures A and B, separately. The total anthocyanin content was calculated via the use of the following formula: $TA = A \times MW \times 5 \times 100 \times V/e$ (Romero et al., 2008). TA represents the total anthocyanin content of the detected sample (mg/100 g, as cyanidin-3-O-glucose equivalent), and V represents the final volume (mL). A = [A510 (pH 1.0)–A700 (pH1.0)]–[A510 (pH 4.5)–A700 (pH 4.5)]. The value of 449.2 represents the molecular mass of cyanidin-3-O-glucose. Furthermore, the value of 26,900 reflects the molar absorptivity (e) at 510 nm [53]. Each biological replicate was repeated in triplicate.

### 2.5. Flavonoid Extraction and MRM

Freeze-dried flowers were crushed via the use of a mixer mill (MM 400, Retsch, Haan, Germany) with a zirconia bead for 1.5 min at 30 Hz. Of the fine powder, 100 mg was extracted overnight at 4 °C with 1.0 mL of 70% aqueous methanol. The extract's supernatant was absorbed (CNWBOND Carbon-GCB SPE Cartridge, 250 mg, 3 mL; ANPEL, Shanghai, China, www.anpel.com.cn/, 25 August 2019) after centrifugation at 10,000× *g* for 10 min before the LC-MS analysis. The supernatant was analyzed using an LC-EMS-MS/MS (HPLC, Shim-pack UFLC SHIMADZU CBM30A system, Kyoto, Japan, www.shimadzu.com.cn/, accessed on 24 April 2023; MS, Applied Biosystems 4500 Q TRAP, Waltham, MA, USA, www.appliedbiosystems.com.cn/, accessed on 24 April 2023). Moreover, the analytical conditions were composed of a column of Waters ACQUITY UPLC HSS T3 C18 (1.8 μm, 2.1 mm × 100 mm). Water and acetonitrile with 0.04% acetic acid were used as solvent systems. The gradient program of HPLC was performed as 100:0 $v/v$ at 0 min, 5:95 $v/v$ at 11.0 min, 5:95 $v/v$ at 12.0 min, 95:5 $v/v$ at 12.1 min, and 95:5 $v/v$ at 15.0 min. The flow rate was 0.40 mL/min with an injection volume of 5 μL, and the column temperature was 40 °C. The effluent was alternatively connected to an ESI-triple quadrupole-linear ion trap (Q TRAP)-MS [6,38].

A triple quadrupole-linear ion trap mass spectrometer (Q TRAP) API 4500 Q TRAP LC/MS/MS System equipped with LIT and triple quadrupole (QQQ) scans were used to obtain the information on the flavonoid metabolites. Moreover, the analysis system was equipped with an ESI Turbo Ion-Spray interface, operating in a positive ion mode and controlled by Analyst 1.6.3 software (AB Sciex). The ESI system index sets were followed by Chen et al. [6] and Zhang et al. [38]. Instrument tuning and mass calibration were performed with 10 and 100 μmol/L polypropylene glycol solutions in QQQ and LIT modes, respectively. QQQ scans were acquired as MRM experiments with the collision gas (nitrogen) set to 5 psi. DP and CE for individual MRM transitions were performed with further DP and CE optimization. MassBank, KNAPSACK, HMDB, MOTDB, and METLIN were used to analyze the structures of metabolites. The KEGG pathway public database was used to identify the specific metabolic pathways. The qualitative and quantitative data of each flavonoid metabolite were identified and quantified based on the MWDB and open public metabolite database. A specific set of MRM transitions was monitored for each period according to the metabolites eluted within this period. The target substance was screened and identified by the parent ion (Q1), fragment ions (Q2), and characteristic fragment ions (Q3) [6,38].

### 2.6. RNA Sample Preparation

Total RNA was prepared by grinding tissues into powder in liquid nitrogen and mixing 0.5 g of said powder with 1 mL of TRIzol reagent (GenStar P124-01). It is processed following the protocols provided by the manufacturer (#124-01, GenStar, Beijing, China, https://www.gene-star.com/pro_cont_10220.html#coming, accessed on 24 April 2023). All samples were replicated three times. Then, 1% agarose gels and NanoDrop (NanoDrop products, Wilmington, NC, USA) were used to detect RNA degradation as well as contamination. The quantity and quality of total RNA were assessed using a Qubit® RNA Assay Kit in a Qubit® 2.0 Fluorometer (Life Technologies, South San Francisco, CA, USA) and an Agilent 2100 Bioanalyzer (Agilent Technologies, Santa Clara, CA, USA), respectively. Two experiments were conducted: One was the different organs (leaf, flower, root, and stem) of the total RNA extracted and mixed with equal amounts to be a pool of *L. chinense* var. *rubrum* RNA. The other was 4 flower cultivars (HJ, XNNY, XF, and XN) with 12 libraries subjected to 2 × 150 paired-end RNA-seq using ILLUMINA novaseq 6000. All of the qualified RNA samples were stored at −20 °C and used for SMRT sequencing in addition to RNA-seq within one week.

### 2.7. PacBio Iso-Seq Library Preparation and SMRT Sequencing

A total of 3.9 μg of equal mixed RNA was sequenced on the PacBio Sequel platform (Pacific Bioscience, Menlo Park, CA, USA) according to the manufacturer's instructions, as previously described (Hoang et al., 2017; Hoang et al., 2019). The Oligo (dT) magnetic beads were used to enrich mRNA for the cDNA libraries. The Iso-Seq library was prepared with a Clontech SMARTer PCR cDNA synthesis kit (#634926; Clontech, Takara Bio Inc., Shiga, Japan) and BluePippin Size Selection System protocol described by Pacific Biosciences (PN 100-092-800-03). Fractions of cDNA with a size over 4 kb were run on the BluePippinTM System to remove the short SMRTbell templates. After size fraction, the optimal library was sequenced on a PacBio RS II instrument at the Novogene technology company (Beijing, China).

The circular consensus sequence was generated from subread BAM files and then classified into non-full-length and full-length FASTA files by using pbclassify.py with ignore polyA false and min Seq Length 200. Next, the isoform-level clustering (ICE) and final arrow polishing were fed into the clustering step via following the parameter configuration of hq_quiver_min_accuracy 0.99, bin_by_primer false, bin_size_kb 1, qv_trim_5p 100, and qv_trim_3q 30. Additionally, the nucleotide errors in consensus reads were corrected via the use of Illumina RNA-seq data with the LoRDEC software [54]. Finally, the final transcripts were obtained in corrected consensus reads by removing the redundant sequences with the CD-HIT software (-c0.95-T6 -G0 -Al 0.00 -aS 0.99) [55]. The raw sequence data reported in this paper have been deposited in the Genome Sequence Archive (Genomics, Proteomics & Bioinformatics 2021) in the National Genomics Data Center (Nucleic Acids Res 2022), China National Center for Bioinformation/Beijing Institute of Genomics, Chinese Academy of Sciences (GSA: CRA009285 and CRA009284), publicly accessible at https://ngdc.cncb.ac.cn/gsa, accessed on 24 April 2023.

### 2.8. Illumina RNA-Seq Library Construction and Sequencing

Twelve libraries of four cultivars of RNA samples were prepared and sequenced, respectively. A total of 2 μg of RNA was used for shot read sequencing on the HiSeq4000 platform, and 150 bp paired-end reads were generated at the Novogene technology company in Beijing, China. The NEBNext® UltraTM RNA Library Prep Kit for Illumina® (NEB, Ipswich, MA, USA) was used for sequencing libraries, following the manufacturer's recommendations.

### 2.9. Gene Functional Annotation, Coding Sequence (CDS) Prediction, Transcription Factor (TF), and Long Non-Coding RNA (lncRNA) Identification

All final transcript annotations were carried out via comparison against the following databases: NCBI nonredundant protein sequences (NR) [56], NCBI nonredundant nucleotide sequences (NT) [57], Protein family (Pfam) [58], Clusters of Orthologous Groups of proteins (KOG/COG) [59], a manually annotated and reviewed protein sequence database (Swiss-Port) [60], KEGG Ortholog database (KO) [61], and Gene Ontology (GO) [62]. The BLAST software with the set of e-values '1e−10' was used for NT database analyses [63]. Diamond Blast with the option of e-value '1e−10' was applied to the NR, KOG, Swiss-Port, and KEGG database analyses [64]. Moreover, the Hmmscan software was used for the Pfam database analyses [65].

The ANGEL pipeline, a long-read ANGLE implementation, was used to determine cDNA protein-coding sequences. We used *L. chinense* var. *rubrum* and closely related species confident protein sequences for ANGEL training and then ran the ANGEL prediction for final transcripts [66]. The iTAK software [67] and Plant Transcription Factor database [68] were used to identify TFs, transcriptional regulators (TRs), and protein kinases (PKs) with final transcripts of *L. chinense* var. *rubrum*. The following four tools were used to predict the lncRNAs of *L. chinense* var. *rubrum*: Coding-Non-Coding-Index (CNCI) [63], Coding Potential Calculator (CPC) [69], Pfam-scan [70], and PLEK [54]. The detected transcripts predicted with coding potential by either/all of the four tools were filtered out, and those without coding potential were our candidate set of LncRNAs.

### 2.10. Quantitation and Differential Expression of Transcripts and Genes Analysis

The twelve samples of raw data were subjected to quality control, including adapters, lower reads, and ploy-N, by fastQC v0.11.2 [71]. Additionally, the clean data were then obtained and mapped onto the *L. chinense* var. *rubrum* full-length transcriptome of final transcripts via the use of Bowtie2 (v2.2.5) [72]. The readcount for each transcript was estimated by the expected number of fragments per kilobase of transcript sequence per millions of base pairs sequenced (FPKM) [73] method through using RESM [74]. A correlation analysis between samples of each cultivar was used to test the repeatability of the samples. The three biological samples with a correlation coefficient value more considerable than 0.92 were performed for differentially expressed gene (DEG) and differentially expressed transcript (DET) analyses. The DEGs and DETs between each paired sample were performed by the DESeq R packages [75]. Genes and transcripts with |log2 fold change| ≥2 and an adjusted $p$-value ≤ 0.05 were considered significant DEGs and DETs. The Gene Ontology

(GO) enrichment of all of the DEGs and DETs was performed via the use of the GO-seq R packages [76]. The KEGG database and KOBAS software were subjected to KEGG pathway enrichment analyses; significantly enriched metabolic or signal transduction pathways with an adjusted $p$ value of $\leq$0.05 were selected.

### 2.11. The Correlation Analysis between DETs and DCMs

The differential content metabolites (DCMs) of flavonoids and anthocyanins in addition to the differentially expressed transcripts (DETs) based on KEGG enrichment analyses of flavonoid biosynthetic pathway and anthocyanin biosynthetic pathway genes were used for integrative analyses. Moreover, Spearman's method was used to analyze the correlation coefficients for transcriptome and metabolome data integration. The connection between DETs and DCMs was shown through a heat plot.

## 3. Results

### 3.1. Petal Color Phenotype and Total Anthocyanins Content among the Four *Loropetalum* Cultivars

Four *Loropetalum* cultivars were selected to elucidate the mechanisms of flavonoid biosynthesis in petals, a white flower petal control cultivar, namely 'Xiangnong Xiangyun' (XX) (Royal Garden Color Card, NN155B), and three *L. chinense* var. *rubrum*, namely 'Huaye Jimu 2' (HJ) (white part: Royal Garden Color Card, NN155C; purple part: Royal Garden Color Card, 61A), 'Xingnong Fenjiao' (XF) (Royal Garden Color Card, 61C), and 'Xiangnong Nichang' (XN) (Royal Garden Color Card, 63A) (Figure 1A). The transverse section of petals showed that the distribution of anthocyanins was significant different in four *Loropetalum* spp. (Figure 1B). The $a^*$ value representing redness was 0.79, 1.5, 4.52, and 2.24 among XX, HJ, XF, and XN, respectively. The $b^*$ value of these four *Loropetalum* cultivars (XX, HJ, XF, and XN) was −5.89, −6.29, −4.36, and −4.74, which represented blueness. The $L^*$ value, representing lightness, showed at 80.43, 100.91, 96.49, and 98.91 in four *Loropetalum* cultivars.

The total anthocyanins in the blooming petals for two days were detected with a spectrophotometric pH differential method. Among these four *Loropetalum* cultivars, the lowest level of anthocyanin accumulation (23.03 mg/g of FW) was detected in XX. Among the three *L. chinense* var. *rubrum* cultivars, the flower color with chimera (Huaye Jimu 2, HJ, 37.71 mg/g of FW) contains more anthocyanins than XX (Figure 1C). The results indicated that the highest level of anthocyanin content was in 'Xiangnong Nichang' (XN, 264.96 mg/g of FW) (Figure 1C). It was shown that the sharp differences in anthocyanin accumulation are due to genetic diversity and specificity. Additionally, a surprising degree of anthocyanin accumulation was observed in the four *Loropetalum* cultivars, especially in XN and XF.

### 3.2. Identification and Qualification of Flavonoid Metabolite Profiles from the Petals of *Loropetalum* Cultivars

To identify flavonoid metabolite profiles in *Loropetalum* cultivars, extracts from the petals of XX, HJ, XF, and XN were analyzed via MRM. A total of 207 flavonoid metabolites were identified from the 4 samples, including 39 flavonol metabolites, 7 isoflavone metabolites, 25 flavonoid metabolites, 79 flavone metabolites, 20 flavanone metabolites, 19 polyphenol metabolites, 15 anthocyanin metabolites, and 3 proanthocyanin metabolites, which were isolated and identified in all 4 of the *Loropetalum* cultivars' petal extracts (Table S1 and Figure 2A). The total content of each flavonoid compound was significantly different. For example, in HJ, XN, and XF the anthocyanin metabolites were the most abundant ones, followed by flavonol or polyphenol metabolites, while flavonol metabolites were the most abundant in XX, followed by polyphenol ones (Figure 2A). In addition, we performed a principal component analysis (PCA), and these 207 flavonoid metabolites could be divided into 4 groups, which is consistent with the flower color phenotypical characteristics of the four *Loropetalum* plants (Figure 2B). To sum up, the accumulation of flavonoid components in flowers is significantly correlated with variety specificity.

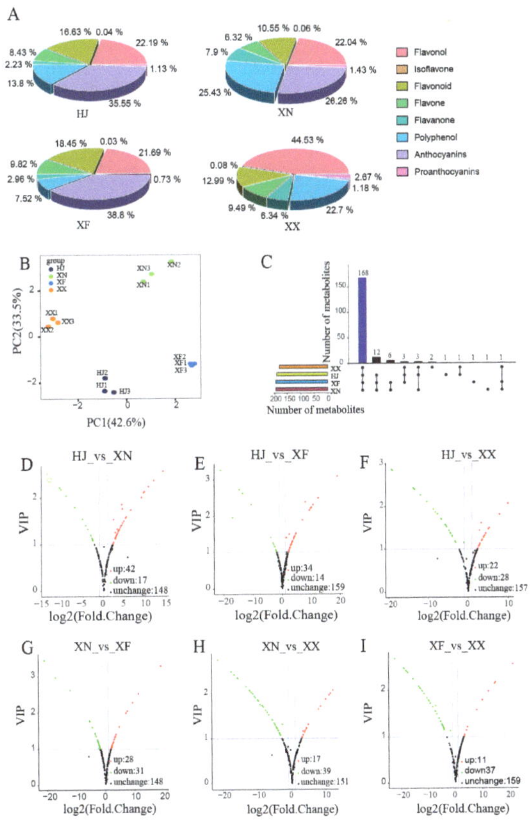


**Figure 2.** Global metabolic changes among the different cultivars of *L. chinense* var. *rubrum*. (**A**) Classification and proportion of 207 flavonoid compounds in different cultivars of XX, XF, XN, and HJ. (**B**) PCA of metabolites in different cultivars of XX, XF, XN, and HJ. (**C**) Upset plot showing overlap of metabolites in different cultivars of XX, XF, XN, and HJ. (**D**) Volcano plot of metabolites between HJ and XN. (**E**) Volcano plot of metabolites between HJ and XF. (**F**) Volcano plot of metabolites between HJ and XX. (**G**) Volcano plot of metabolites between XN and XF. (**H**) Volcano plot of metabolites between XN and XX. (**I**) Volcano plot of metabolites between XF and XX.

### *3.3. Differential Accumulation of Flavonoid Metabolites in Loropetalum Plant's Petals*

To investigate the different metabolites' changes and the expression levels of flavonoids involved in *L.* spps., the types and relative accumulation levels of flavonoid compounds were analyzed among the four selected cultivars (Supplementary Tables S1–S7). The results shown in the upset plot (Figure 2C) and volcano plot (Figure 2D–I) demonstrate that metabolites significantly differed among them. In total, 168 flavonoid metabolites were commonly found in the 4 samples. Notably, eight flavonols (Kaempferide, Ayanin, Chrysoeriol 6-*C*-hexoside, 'di-*C*, *C*-hexosyl-apigenin', 8-*C*-hexosyl chrysoeriol *O*-hexoside, Tricin 7-*O*-hexosyl-*O*-hexoside, Acacetin *O*-glucuronic acid, and Apigenin 6,8-*C*-diglucoside), one flavanone (Hesperetin *O*-Glucuronic acid), and three anthocyanins (Peonidin, Peonidin 3-sophoroside-5-glucoside, and Petunidin 3-*O*-glucoside) were coexisting in the petals of HJ, XF, and XN (Figure 2C, Supplementary Table S8). Furthermore, five flavones (Apiin, Tricin *O*-rhamnosyl-*O*-malonylhexoside, *C*-hexosyl-chrysin *O*-feruloylhexoside, Chrysoeriol *O*-sinapoylhexoside, and Apigenin 7-rutinoside (Isorhoifolin)) and one anthocyanin (Malvidin 3-acetyl-5-diglucoside) commonly existed in XF and XN (Figure 2C, Supplementary Table S8). Additionally, three anthocyanidins (Peonidin, Petunidin 3-*O*-

glucoside, and Peonidin 3-sophoroside-5-glucoside), one flavanone (Hesperetin *O*-Glucuronic acid), six flavones (Chrysoeriol 6-*C*-hexoside, di-*C*,*C*-hexosyl-apigenin, 8-*C*-hexosyl chrysoeriol *O*-hexoside, Tricin 7-*O*-hexosyl-*O*-hexoside, Acacetin *O*-glucuronic acid, and Apigenin 6,8-*C*-diglucoside), and two flavonols (Kaempferide, Ayanin) were found in HJ, XN, and XF, respectively. Moreover, one flavone (*C*-hexosyl-luteolin *O*-hexoside) and one isoflavone (Glycitin) were detected in XX (Figure 2C, Supplementary Table S8). In addition, Cyanidin 3-*O*-glucoside was found in XF, XN, XX, and HJ, significantly differing in the four samples (Figure 2C, Supplementary Table S8).

The flavonoid metabolites' distribution could be divided into up- and downregulated types. Based on the fold changes and VIP values of the OPLS-DA of flavonoid metabolites in the four samples, 59, 48, 50, 59, 56, and 48 differential concentrations of flavonoid metabolites (DCMs) (fold change $\geq$ 2 or fold change $\leq$ 0.5, and VIP $\geq$ 1) were detected in HJ VS XN (Figure 2D, Supplementary Table S9), HJ VS XF (Figure 2E, Supplementary Table S10), HJ VS XX (Figure 2F, Supplementary Table S11), XN VS XF (Figure 2G, Supplementary Table S12), XN VS XX (Figure 2H, Supplementary Table S13), and XF VS XX (Figure 2I, Supplementary Table S14), respectively. These DCMs were then introduced into the Kyoto Encyclopedia of Genes and Genomes (KEGG) database and KEGG enrichment analyses to explore the potential metabolic pathways affected by the different colored petals. The 'Biosynthesis of secondary metabolites' (ko01110), 'Flavonoid biosynthesis' (ko00941), 'Anthocyanin biosynthesis' (ko00942), 'Isoflavonoid biosynthesis' (ko00943), and 'Flavone and flavonol biosynthesis' (ko00944) were screened out to be the most relative metabolic pathways related to the petals' coloring (Figure S1, Supplementary Tables S15–S20). Notably, the different relative contents of Kaempferide (Flavonol), Glycitin (Isoflavone), Apiin (Flavone), Tricetin (Flavone), Naringenin 7-*O*-glucoside (Flavanone), Cyanidin 3,5-*O*-diglucoside (Anthocyanin), Cyanidin 3-*O*-glucoside (Anthocyanin), Delphinidin (Anthocyanin), Delphinidin 3-*O*-glucoside (Anthocyanin), Petunidin 3-*O*-glucoside (Anthocyanin), and Pelargonidin (Anthocyanin) were significantly different in the four cultivars of petals. Additionally, the Kaempferide, Malvidin 3-acetyl-5-diglucoside, Peonidin, Peonidin 3-sophoroside-5-glucoside, Petunidin 3-*O*-glucoside, Cyanidin 3,5-*O*-diglucoside, Cyanidin 3-*O*-glucoside, Delphinidin 3-*O*-glucoside, and Pelargonidin had much higher contents in XN, XF, and HJ, while the content of Delphinidin was the highest in XX (Figure S2, Supplementary Tables S15–S20). The phenomenon mentioned above, flavonoid compounds from the three purple petals, indicated that these purple cultivars might differ in the expression of anthocyanin biosynthetic or regulatory gene expression from the white one.

### 3.4. Combined Sequencing Approach to Tissues of *L. chinensis var. rubrum*

To identify and differentiate the flower transcriptome of *L. chinense*, two sequencing strategies were adopted by using the NGS and SMRT sequencing platforms. First, the twelve mRNA samples from four flower petals (HJ, XN, XF, and XX, each in triplicate) were performed via 2 $\times$ 150 paired-end sequencing by using a HiSeq 4000 platform, with each sample yielding 8.14 Gb of clean data on average, and the Q30 values of all of the 12 samples were higher than 93.82%. The average coverage of each Illumina transcriptome data mapping to the *L. chinense* var. *rubrum* full-length transcriptome was 82.92% (80.36~86.77%), which indicated that the data were comparable (Supplementary Table S21). The second strategy was that full-length cDNAs from 12 pooled total RNA samples were normalized and SMRT sequencing was performed by using the PacBio platform (Supplementary Table S22). In total, 46.16 Gb of polymerase read bases (1,562,962 polymerase reads) were generated by the PacBio sequence, with an average read length of 29,532 bp and N50 of 51,440 bp. After filtering using the RS Subreads.1 of the PacBio RS, 35,173,249 subreads representing 43.50 Gb were obtained. After clustering and polishing with ICE and arrow, 723,655 full-length non-chimeric reads were obtained. Finally, the clean Illumina reads were used to correct the SMRT reads of the polished consensus via the use of the LoRDEC software, the redundant sequences were removed by CD-HIT software, and 171,783 high-quality nonredundant transcripts were produced, with a total length of 406,454,922 bp and N50 length of 2935 bp; their lengths were in the range of 132 to 14,258 bp. High-quality full-length transcriptome data were obtained.

The final transcripts were annotated with the NR, KOG, GO, NT, Pfam, Swiss-Port, and KEGG databases, and they were also analyzed in terms of metabolic pathways and functional classifications. In total, 144,893 transcripts (84.35% of the total) were functionally annotated, and 52,851 transcripts (24.94% of the total) were subjected to all databases (Figure 3A). Furthermore, 5 public protein databases shared 43,776 transcripts (Figure 3B), with 131,741 transcripts (76.69% of the total) being assigned to the NR database. The NR databases had the most matched sequences, and the top five species, including *Vitis vinifera*, *Juglans regia*, *Theobroma cacao*, *Nelumbo nucifera*, and *Ziziphus jujuba*, had the best Blast hits ratios of over 50% (Figure S3). In the KOG database, 85,328 transcripts were subjected to 26 functional categories (Figure S4), and the largest group was general function prediction only, with 19,255 transcripts (22.57% of the total). Especially in the Q group, 3830 transcripts were assigned to the category of 'Secondary metabolites biosynthesis, transport and catabolism'. A total of 76,974 transcripts were grouped into three GO categories of biological processes (BPs), cellular components (CCs), and molecular functions (MFs). All of the hit transcripts were divided into 54 subgroups, with many transcripts related to 'metabolic process' (35,888 of the total) and 'biological regulation' (9532 of the total) (Figure S5). A total of 129,856 transcripts were annotated into 6 categories subjected to the KEGG database (Figure S6). The most enriched was 'Metabolism' (40,595 of the total), and these transcripts were associated with the biosynthesis of carbohydrate metabolism (5657 of the total), phenylpropanoid biosynthesis (641 of the total), flavonoid biosynthesis (267 of the total), isoflavonoid biosynthesis (5 of the total), and flavone as well as flavonol biosynthesis (62 of the total) pathways.

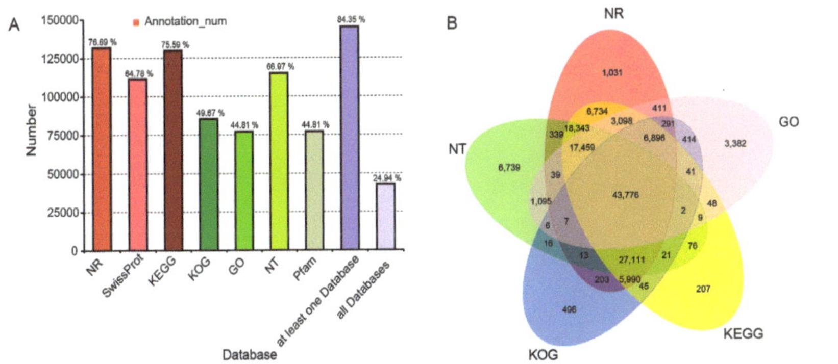


**Figure 3.** The annotation results of full-length nonredundant final transcripts to public databases. (**A**) The static annotation results of full-length nonredundant final transcripts to seven databases. (**B**) Annotation of full-length nonredundant final transcripts to NR, NT, KOG, KEGG, and GO.

The protein-coding and long non-coding RNA were obtained via full-length transcriptome sequencing. A total of 16,1387 CDSs were predicted from the transcripts (Figure S7A). Furthermore, these CDSs were blasted and hit with PlnTFDB and iTAK software for TF annotation. In total, 5487 TFs were identified and divided into more than 28 families (Figure S7B). The most abundant family was *C2H2* (397 of the total), followed by *C3H* (341 of the total), *bHLH* (324 of the total), *FAR1* (291 of the total), and *SNF2* (254 of the total). In addition, 26,590 *LncRNAs* were predicted from the final transcripts, ranging from 200 to 11,293 bp (Figure S7C,D). These data provided an abundant gene and transcript pool for the further study of metabolic processes.

### 3.5. Global Transcriptomic Characteristics of Flowers during the Full-Bloom Stage

To acquire insight into the molecular mechanisms of *L. chinense* var. *rubrum* flowers, RNA samples from the four cultivars with dissimilar flower colorings were used for transcript quantification. The box histogram showed that all of the transcripts' expression levels were significantly different in the four samples, and that the lowest level was in XN (Figure 4A); however, a correlation analysis indicated that all of the samples in each group had good reproducibility (Figure 4B). Moreover, the cluster dendrogram analysis and PCA analysis revealed that all of the samples within each group had high consistency (Figure 4C,D). A total of 8713, 17,085, 11,818, 17,652, 9803, 17,687, 13,412, 15,772, 1412, 16,058, 7854, and 7521 differential expression transcripts (DETs) were identified in the XF vs. HJ, XF vs. XN, XF vs. XX, XN vs. HJ, XN vs. XX, and XX vs. HJ compared combinations, respectively (Figure 4E). In particular, focusing on the pairwise comparisons group content of XF and XN (petals with a pearl coloring), the number of upregulated DETs was higher than that of downregulated ones, while the downregulated DETs were higher than upregulated ones in the XX vs. HJ group (petals with white and mosaic coloring). Additionally, the number of upregulated differentials expressing LncRNA was higher than downregulated ones in the groups of XF vs. HJ and XF vs. XX (Figure S8).

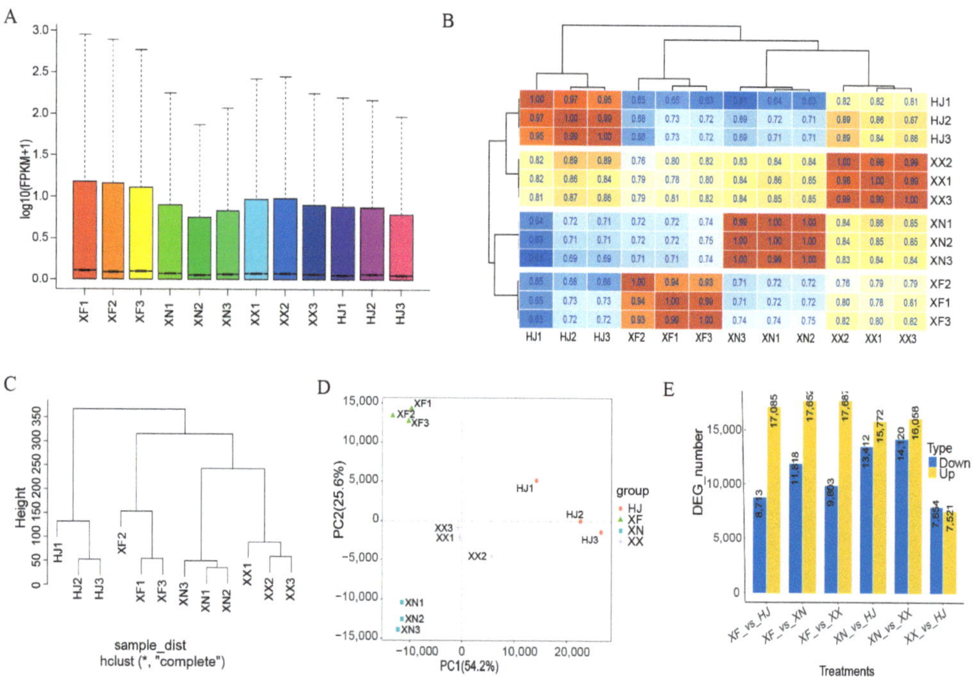


**Figure 4.** RNA-seq data profiles of *L. chinense* var. *rubrum*. Note: XF1–XF3 belong to XF, XN1–XN3 belong to XN, XX1–XX3 belong to XX, and HJ1–HJ3 belong to HJ. (**A**) Box histogram of all transcripts' expressions among all of the samples. Note: FPKM value means fragments per kilobase of transcript per million mapped reads. (**B**) Heatmap of Pearson's correlation coefficient of all transcripts' expressions among all of the samples. (**C**) Clustering tree of transcripts' expressions of all of the samples, which shows the distance between samples. (**D**) PCA of transcripts' expressions of all of the samples, which shows that principal components 1 and 2 represent high cohesion within groups and good separation among different cultivars. (**E**). The number of up- and downregulated different expression transcripts in the different compared combinations. Note: DEG_number is the number of differentially expressed genes and transcripts.

Differences in transcripts were analyzed in order to determine genes that may be involved in flower coloring formation among colorful petal cultivars. Furthermore, GO enrichment analysis was used to classify the function of the DETs. The results showed that the main enriched GO terms of DETs in the selected group belong to the 'biological process' (BP), 'cellar component' (CC), and 'molecular function' (MF) GO categories. In the groups of XN vs. HJ (Figure S9, Supplementary Table S23), XN vs. XX (Figure S10, Supplementary Table S24), XF vs. HJ (Figure S11, Supplementary Table S25), XF vs. XN (Figure S12, Supplementary Table S26), XF vs. XX (Figure S13, Supplementary Table S27), and XX vs. HJ (Figure S14, Supplementary Table S28) paired comparisons, the BP had 15, 15, 15, 15, 15, and 15 significantly enriched GO terms, respectively; the CC had 11, 12, 3, 6, 8, and 1 significantly enriched GO terms, respectively; and the MF had 15, 15, 16, 15, 15, and 15 significantly enriched GO terms, respectively. Moreover, the 'biding', 'catalytic activity', 'metabolic process', 'cellular process', and 'organic substance metabolic process' were the top five processes and molecular functions of significantly enriched GO terms.

The KEGG enrichment analysis of the differential expressed transcripts was analyzed for their biological functions (Supplementary Tables S23–S28, Figure S15). The results indicated that the 'Phenylpropanoid biosynthesis', 'Flavonoid biosynthesis', 'Carotenoid biosynthesis', 'Monoterpenoid biosynthesis', 'Phenylalanine, tyrosine, and tryptophan biosynthesis', 'Photosynthesis-antenna proteins', and 'Plant hormone signal transduction' enrichment pathways were obtained in the six paired comparisons, which demonstrated that those DETs had significant effects on petal coloring. Furthermore, phenylpropanoids and flavonoids are the most significant contributors to flower and leaf coloring. Therefore, the DETs involved in 'Phenylpropanoid biosynthesis' and 'Flavonoid biosynthesis' might play critical roles in different petal coloring formations within *L. chinenses* var. *rubrum* and *L. chinenses*.

### 3.6. Expression Analysis Indicates Flavonoid Compounds' Biosynthesis and Accumulation

Flavonoids are the primary pigments in plants, which can be classified into several subgroups, such as flavones, flavonols, anthocyanins, et al. In total, 433 DETs were detected in 6 compared subgroups (Figure S16). The KEGG enrichment analysis indicated that the "flavonoid biosynthesis" pathway was one of the most representative pathways in XN vs. HJ (Supplementary Table S29), XN vs. XX (Supplementary Table S30), XF vs. HJ (Supplementary Table S31), XF vs. XN (Supplementary Table S32), XF vs. XX (Supplementary Table S33), and XX vs. HJ (Supplementary Table S34). Based on the KEGG enrichment analysis of DCMs and DETs, the flavonoid biosynthetic pathway of *L. chinense* var. *rubrum* was constructed according to the KEGG public database (Figure 5A). Eight flavonoid compounds were significantly different among the four samples (Figure 5A). The XX had the highest relative contents of myricetin, Naringenin-7-*O*-glucoside, Glycitin, Tricetin, and Delphinidin, while the XN and XF had higher relative contents of Apiin, Kaempferol, and Pelargonidin. Furthermore, 73 DETs were related to the biosynthesis of flavonoid compounds, including eight *LcCYP73A* (encoding trans-cinnamate 4-monooxygenase), twenty *LcCHS* (encoding chalcone synthase), three *LcCHI* (encoding chalcone isomerase), nineteen *LcF3H* (encoding naringenin 3-dioxygenase), eight *LcDFR* (encoding bifunctional dihydroflavonol 4-reductase or flavanone 4-reductase), and four *LcANS* (encoding anthocyanidin reductase) (Figure 5B). Additionally, all of the differentially expressed *LcDFR* as well as *LcANS* genes were highly expressed in XF or XN. Moreover, there was a strong correlation between the expression of flavonoid-biosynthesis-related gene DETs and the KEGG enrichment DCMs of flavonoids (Figure S17).

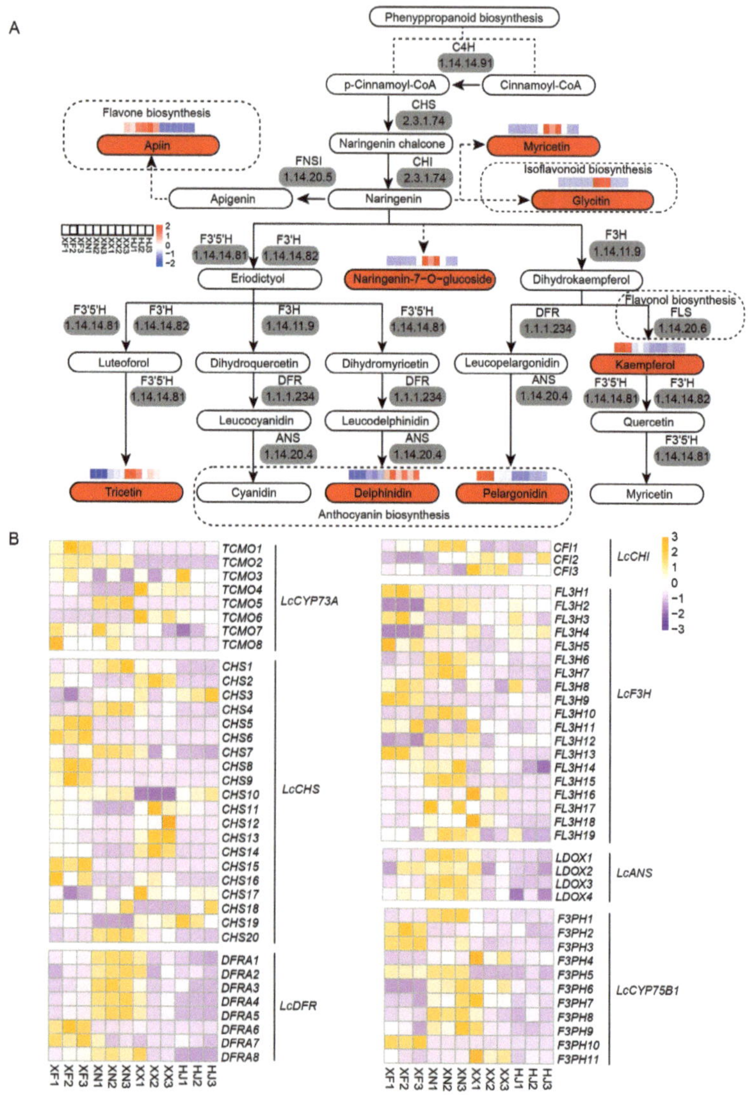


**Figure 5.** Simplified representation of flavonoid metabolism and heat map produced by significantly different genes related to the flavonoid synthetic of *L. chinense* var. *rubrum*. (**A**) Simplified representation of flavonoid metabolism. Red rectangular boxes represent significantly changed metabolism according to a KEGG enrichment analysis. (**B**) Heat map produced by significantly different genes related to the flavonoid synthetic. Additionally, the gray rectangular boxes represent the enzyme coded by related genes. The orange square represent upregulated, and the purple square represent downregulated. The white square represent the gene related to the flavonoid synthetic that had not significant changed.

### 3.7. Co-Expression Analysis for the Investigation of Anthocyanin Biosynthesis

Anthocyanins were the primary coloring pigments that provided the orange, red, blue, and purple colors in flowers, leaves, fruits, seeds, and other tissues (Bueno et al., 2012). Anthocyanins were the essential products in the flavonoid pathway biosynthesis. Moreover, the colored unstable anthocyanidins (pelargonidin, cyanidin, and delphinidin) were converted into stable anthocyanins (pelargonidin-3-*O*-glucoside, cyanidin-3-*O*-glucoside, and delphinidin-3-glucoside) and their further modifications (such as acylation, glycosylation, and methylation) (Liu et al., 2021b). The KEGG enrichment analysis indicated that eight anthocyanins, including pelargonidin, delphinidin, cyanidin-3-*O*-glucoside, cyanidin-3,5-*O*-glucoside, petunidin-3-*O*-glucoside, peonidin-3-*O*-sophoroside-5-*O*-glucoside, malvidin-3-acetyl-5-*O*-diglucoside, and Delphinidin-3-*O*-glucoside, were significantly different among the four samples (Supplementary Table S1, Figure 6A). Furthermore, the relative concentrations of pelargonidin, peonidin-*O*-3-sophoroside-5-*O*-glucoside, and Malvidin-3-acetyl-5-*O*-diglucoside were consistent with the petal coloring of all of the samples (Supplementary Table S1, Figure 6A). Twenty DET-related genes of anthocyanin biosynthetic were identified (Figure 6B). The results indicated that the genes *UFOG1*, *UFOG7*, *UFOG9*, *UFOG12*, and *3AT1_2* were relatively higher expressed in the darker petals of the samples.

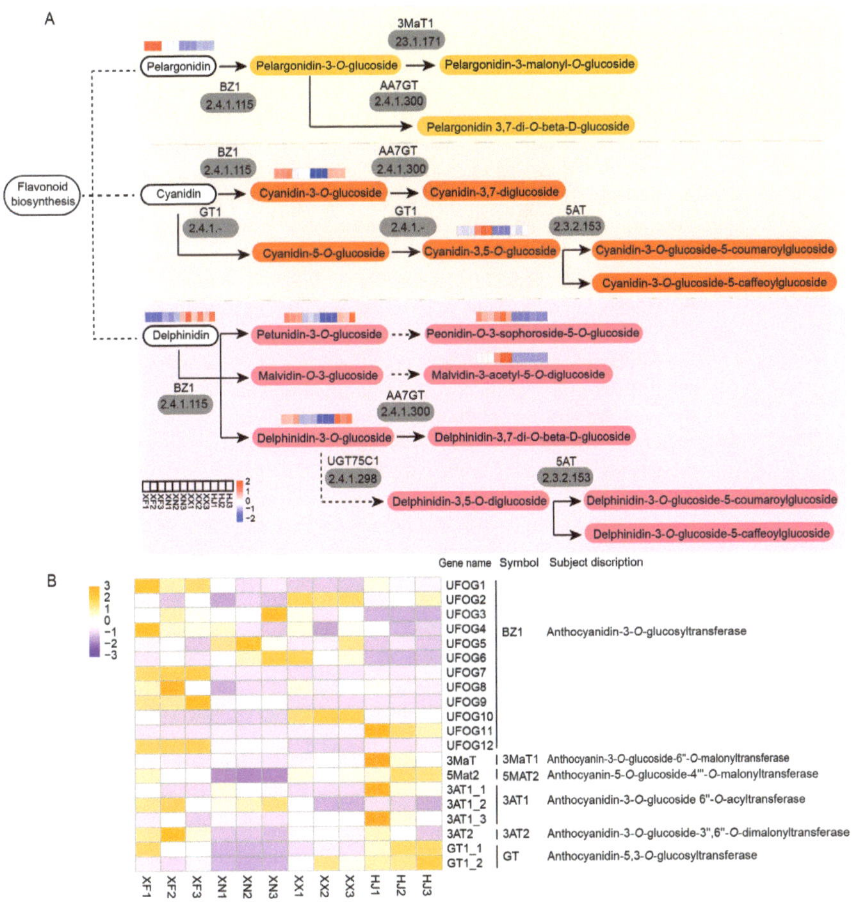


**Figure 6.** Simplified representation of anthocyanins' metabolism and heat map of genes related to *L. chinense* var. *rubrum*. (**A**) Simplified representation of flavonoid metabolism. (**B**) Heat map produced by significantly different genes related to flavonoid synthetic. Additionally, the gray rectangular boxes represent the enzymes coded by related genes. The orange square represents upregulated, and the purple square represents downregulated. The white square represents the gene related to flavonoid synthetic that had not significant changed.

To explore the roles of the DETs involved in the accumulation of anthocyanins, a correlation analysis of the relative expression profiles of the DETs and the DCMs of anthocyanins mentioned above was performed (Figure S18). As a result, the expressions of *UFOG1, UFOG7, UFOG9, UFOG12*, and *3AT1_2* were positively correlated with the contents of cyanindin-3-*O*-glucoside, delphinidin-3-*O*-glucoside, petunidin-3-*O*-glucoside, pelargonidin, malvidin-3-acetyl-5-*O*-diglucoside, peonidin, and peonidin-3-sophoroside-5-*O*-glucoside, and negatively correlated with the content of delphinidin. In addition, *UFOG2, UFOG10*, and *GT1_2* positively correlated with the content of delphinidin.

## 4. Discussion

What flowers do we like? Flower color is one of the most important factors influencing flowers' beauty [12]. Flavonoids are common pigment components that present a broad spectrum of colors, from pale yellow to blue to purple, especially the case in petals. Flavonoids are a large group of plant biosynthetic compounds that can be classified as anthocyanins, flavones, flavanols, flavanones, isoflavones, or other flavonoid compounds [77]. The anthocyanin in leaves is the primary coloring pigment in *L. chinense* var. *rubrum* [6]; the total anthocyanins of PL are much higher than ML and GL, and the lowest is GL [7]. In the present study, we found that the anthocyanin contents of red *L. chinense* var. *rubrum* flowers are higher than those of white *L. chinense* flowers. Moreover, the content of anthocyanins is a vital factor influencing their flower coloring. XN had the highest total anthocyanins (XNNC, 264.96 mg/g of FW) and the redness color of $a^*$ value ($a^*$, 4.52), while XX had the lowest total anthocyanins (23.03 mg/g of FW) and lightness ($L^*$, 80.43). It might be the main reason for the transformation from white to pink to purple. These results also showed that *L. chinense* var. *rubrum* was a critical resource of anthocyanins compared to *Asparagus officinalis* (22.04 mg/g of FW) [52], *Vaccinium corymbosum* (about 1.8 mg/g of FW) [78], *Lycoris longituba* (1.25 mg/g of FW) [79], and other horticultural plants.

Flavonoids are some of the essential pigments in many ornamental plant petals, and the composition of flavonoids' composition may vary among different color petals of the same species. The white chrysanthemum flowers only contain flavones and flavonols, and the pink ones mainly contain anthocyanins, flavones, and flavonols [80]. The derivatives of delphinidin and cyanidin were more complicated in the red group of water lily than in others [81]. In total, 15 anthocyanins and 20 flavonols were identified in primula vulgaris cultivars, while peonidin-type anthocyanins, cyanindi-3-*O*-glucoside, and dephininidin-3,5-di-*O*-glucoside-3′-caffeic ester were accumulated in pink flowers [36]. We found that 207 flavonoid metabolites were identified in the 4 cultivars (Table S1 and Figure 2A), and they were mainly involved in 39 flavonols, 7 isoflavones, 25 flavonoids, 79 flavones, 20 flavanones, 19 polyphenols, 15 anthocyanins, and 3 proanthocyanins in *L. chinense* var. *rubrum* and *L. chinense*; it is similarly the case with the leaves of *L. chinense* var. *rubrum* [7]. About 168 flavonoids were common in the 4 samples, indicating no apparent alternatives of flavonoids among the samples of *Loropetalum* species. The same result was found in Nicotiana species [82], which might be the same species' close genetic background. Eight flavonols, one flavone, and three anthocyanins, reported for the first time in *Loropetalum* species, were identified in the petals of HJ, XF, and XN. These DCMs had a close correlation with the petal colorings of *Loropetalum* species. Furthermore, the enrichment analysis showed that these DCMs were identified in the metabolic pathways of 'Biosynthesis of secondary metabolites' (ko01110), 'Flavonoid biosynthesis' (ko00941), 'Anthocyanin biosynthesis' (ko00942), 'Isoflavonoid biosynthesis' (ko00943), and 'Flavone and flavonol biosynthesis' (ko00944).

The structural and functional genomic analyses provide good insights into discovering the metabolic processing pathways in plants. The full-length transcripts, based on the SMRT sequencing stratagem that improves genome and transcriptome assembly, are essential for structural, functional, and comparative genomic studies [83–85]. The NGS strategies for transcriptomes were a convenient tool for correcting the SMRT sequencing errors reported in many plants [43,86]. In this study, we obtained 46.16 GB of raw data via

PacBio sequencing, with an average length of 29,523 bp and N50 of 51,440 bp. After correcting the SMRT reads with NGS data and CD-HIT software, 171,783 high-quality nonredundant transcripts were produced. This is a much larger number than *Carthamus tincorius* (79,926 transcripts) [87], *Litchi chinense* (50,808 transcripts) [88], *Camellia oleifera* (40,143 transcripts) [89], and *Lolium multiflorum* (72,722 transcripts) (Chen et al., 2018). These transcripts were then annotated with the NR, KOG, GO, NT, Pfam, Swiss-Port, and KEGG databases for gene function annotation and metabolic pathway analysis. As a result, 144,893 transcripts were functionally annotated, and the most enriched were 'Metabolism', including phenylpropanoid biosynthesis, flavonoid biosynthesis, isoflavonoid biosynthesis, and flavone as well as flavonol biosynthesis pathways (Figures 3A,B and S5–S8). Thus, SMRT sequencing, combined with an NGS strategy, established an accurate abundant reference transcriptome of *L. chinense* var. *rubrum* for the first time.

Cultivar-specific flavonoids in *L. chinense* var. *rubrum* might be associated with the differential expression of essential biosynthetic genes. Flavonoid biosynthesis and anthocyanin biosynthesis genes have been well summarized in some reviews [22,42,44,85]. The transcriptomes are based on an NGS strategy as a convenient tool for quantitative gene expression data in plants. This study used four cultivars with twelve samples for transcript qualification. All of the differentially expressed genes were annotated with high-quality full-length nonredundant reference transcriptomes. The DETs' enrichment, analyzed with KEGG annotation, demonstrated that they were involved in 'Phenylpropanoid biosynthesis' and 'Flavonoid biosynthesis'. Combined with the DCMs analyzed, eight *LcCYP73A* (encoding trans-cinnamate 4-monooxygenase), twenty *LcCHS* (encoding chalcone synthase), three *LcCHI* (encoding chalcone isomerase), nineteen *LcF3H* (encoding naringenin 3-dioxygenase), eight *LcDFR* (encoding bifunctional dihydroflavonol 4-reductase or flavanone 4-reductase), four *LcANS* (encoding anthocyanidin reductase), and twenty genes of anthocyanin biosynthetic were identified in our study (Figures 5A,B and 6A,B). Moreover, the correlation of DCMs and flavonoid-pathway (including the anthocyanin pathway)-related genes were also discovered. The genes *UFOG1*, *UFOG7*, *UFOG9*, *UFOG12*, and *3AT1_2* were expressed relatively higher in the darker petals of samples. Furthermore, *UFOG2*, *UFOG10*, and *GT1_2* were positively correlated with delphinidin content.

## 5. Conclusions

For the first time, this study reports on the intergrade application of phenotypes, metabolomics, and transcriptomics to elucidate the flower coloring mechanisms of *L. chinense* var. *rubrum*. The different pigment combinations and their accumulations were the crucial factors caused by the different petal coloration in *L. chinense* var. *rubrum*. Most flavonoids had significantly different contents among the four cultivars for close genetic backgrounds. Additionally, the relative contents of Kaempferide, Apiin, Glycitin, Tricetin, Naringenin 7-*O*-glucoside, pelargonidin, delphinidin, cyanidin-3-*O*-glucoside, cyanidin-3,5-*O*-glucoside, petunidin-3-*O*-glucoside, peonidin-3-*O*-sophoroside-5-*O*-glucoside, malvidin-3-acetyl-5-*O*-diglucoside, and Delphinidin-3-*O*-glucoside were the main differential contents of components in *L. chinense* var. *rubrum*. An accurate abundant referent transcriptome was established by combining SMRT with an NGS strategy. The flavonoid biosynthesis genes, such as *LcCYP73A*, twenty *LcCHS*, three *LcCHI*, nineteen *LcF3H*, eight *LcDFR*, four *LcANS*, four UFOGs, and one *3AT1_2*, correlated with the specific flavonoids' high determinations of the flower color of *L. chinense* var. *rubrum*. Together, these findings offer novel insights into the molecular basis for flavonoid biosynthesis in *L. chinense* var. *rubrum*, and could serve as the basis for future research on the selective breeding of colorful ornamental plants.

**Supplementary Materials:** The following supporting information can be downloaded at: https://www.mdpi.com/article/10.3390/agronomy13051296/s1, Figure S1: The metabolism view map of the significant metabolism pathways of the four different cultivars of *L. chinense* and *L. chinense* var. *rubrum*. (A) Significantly changed pathways on the enrichment of HJ VS XN. (B) Significantly changed pathways on the enrichment of HJ vs. XF. (C) Significantly changed pathways on the enrichment of HJ vs. XX. (D) Significantly changed pathways on the enrichment of XN vs. XF. (E) Significantly changed pathways on the enrichment of XN vs. XX. (F) Significantly changed pathways on the enrichment of XN vs. XX. (G) Significantly changed pathways on the enrichment of XF vs. XX.; Figure S2: The top 20 differential metabolites among the different cultivars of *L. chinense* var. *rubrum*. (A) The top 20 changed metabolites of HJ vs. XN. (B) The top 20 changed metabolites of HJ vs. XF. (C) The top 20 changed metabolites of HJ vs. XX. (D) The top 20 changed metabolites of XN vs. XF. (E) The top 20 changed metabolites of XN VS XX. (F) The top 20 changed metabolites of XF vs. XX; Figure S3: The annotation static results of full-length nonredundant final transcripts to NR databases in different species; Figure S4: The classification map of full-length nonredundant final transcripts to the GO database; Figure S5: The classification map of full-length nonredundant final transcripts to the GO database; Figure S6: The classification chart of the annotation of full-length nonredundant final transcripts to the KEGG metabolic pathway; Figure S7: The annotation and prediction of the CDS, transcription factors, and LncRNAs of *L. chinense* var. *rubrum*. (A) The number, percentage, and length of CDS predicted by the final transcripts of *L. chinense* var. *rubrum*. (B) The top 30 transcription factor families predicted by the final transcripts of *L. chinense* var. *rubrum*. (C) Scatter diagram of LncRNAs and mRNA length distributions of *L. chinense* var. *rubrum*. (D) Annotation of full-length noncoding final transcripts predicted by cnci, pfam, plek and cpc software; Figure S8: The number of up- and downregulated different expression LncRNAs in the different compared combinations; Figure S9: GO enrichment analysis of DETs between the XN and HJ groups; Figure S10: GO enrichment analysis of DETs between the XN and XX groups; Figure S11: GO enrichment analysis of DETs between the XF and HJ groups; Figure S12: GO enrichment analysis of DETs between the XF and XN groups; Figure S13: GO enrichment analysis of DETs between the XF and XX groups; Figure S14: GO enrichment analysis of DETs between the XX and HJ groups; Figure S15: The top 20 enriched KEGG pathways of the differentially expressed genes (DEGs) among the different cultivars of *L. chinense* var. *rubrum*. (A) The enrichment pathway of XN vs. HJ. (B) The enrichment pathway of XN vs. XX. (C) The enrichment pathway of XF vs. HJ. (D) The enrichment pathway of XF vs. XN. (E) The enrichment pathway of XF vs. XX. (F) The enrichment pathway of XX vs. HJ; Figure S16: The Venn diagram of DETs between different comparisons; Figure S17: Correlation analyses of differentially expressed transcripts (DETs) involved in flavonoids; Figure S18: Correlation analyses of differentially expressed transcripts (DETs) involved in anthocyanins; Table S1: Type and content of flavonoids in four *Loropetalum* cultivars; Table S2: The differentially accumulated metabolites in the group of HJ vs. XN; Table S3: The differentially accumulated metabolites in the group of HJ vs. XF; Table S4: The differentially accumulated metabolites in the group of HJ vs. XF; Table S5: The differentially accumulated metabolites in the group of XN vs. XF; Table S6: The differentially accumulated metabolites in the group of XN vs. XX; Table S7: The differentially accumulated metabolites in the group of XF vs. XX; Table S8: The differentially accumulated flavonoid metabolites in all groups of *L. chinense* and *L. chinense* var. *rubrum*; Table S9: The differential concentration of flavonoid metabolites analysis in the group of HJ vs. XN; Table S10: The differential concentration of flavonoid metabolites analysis in the group of HJ vs. XF; Table S11: The differential concentration of flavonoid metabolites analysis in the group of HJ vs. XX; Table S12: The differential concentration of flavonoid metabolites analysis in the group of XN vs. XF; Table S13: The differential concentration of flavonoid metabolites analysis in the group of XN vs. XX; Table S14: The differential concentration of flavonoid metabolites analysis in the group of XF vs. XX; Table S15: The compared transcriptome of HJ vs. XN on KEGG metabolic pathway annotation; Table S16: The compared transcriptome of HJ vs. XN on KEGG metabolic pathway annotation; Table S17: The compared transcriptome of HJ vs. XX on KEGG metabolic pathway annotation; Table S18: The compared transcriptome of XN vs. XF on KEGG metabolic pathway annotation; Table S19: The compared transcriptome of XN vs. XX on KEGG metabolic pathway annotation; Table S20: The compared transcriptome of XF vs. XX on KEGG metabolic pathway annotation; Table S21: Quality statistics of filtered reads; Table S22: Major indicators of the full-length transcriptome of *Loropetalum chinense* var. *rubrum*; Table S23: The GO enrichment analysis of DETs between the XN and HJ groups; Table S24: The GO enrichment analysis of DETs between the XN and XX groups; Table S25: The GO enrichment analysis of DETs between the XF and HJ groups; Table S26: The GO enrichment analysis of DETs between the XF and XN groups; Table S27: The GO enrichment analysis of DETs between the XF and XX groups; Table S28: The GO enrichment analysis of DETs between the XX and HJ groups; Table S29: The KEGG enrichment analysis of DETs between the XN and HJ groups; Table S30: The KEGG enrichment analysis of DETs between the XN and XX groups; Table S31: The KEGG enrichment analysis of DETs between the XF and HJ groups; Table S32: The KEGG enrichment analysis of DETs between the XF and XN groups; Table S33: The KEGG enrichment analysis of DETs between the XF and XX groups; Table S34: The KEGG enrichment analysis of DETs between the XX and HJ groups.

**Author Contributions:** Conceptualization, M.C. and X.Y.; methodology, X.Z.; software, D.Z. (Damao Zhang); validation, Y.L. (Yang Liu), L.L. and L.Z.; formal analysis, L.Z.; investigation, X.Y.; resources, D.Z. (Donglin Zhang); data curation, M.S.; writing—original draft preparation, Y.L. (Yanlin Li); writing—review and editing, Y.L. (Yanlin Li); visualization, M.C.; supervision, Y.L. (Yanlin Li); project administration, X.X.; funding acquisition, X.Y. All authors have read and agreed to the published version of the manuscript.

**Funding:** This research was funded by the Forestry Science and Technology Innovation Foundation of Hunan Province for Distinguished Young Scholarship, grant number XLKJ202205; the Open Project of Horticulture Discipline of Hunan Agricultural University, grant number 2021YYXK001; the fund of the Changsha Municipal Science and Technology Bureau, grant number KQ2202227; the Key Project of the Hunan Provincial Education Department, grant number 22A0155; the Forestry Bureau for Industrialization management of Hunan Province, grant number 2130221; Hunan Provincial Education Department Teaching Reform Project, grant number 2021JGYB101; and Hunan Agricultural University Teaching Reform Research Project, grant number XJJG-2020-071.

**Data Availability Statement:** Data are contained within the article and the Supplementary Materials.

**Acknowledgments:** Lili Xiang, Yujie Yang, and Lu Xu were acknowledged to provide technical supports. Yong Song was acknowledged for funding support.

**Conflicts of Interest:** The authors declare no conflict of interest.

## References

1. Zhang, X.; Zhang, D.; Zhang, L.; Wang, X.; Xiong, X.; Gan, D.; Yu, X.; Li, Y. The Whole Genome Analysis of *Loropetalum chinense* var. *Rubrum*. *Mol. Plant Breed.* **2020**, *18*, 7023–7029.
2. Shi, F. Comparison and Application of *Lorpetalum chinense* Var. *Rubrum* and *Photinia Fraseryin* Landscape. *J. Hengyang Norm. Univ.* **2016**, *37*, 106–111.
3. Shao, Z.; Hou, W.; Long, X.; Yang, G.; Chen, C.; Zeng, X.; Hou, B.; Yu, G.; Wu, W. The formation and development of the geography symbol product *Loropetalum chinense* var. *rubrum* in Liuyang City. *Hunan For. Sci. Technol.* **2007**, *2*, 71–73.
4. Zhang, D.; Cai, W.; Zhang, X.; Li, W.; Zhou, Y.; Chen, Y.; Mi, Q.; Jin, L.; Xu, L.; Yu, X.; et al. Different pruning level effects on flowering period and chlorophyll fluorescence parameters of *Loropetalum chinense* var. *rubrum*. *PeerJ* **2022**, *10*, 1–17. [CrossRef]
5. Zhang, L.; Yu, X.; Zhang, X.; Zhang, D.; Li, W.; Xiang, L.; Yang, Y.; Li, Y.; Xu, L. Phenotypic Diversity Analysis of the Progeny Variation of a 'Mosaic Leaf' *Loropetalum chinense* var. *rubrum* Based on Flower Organ Characteristics. *Diversity* **2022**, *14*, 913. [CrossRef]
6. Chen, Q.; Cai, W.; Zhang, X.; Zhang, D.; Li, W.; Xu, L.; Yu, X.; Li, Y. The Comparative Studies on Phytochemicals of Leaf Coloration of *Loropetalum chinense* var. *rubrum*. *Acta Hortic. Sin.* **2021**, *48*, 1969–1982. [CrossRef]
7. Zhang, X.; Zhang, L.; Zhang, D.; Su, D.; Li, W.; Wang, X.; Chen, Q.; Cai, W.; Xu, L.; Cao, F.; et al. Comprehensive analysis of metabolome and transcriptome reveals the mechanism of color formation in different leave of *Loropetalum chinense* var. *rubrum*. *BMC Plant Biol.* **2023**, *23*, 133. [CrossRef]
8. Iwashina, T. Contribution to Flower Colors of Flavonoids Including Anthocyanins: A Review. *Nat. Prod. Commun.* **2015**, *10*, 529–544. [CrossRef]
9. Tanaka, Y.; Brugliera, F. Flower Colour. In *Annual Plant Reviews*; Wiley: Hoboken, NJ, USA, 2018; Volume 20, pp. 201–239.
10. Goto, T. Structure, stability and color variation of natural anthocyanins. In *Fortschritte der Chemie Organischer Naturstoffe/Progress in the Chemistry of Organic Natural Products*; Springer: Berlin/Heidelberg, Germany, 1987; Volume 52, pp. 113–158.
11. Koes, R.E.; Van Blokland, R.; Quattrocchio, F.; Van Tunen, A.J.; Mol, J.N. Chalcone synthase promoters in petunia are active in pigmented and unpigmented cell types. *Plant Cell* **1990**, *2*, 379–392. [CrossRef]
12. Hůla, M.; Flegr, J. What flowers do we like? The influence of shape and color on the rating of flower beauty. *PeerJ* **2016**, *4*, e2106. [CrossRef]
13. Yin, X.; Wang, T.; Zhang, M.; Zhang, Y.; Irfan, M.; Chen, L.; Zhang, L. Role of core structural genes for flavonoid biosynthesis and transcriptional factors in flower color of plants. *Biotechnol. Biotechnol. Equip.* **2021**, *35*, 1214–1229. [CrossRef]
14. Tanaka, Y.; Brugliera, F.; Chandler, S. Recent progress of flower colour modification by biotechnology. *Int. J. Mol. Sci.* **2009**, *10*, 5350–5369. [CrossRef]
15. Tanaka, Y.; Brugliera, F. Flower colour and cytochromes P450. *Philos. Trans. R. Soc. Lond. B Biol. Sci.* **2013**, *368*, 20120432. [CrossRef]
16. Tanaka, Y.; Sasaki, N.; Ohmiya, A. Biosynthesis of plant pigments: Anthocyanins, betalains and carotenoids. *Plant J.* **2008**, *54*, 733–749. [CrossRef]

17. Liang, C.-Y.; Rengasamy, K.P.; Huang, L.-M.; Hsu, C.-C.; Jeng, M.-F.; Chen, W.-H.; Chen, H.-H. Assessment of violet-blue color formation in Phalaenopsis orchids. *BMC Plant Biol.* **2020**, *20*, 212. [CrossRef]
18. Diretto, G.; Jin, X.; Capell, T.; Zhu, C.; Gomez-Gomez, L. Differential accumulation of pelargonidin glycosides in petals at three different developmental stages of the orange-flowered gentian (*Gentiana lutea* L. var. *aurantiaca*). *PLoS ONE* **2019**, *14*, e0212062. [CrossRef]
19. Liu, W.; Feng, Y.; Yu, S.; Fan, Z.; Li, X.; Li, J.; Yin, H. The flavonoid biosynthesis network in plants. *Int. J. Mol. Sci.* **2021**, *22*, 12824. [CrossRef]
20. Haslam, E. *Practical Polyphenolics: From Structure to Molecular Recognition and Physiological Action*; Cambridge University Press: Cambridge, UK, 1998.
21. Winkel-Shirley, B. Molecular genetics and control of anthocyanin expression. *Adv. Bot. Res.* **2002**, *37*, 75–94.
22. Fraser, C.M.; Chapple, C. The phenylpropanoid pathway in Arabidopsis. *Arab. Book* **2011**, *9*, e0152. [CrossRef]
23. Schröder, G. Stilbene and chalcone synthases: Related enzymes with key functions in plant-specific pathways. *Z. Für Nat. C* **1990**, *45*, 1–8. [CrossRef]
24. Shirley, B.W.; Kubasek, W.L.; Storz, G.; Bruggemann, E.; Koornneef, M.; Ausubel, F.M.; Goodman, H.M. Analysis of Arabidopsis mutants deficient in flavonoid biosynthesis. *Plant J.* **1995**, *8*, 659–671. [CrossRef] [PubMed]
25. van Tunen, A.J.; Mur, L.A.; Recourt, K.; Gerats, A.; Mol, J. Regulation and manipulation of flavonoid gene expression in anthers of petunia: The molecular basis of the Po mutation. *Plant Cell* **1991**, *3*, 39–48. [PubMed]
26. Wisman, E.; Hartmann, U.; Sagasser, M.; Baumann, E.; Palme, K.; Hahlbrock, K.; Saedler, H.; Weisshaar, B. Knock-out mutants from an En-1 mutagenized Arabidopsis thaliana population generate phenylpropanoid biosynthesis phenotypes. *Proc. Natl. Acad. Sci.* **1998**, *95*, 12432–12437. [CrossRef] [PubMed]
27. Holton, T.A.; Brugliera, F.; Lester, D.R.; Tanaka, Y.; Hyland, C.D.; Menting, J.G.; Lu, C.-Y.; Farcy, E.; Stevenson, T.W.; Cornish, E.C. Cloning and expression of cytochrome P450 genes controlling flower colour. *Nature* **1993**, *366*, 276–279. [CrossRef] [PubMed]
28. Winkel-Shirley, B. Flavonoid biosynthesis. A colorful model for genetics, biochemistry, cell biology, and biotechnology. *Plant Physiol.* **2001**, *126*, 485–493. [CrossRef]
29. Schoenbohm, C.; Martens, S.; Eder, C.; Forkmann, G.; Weisshaar, B. Identification of the Arabidopsis thaliana flavonoid 3′-hydroxylase gene and functional expression of the encoded P450 enzyme. *Bio. Chem.* **2000**, *381*, 749–753. [CrossRef]
30. Shirley, B.W.; Hanley, S.; Goodman, H.M. Effects of ionizing radiation on a plant genome: Analysis of two Arabidopsis transparent testa mutations. *Plant Cell* **1992**, *4*, 333–347.
31. Hichri, I.; Barrieu, F.; Bogs, J.; Kappel, C.; Delrot, S.; Lauvergeat, V. Recent advances in the transcriptional regulation of the flavonoid biosynthetic pathway. *J. Exp. Bot.* **2011**, *62*, 2465–2483. [CrossRef]
32. Zhang, Y.; Chu, G.; Hu, Z.; Gao, Q.; Cui, B.; Tian, S.; Wang, B.; Chen, G. Genetically engineered anthocyanin pathway for high health-promoting pigment production in eggplant. *Mol. Breed.* **2016**, *36*, 1–14. [CrossRef]
33. Tohge, T.; Nishiyama, Y.; Hirai, M.Y.; Yano, M.; Nakajima, J.; Awazuhara, M.; Inoue, E.; Takahashi, H.; Goodenowe, D.B.; Kitayama, M. Functional genomics by integrated analysis of metabolome and transcriptome of Arabidopsis plants over-expressing an MYB transcription factor. *Plant J.* **2005**, *42*, 218–235. [CrossRef]
34. Veitch, N.C.; Grayer, R.J. Flavonoids and their glycosides, including anthocyanins. *Nat. Prod. Rep.* **2008**, *25*, 555–611. [CrossRef]
35. Wei, X.; Yu, X.; Chen, J.; Fu, H.; Hu, B.; Chen, Y.; Da, L. Cloning and Sequence Analyzing of Chalcone Synthase Gene in *Loropetalum chinense* var. *rubrum*. *Chin. Agric. Sci. Bull.* **2013**, *29*, 24–28.
36. Li, C.H.; Duo, D.Y.; Liao, X.S.; Zhang, Y.B. Cloning and Bioinformatic Analysis of Two Chalcone Synthases from *Loropetalum chinense* var. *Rubrum*. *J. Hunan Univ. Technol.* **2020**, *34*, 92–98.
37. Rong, D.; Zhang, X.; Pan, T.; Wang, J.; Yang, G.; Zhang, B. Cloning, Expression and Transformation of LcFLS1 Gene from *Loropetalum chinense* var. *rubrum*. *Acta Bot. Boreali-Occident. Sin.* **2019**, *28*, 1877–1887.
38. Zhang, B.Y.; Li, C.H.; Liu, X.; Liao, X.S.; Rong, D.Y. Cloning and subcellular localization analysis of LcDFR1 and LcDFR2 in *Loropetalum chinense* var. *rubrum*. *J. South. Agric.* **2020**, *51*, 2865–2874.
39. Dong, N.Q.; Lin, H.X. Contribution of phenylpropanoid metabolism to plant development and plant–environment interactions. *J. Integr. Plant Biol.* **2021**, *63*, 180–209. [CrossRef]
40. Ma, D.; Constabel, C.P. MYB repressors as regulators of phenylpropanoid metabolism in plants. *Trends Plant Sci.* **2019**, *24*, 275–289. [CrossRef]
41. Ohtani, M.; Demura, T. The quest for transcriptional hubs of lignin biosynthesis: Beyond the NAC-MYB-gene regulatory network model. *Curr. Opin. Biotechnol.* **2019**, *56*, 82–87. [CrossRef]
42. Nabavi, S.M.; Šamec, D.; Tomczyk, M.; Milella, L.; Russo, D.; Habtemariam, S.; Suntar, I.; Rastrelli, L.; Daglia, M.; Xiao, J. Flavonoid biosynthetic pathways in plants: Versatile targets for metabolic engineering. *Biotechnol. Adv.* **2020**, *38*, 107316. [CrossRef]
43. Xu, W.; Dubos, C.; Lepiniec, L. Transcriptional control of flavonoid biosynthesis by MYB–bHLH–WDR complexes. *Trends Plant Sci.* **2015**, *20*, 176–185. [CrossRef]
44. Sunil, L.; Shetty, N.P. Biosynthesis and regulation of anthocyanin pathway genes. *Appl. Microbiol. Biotechnol.* **2022**, *106*, 1783–1798. [CrossRef] [PubMed]
45. Kiferle, C.; Fantini, E.; Bassolino, L.; Povero, G.; Spelt, C.; Buti, S.; Giuliano, G.; Quattrocchio, F.; Koes, R.; Perata, P. Tomato R2R3-MYB proteins SlANT1 and SlAN2: Same protein activity, different roles. *PLoS ONE* **2015**, *10*, e0136365. [CrossRef] [PubMed]

46. Cao, Y.; Jia, H.; Xing, M.; Jin, R.; Grierson, D.; Gao, Z.; Sun, C.; Chen, K.; Xu, C.; Li, X. Genome-wide analysis of MYB gene family in Chinese bayberry (*Morella rubra*) and identification of members regulating flavonoid biosynthesis. *Front. Plant Sci.* **2021**, *12*, 1244. [CrossRef] [PubMed]
47. Zhang, B.; Xu, X.; Huang, R.; Yang, S.; Li, M.; Guo, Y. CRISPR/Cas9-mediated targeted mutation reveals a role for AN4 rather than DPL in regulating venation formation in the corolla tube of *Petunia hybrida*. *Hortic. Res.* **2021**, *8*, 116. [CrossRef]
48. Lai, Y.-S.; Shimoyamada, Y.; Nakayama, M.; Yamagishi, M. Pigment accumulation and transcription of LhMYB12 and anthocyanin biosynthesis genes during flower development in the Asiatic hybrid lily (*Lilium* spp.). *Plant Sci.* **2012**, *193*, 136–147. [CrossRef]
49. Nesi, N.; Debeaujon, I.; Jond, C.; Pelletier, G.; Caboche, M.; Lepiniec, L. The TT8 gene encodes a basic helix-loop-helix domain protein required for expression of DFR and BAN genes in Arabidopsis siliques. *Plant Cell* **2000**, *12*, 1863–1878. [CrossRef]
50. Qi, F.; Liu, Y.; Luo, Y.; Cui, Y.; Lu, C.; Li, H.; Huang, H.; Dai, S. Functional analysis of the ScAG and ScAGL11 MADS-box transcription factors for anthocyanin biosynthesis and bicolour pattern formation in *Senecio cruentus* ray florets. *Hortic. Res.* **2022**, *9*, uhac071. [CrossRef]
51. Zhang, Y.-L.; Fang, Z.-Z.; Ye, X.-F.; Pan, S.-L. Identification of candidate genes involved in anthocyanin accumulation in the peel of jaboticaba (*Myrciaria cauliflora*) fruits by transcriptomic analysis. *Gene* **2018**, *676*, 202–213. [CrossRef]
52. Dong, T.; Han, R.; Yu, J.; Zhu, M.; Li, Z. Anthocyanins Accumulation and Molecular Analysis of Correlated Genes by Metabolome and Transcriptome in Green and Purple Asparaguses (*Asparagus officinalis* L.). *Food Chem.* **2018**, *271*, 18–28. [CrossRef]
53. Wrolstad, R.E.; Culbertson, J.D.; Cornwell, C.J.; Mattick, L.R. Detection of adulteration in blackberry juice concentrates and wines. *J.-Assoc. Off. Anal. Chem.* **1982**, *65*, 1417–1423. [CrossRef]
54. Leena, S.; Eric, R. LoRDEC: Accurate and efficient long read error correction. *Bioinformatics* **2014**, *30*, 3506–3514. [CrossRef]
55. Fu, L.; Niu, B.; Zhu, Z.; Wu, S.; Li, W. CD-HIT: Accelerated for clustering the next-generation sequencing data. *Bioinform. Oxf.* **2012**, *28*, 3150–3152. [CrossRef]
56. Deng, Y.; Li, J.; Wu, S.; Zhu, Y.; Chen, Y.; He, F. Integrated nr database in protein annotation system and its localization. *Comput. Eng.* **2006**, *32*, 71–73.
57. Li, W.; Jaroszewski, L.; Godzik, A. Tolerating some redundancy significantly speeds up clustering of large protein databases. *Bioinformatics* **2002**, *18*, 77–82. [CrossRef]
58. Finn, R.D.; Alex, B.; Jody, C.; Penelope, C.; Eberhardt, R.Y.; Eddy, S.R.; Andreas, H.; Kirstie, H.; Liisa, H.; Jaina, M. Pfam: The protein families database. *Nucleic Acids Res.* **2014**, *42*, D222–D230. [CrossRef]
59. Tatusov, R.L.; Fedorova, N.D.; Jackson, J.D.; Jacobs, A.R.; Kiryutin, B.; Koonin, E.V.; Krylov, D.M.; Mazumder, R.; Mekhedov, S.L.; Nikolskaya, A.N.; et al. The COG database: An updated version includes eukaryotes. *BMC Bioinform.* **2003**, *4*, 41. [CrossRef]
60. Bairoch, A.; Apweiler, R. The SWISS-PROT protein sequence database and its supplement TrEMBL in 2000. *Nucleic Acids Res.* **2000**, *28*, 45–48. [CrossRef]
61. Minoru, K.; Susumu, G.; Shuichi, K.; Yasushi, O.; Masahiro, H. The KEGG resource for deciphering the genome. *Nucleic Acids Res.* **2004**, *32*, D277.
62. Ashburner, M.; Ball, C.A.; Blake, J.A.; Botstein, D.; Butler, H.; Cherry, J.M.; Davis, A.P.; Dolinski, K.; Dwight, S.S.; Eppig, J.T.; et al. Gene Ontology: Tool for the unification of biology. *Nat. Genet.* **2000**, *25*, 25–29. [CrossRef]
63. Altschul, S.F.; Madden, T.L.; Schffer, A.A.; Zhang, J.; Zhang, Z.; Webb, M.; Lipman, D.J. Gapped BLAST and PSI-BLAST: A new generation of protein database search programs. *Nucleic Acids Res.* **1997**, *25*, 3389. [CrossRef]
64. Buchfink, B.; Xie, C.; Huson, D. Fast and sensitive protein alignment using DIAMOND. *Nat. Methods* **2015**, *12*, 59–60. [CrossRef] [PubMed]
65. Mistry, J.; Finn, R.D.; Eddy, S.R.; Bateman, A.; Punta, M. Challenges in homology search: HMMER3 and convergent evolution of coiled-coil regions. *Nucleic Acids Res.* **2013**, *41*, e121. [CrossRef] [PubMed]
66. Shimizu, K.; Adachi, J.; Muraoka, Y. ANGLE: A sequencing errors resistant program for predicting protein coding regions in unfinished cDNA. *J. Bioinform. Comput. Biol.* **2006**, *4*, 649–664. [CrossRef] [PubMed]
67. Zheng, J.; Sun, H.H.; Rosli, H.G.; Pombo, M.A.; Zhang, P.F. iTAK: A Program for Genome-wide Prediction and Classification of Plant Transcription Factors, Transcriptional Regulators, and Protein Kinases. *Mol. Plant* **2016**, *9*, 1667–1670. [CrossRef]
68. Jin, J.; Tian, F.; Yang, D.-C.; Meng, Y.-Q.; Kong, L.; Luo, J.; Gao, G. PlantTFDB 4.0: Toward a central hub for transcription factors and regulatory interactions in plants. *Nucleic Acids Res.* **2017**, *45*, D1040–D1045. [CrossRef]
69. Kong, L.; Zhang, Y.; Ye, Z.Q.; Liu, X.Q.; Gao, G. CPC: Assess the protein-coding potential of transcripts using sequence features and support vector machine. *Nucleic Acids Res.* **2007**, *35*, W345–W349. [CrossRef]
70. Finn, R.; Coggill, P.; Eberhardt, R.; Eddy, S.; Mistry, J.; Mitchell, A.; Potter, S.; Punta, M.; Qureshi, M.; Sangrador-Vegas, A. The Pfam protein families database: Towards a more sustainable future. *Nucleic Acids Res.* **2016**, *44*, D279–D285. [CrossRef]
71. Freire-Pritchett, P.; Ray-Jones, H.; Della Rosa, M.; Eijsbouts, C.Q.; Orchard, W.R.; Wingett, S.W.; Wallace, C.; Cairns, J.; Spivakov, M.; Malysheva, V. Detecting chromosomal interactions in Capture Hi-C data with CHiCAGO and companion tools. *Nat. Protoc.* **2021**, *16*, 4144–4176. [CrossRef]
72. Langmead, B.; Salzberg, S.L. Fast gapped-read alignment with Bowtie 2. *Nat. Methods* **2012**, *9*, 357–359. [CrossRef]
73. Trapnell, C.; Williams, B.A.; Pertea, G.; Mortazavi, A.; Kwan, G.; van Baren, M.J.; Salzberg, S.L.; Wold, B.J.; Pachter, L. Transcript assembly and quantification by RNA-Seq reveals unannotated transcripts and isoform switching during cell differentiation. *Nat. Biotechnol.* **2010**, *28*, 511–515. [CrossRef]

74. Li, B.; Dewey, C.N. RSEM: Accurate transcript quantification from RNA-Seq data with or without a reference genome. *BMC Bioinform.* **2011**, *12*, 323. [CrossRef]
75. Robinson, M.D.; Mccarthy, D.J.; Smyth, G.K. edgeR: A Bioconductor package for differential expression analysis of digital gene expression data. *Bioinformatics* **2010**, *26*, 139–140. [CrossRef]
76. Young, M.D.; Wakefield, M.J.; Smyth, G.K.; Oshlack, A. Gene ontology analysis for RNA-seq: Accounting for selection bias. *Genome Biol.* **2010**, *11*, R14. [CrossRef]
77. Qiu, X.H.; Zhang, Y.J.; Yin, G.F.; Shi, C.Y.; Yu, X.Y.; Zhao, N.J.; Liu, W.Q. Photosynthetic Parameters Inversion Algorithm Study Based on Chlorophyll Fluorescence Induction Kinetics Curve. *Guang Pu Xue Yu Guang Pu Fen Xi = Guang Pu* **2015**, *35*, 2194–2197.
78. Spinardi, A.; Cola, G.; Gardana, C.S.; Mignani, I. Variation of Anthocyanin Content and Profile Throughout Fruit Development and Ripening of Highbush Blueberry Cultivars Grown at Two Different Altitudes. *Front. Plant Sci.* **2019**, *10*, 1045. [CrossRef]
79. Yue, Y.; Liu, J.; Shi, T.; Chen, M.; Wang, L. Integrating Transcriptomic and GC-MS Metabolomic Analysis to Characterize Color and Aroma Formation during Tepal Development in *Lycoris longituba*. *Plants* **2019**, *8*, 53. [CrossRef]
80. Chen, S.M.; Li, C.H.; Zhu, X.R.; Deng, Y.M.; Sun, W.; Wang, L.S.; Chen, F.D.; Zhang, Z. The identification of flavonoids and the expression of genes of anthocyanin biosynthesis in the chrysanthemum flowers. *Biol. Plant.* **2012**, *56*, 458–464. [CrossRef]
81. Zhu, M.; Zheng, X.; Shu, Q.; Li, H.; Zhong, P.; Zhang, H.; Xu, Y.; Wang, L.; Wang, L. Relationship between the composition of flavonoids and flower colors variation in tropical water lily (*Nymphaea*) cultivars. *PLoS ONE* **2012**, *7*, e34335. [CrossRef]
82. Xiao, Q.; Zhu, Y.; Cui, G.; Zhang, X.; Hu, R.; Deng, Z.; Lei, L.; Wu, L.; Mei, L. A Comparative Study of Flavonoids and Carotenoids Revealed Metabolite Responses for Various Flower Colorations between *Nicotiana tabacum* L. and *Nicotiana rustica* L. *Front Plant Sci.* **2022**, *13*, 828042. [CrossRef]
83. Wang, B.; Tseng, E.; Regulski, M.; Clark, T.A.; Hon, T.; Jiao, Y.; Lu, Z.; Olson, A.; Stein, J.C.; Ware, D. Unveiling the complexity of the maize transcriptome by single-molecule long-read sequencing. *Nat. Commun.* **2016**, *7*, 11708. [CrossRef]
84. Abdel-Ghany, S.E.; Hamilton, M.; Jacobi, J.L.; Ngam, P.; Devitt, N.; Schilkey, F.; Ben-Hur, A.; Reddy, A.S. A survey of the sorghum transcriptome using single-molecule long reads. *Nat. Commun.* **2016**, *7*, 11706. [CrossRef] [PubMed]
85. Liu, D.; Chen, L.; Chen, C.; An, X.; Zhang, Y.; Wang, Y.; Li, Q. Full-length transcriptome analysis of *Phytolacca americana* and its congener *P. icosandra* and gene expression normalization in three *Phytolaccaceae* species. *BMC Plant Biol.* **2020**, *20*, 396. [CrossRef] [PubMed]
86. Zhou, C.; Zhu, C.; Tian, C.; Xu, K.; Huang, L.; Shi, B.; Lai, Z.; Lin, Y.; Guo, Y. Integrated volatile metabolome, multi-flux full-length sequencing, and transcriptome analyses provide insights into the aroma formation of postharvest jasmine (*Jasminum sambac*) during flowering. *Postharvest Biol. Technol.* **2022**, *183*, 111726. [CrossRef]
87. Chen, J.; Tang, X.; Ren, C.; Wei, B.; Wu, Y.; Wu, Q.; Pei, J. Full-length transcriptome sequences and the identification of putative genes for flavonoid biosynthesis in safflower. *BMC Genom.* **2018**, *19*, 548. [CrossRef]
88. Zhou, Y.; Chen, Z.; He, M.; Gao, H.; Zhu, H.; Yun, Z.; Qu, H.; Jiang, Y. Unveiling the complexity of the litchi transcriptome and pericarp browning by single-molecule long-read sequencing. *Postharvest Biol. Technol.* **2020**, *168*, 111252. [CrossRef]
89. Gong, W.; Qiling, S.; Ji, K.; Gong, S.; Wang, L.; Chen, L.; Zhang, J.; Yuan, D. Full-Length Transcriptome from *Camellia oleifera* Seed Provides Insight into the Transcript Variants Involved in Oil Biosynthesis. *J. Agric. Food Chem.* **2020**, *68*, 14670–14683. [CrossRef]

*agronomy*

MDPI

*Article*

# Petal Morphology Is Correlated with Floral Longevity in *Paeonia suffruticosa*

**Yingzi Guo [1,2,†], Yongjie Qiu [1,†], Huan Hu [1], Yanli Wang [1], Zhaorong Mi [1], Shulin Zhang [1], Songlin He [1,2,*] and Wenqing Jia [1,*]**

1 College of Horticulture and Landscape Architecture, Henan Institute of Science and Technology, Xinxiang 453003, China
2 College of Landscape Architecture and Art, Henan Agricultural University, Zhengzhou 450002, China
* Correspondence: hsl213@yeah.net (S.H.); jiawq2022@hist.edu.cn (W.J.); Tel.: +86-373-3040989 (S.H.); +86-15936519769 (W.J.)
† These authors contributed equally to this work and are co-first authors.

**Abstract:** Floral longevity (FL) is an important floral functional trait which is critical for flowering plants. FL shows great diversity among angiosperms; however, there is limited information on the mechanisms that influence differences in floral longevity, especially the relationship between petal anatomical traits and floral longevity. We aimed to examine (1) the relationships between petal anatomical traits and FL in tree peony cultivars and (2) the petal anatomical characteristics of longer FL cultivars. Eleven traits of six tree peony cultivars with different FL were investigated, including six water conservation traits (petal thickness, cuticle thickness, number of cell layers, mesophyll thickness, adaxial epidermis thickness and abaxial epidermis thickness), three water supply traits (vein density, number of xlylem vessels and xylem vessel diameter), petal fresh mass and petal dry mass across cultivars. There are significant differences in traits related to water conservation and water supply ability of tree peonies with different FL. Tree peony cultivars with long FL were characterized by the thicker Mesophyll, cuticles, adaxial and abaxial epidermis of the petals. There was a positive correlation between FL and vessel number and vessel diameter. These results suggest that the ability to retain water in flowers is associated with floral longevity. Petal traits related to water conservation and supply, including vein densities, mesophyll thickness, and epidermis thickness, are beneficial for prolonging the flower longevity in tree peonies.

**Keywords:** *Paeonia suffruticosa*; cuticle thickness; petal thickness; vein density; dry mass

**Citation:** Guo, Y.; Qiu, Y.; Hu, H.; Wang, Y.; Mi, Z.; Zhang, S.; He, S.; Jia, W. Petal Morphology Is Correlated with Floral Longevity in *Paeonia suffruticosa*. *Agronomy* **2023**, *13*, 1372. https://doi.org/10.3390/agronomy13051372

Academic Editors: Jinzhi Zhang and Avi Sadka

Received: 28 March 2023
Revised: 10 May 2023
Accepted: 11 May 2023
Published: 14 May 2023



## 1. Introduction

Floral longevity (FL), the period of time between anthesis and floral senescence, plays an important role in the ornamental value of flowering plants. A longer duration of the flowering not only enhances the ability of plants to attract pollinators, but also significantly increases their ornamental appearance, which is critical for ornamental crops. The FL trait shows great diversity in angiosperms [1]. For example, the flowers of *Epiphyllum oxypetalum* and *Oenothera* spp. remain open for only a few hours, while other species, such as *Phalaenopsis aphrodite* and *Cattleya* × *hybrida*, can last for weeks or even several months [1–3]. The factors that affect floral longevity have been a focus of recent theoretical and empirical research [1–4].

Previous studies have found that FL is closely related to biotic and abiotic factors [5–12]. In many flowering plants, flowering senescence is triggered by pollination. Premature artificial or insect pollination shortens FL [11–13], which may be an adaptive trait that minimizes water loss and resource expense incurred by maintaining open flowers [14,15].

FL depends on temperature and water balance and is affected by floral respiration and transpiration [2,16]. A previous study on cut flower preservation found that preservative application can improve the water balance of cut flowers, maintain petal swelling pressure, and extend the life of cut flowers [17]. Oblique trimming or recutting of the stems can increase the water absorption efficiency of cut flowers and prolong FL [18]. Reducing temperature and the content of carbon dioxide and oxygen in the air can reduce plant respiration, thereby slowing down the rate of energy and water loss [19]. For example, hemerocallis (daylily) flowers, affected by temperature, usually last one day and bloom for two days in September and three days in late October [20].

The balance of water in petals is closely related to the epidermal features and anatomical traits of plant petals. Long FL is basically accompanied by thicker adaxial and abaxial epidermis of the flower, and tightly coupled with water conservation in flowers [21]. Cuticle as one of epidermal features not only minimizes water loss and provides protection against desiccation, but also has a role in plant development and environmental interactions [22]. For example, Kirsch et al. [23] reported that species with short leaf longevity have relatively thin cuticles, while species where foliage has a longer leaf longevity tend to possess relatively thicker/denser cuticles with a more robust leaf to maximize net photosynthesis [24]. Flower dry mass is also strongly and positively related to drought tolerance of the flowers and FL [25]. Zhang et al. conducted a systematic study and concluded that there was a positive relationship between FL and floral dry mass per unit area in 11 species of *Paphiopedilum* (slipper orchids) [26]. An adequate water supply is needed during all periods of floral display [27–29]. Therefore, understanding the relationship between the floral water transport system (xylem vessels in veins) and FL may provide new insights into the evolution of flowers [30–32]. At present, studies on longevity in plants focus only on the functional link between leaf longevity and morphology, whereas the mechanisms that influence the FL difference between longer-FL and shorter-FL plants are still unclear.

Tree peony (*Paeonia* sect. *Moutan*) is well known for its high ornamental value, with approximately nine species, two subspecies, and more than 1500 cultivars [33]. Our long-term observation showed that most cultivars of tree peonies with very thin and soft petals have a flowering period of only 3–5 days, but some cultivars with hard petals have a flowering period of 8–10 days. The differences in FL, morphology, habitat, and physiology between long FL cultivars and shorter FL cultivars make them ideal candidates for exploring the association between FL and petal structure in tree peonies.

The current study explores the relationship between epidermal characteristics, anatomical structures and tree peony FL and examines three questions regarding tree peony FL and petal characteristics: (i) Could epidermal cuticle and thickness be associated with FL? It can be hypothesized that that the presence of the epidermal cuticle is associated with a long floral life. (ii) Do cultivars with a higher number of petal cell layers have longer FL? One may expect that petals with a high number of cell layers, which have an enhanced water retention capacity, have a longer FL than petals of cultivars with a low number of cell layers. (iii) Do petal flowers with dense veins and well-developed vascular bundles live longer? It can be hypothesized that petals with denser vasculatures and well-developed vascular bundles have increased FL due to their enhanced ability to transport water, which helps maintain water balance within the petals. The aim of this paper was to investigate the floral structural traits related to water balance and six tree peony FL and to explore the related mechanism that affects the differences in FL between six tree peonies.

## 2. Materials and Methods

### 2.1. Study Site and Species

FL and other floral traits (Table 1) were observed for three cultivars with longer FL ('Souvenir de Maxime Cornu', 'Haihuang', 'Changshoule') and three cultivars with shorter FL ('Sihelian', 'Zhihong' and 'Shanhutai'). Six decade-old tree peony cultivars with different FLs (Figure 1) were planted in the National Peony Garden (112°41″ E, 34°71″ N) in Luoyang city. The weather conditions during the observation and sampling periods were as follows: the highest daily average temperature was 19 °C and the lowest was 10.4 °C, wind speed < 10 km/h with no precipitation.

**Table 1.** Floral traits and floral longevity of six *Paeonia suffruticosa* cultivars.

| Code | Name | Flower Color | Flower Type | FL (Days) |
|---|---|---|---|---|
| 1 | 'Souvenir de Maxime' Cornu' | Yellowish orange | double-petaled | 8.5 ± 1.5 [a] |
| 2 | 'Haihuang' | Yellow | single-petaled | 9.0 ± 1.2 [a] |
| 3 | 'Changshoule' | Light pink | double-petaled | 8.9 ± 1.1 [a] |
| 4 | 'Sihelian' | Pink | single-petaled | 4.0 ± 1.1 [b] |
| 5 | 'Zhihong' | Rouge red | double-petaled | 3.5 ± 1.5 [b] |
| 6 | 'Shanhutai' | Pink | double-petaled | 3.7 ± 1.3 [b] |

Note: Data shown are average ± standard deviation. The data with different lowercase letter indicate significant differences at the 0.05 level, the same below.

**Figure 1.** Flowers of six tree peony cultivars. (**A**) 'Souvenir de Maxime Cornu'; (**B**) 'Hahuang'; (**C**) 'Changshoule'; (**D**) 'Sihelian'; (**E**) 'Zhihong'; (**F**) 'Shanhutai'.

The experiments were carried out at the Henan Horticultural Plant Resources Innovation and Utilization Engineering Research Center in April 2021. The surfaces of epidermal cells on the petals were examined by scanning electron microscopy (SEM). The cross-sectional anatomy of the petals was prepared following the method detailed by Chavarria et al. [34] and Olsen et al. [35] and visualized under a light microscope (LM) [36–38]. Several anatomical characteristics including petal thickness, number of petal cell layers, cuticle thickness, mesophyll thickness, adaxial epidermal thickness, abaxial epidermal thickness, and vessel diameter were measured with the help of the Olympus cellSens standard software [37].

### *2.2. Flower Longevity*

The floral longevity was defined from the beginning of the petals opening slightly and exposing the stamens to the wilting or abscission of the petals. To investigate the FL of a single flower for each cultivar, 8–10 newly emerging floral buds we randomly selected of 3–5 plants per cultivar over three consecutive years (2018–2020). Their individual opening and wilting dates were recorded throughout the flowering season.

### *2.3. Fresh Weight and Dry Weight of Petals*

A 4 cm × 4 cm sample was taken from midway between the midrib and the margin of five petals per species, and the fresh mass of the samples was recorded. Subsequently, these samples were oven-dried at 70 °C for 48 h to obtain petal dry mass [26,39].

### 2.4. Scanning Electron Microscope

For scanning electron microscope observation, the fragments of about 5 mm × 5 mm were taken from the central part of each tree peony petal and glued to the sample table with double-sided black conductive tape without any treatment. The samples were sputtered with gold–palladium. Then, the micromorphological details (stomata, sculptures, and epidermal cells) of samples were observed and photographed under a scanning electron microscope (Quanta 200, FEI Corporation, Hillsboro, OR, USA). The analysis of epidermis sculptures involved the observation and description of the epidermal cells, calculated following the method of Zhou et al. [40].

### 2.5. Light Microscope

Fragments of 1 cm$^2$ were cut from the central part of the outer whorl petals and soaked in distilled water to avoid rapid dehydration. Subsequently, the fragments were cut into 0.1–0.3 mm fragments quickly along the direction perpendicular to the texture of the petal epidermal with razor blade. These fragments were placed in a drop of water on a slide, covered with cover slides, and later observed under a microscope (BX53, Olympus, Tokyo, Japan).

The fragments in the same batch were fixed in formalin–acetic acid–alcohol (FAA) fixing solution (5% formaldehyde, 5% acetic acid and 70% alcohol) at 4 °C for three days, and then rinsed with distilled water for 24 h. The samples were then dehydrated in 50%, 70%, 90%, 95% and 100% graded ethanol solutions, infiltrated in the mixtures of anhydrous ethanol and dichloromethane (3:1, 1:1, 1:3 [V]: [V]) and embedded in a mixed solution of dichloromethane and paraffin wax (3:1, 1:1, 1:3 [V]: [V]). For observations of the anatomical structure of petals, thin (10 μm) sections were obtained using an RM2245 semiautomatic rotary microtome (Leica Microsystems, Nussloch, Germany). These paraffin sections were stained with safranin and fast green dye [41], and observed and photographed under a fluorescence microscope (BX53, Olympus, Tokyo, Japan). Petal thickness, cuticle thickness, adaxial epidermal thickness, abaxial epidermal thickness, and mesophyll thickness of the petals were determined with the help of the Olympus cellSens standard software 1.12 [37]. Vein density was measured as the total length of the vascular tissue per mm$^2$ of petal surface area. The xylem vessel number of each cultivar was the mean of the xylem vessels in 18 veins. Xylem vessel diameter was determined as the average of 18 randomly selected xylem vessel diameters.

### 2.6. Statistical Analysis

To assess the stability of micromorphological details, three petals per flower, two flowers per bush and three bushes per cultivar were analyzed. Petal micromorphology data are the average of at least 18 replicates. The correlation among variables was statistically evaluated using Pearson's correlation coefficient (r) at a $p$ value of 0.05. Duncan's multiple comparisons were used to evaluate the differences in petal micromorphology of six cultivars of tree peonies at a 95% confidence interval ($p < 0.05$). The classification of six tree peony cultivars with different FL was determined based on petal traits using the Principal Component Analysis (PCA) method. Differences in floral traits among six tree peony cultivars were determined by the Mann–Whitney U test. All statistical analyzes for petal traits were performed with the SPSS 19.0 software (SPSS Inc., Chicago, IL, USA). All figures were drawn using the SigmaPlot 14.0 software.

## 3. Results

### 3.1. Flower Longevity

The length of FLs in different cultivars of tree peonies was significantly different ($p < 0.05$) (Table 1). The longest floral longevity was recorded in 'Haihuang' and the shortest was recorded in 'Zhihong'.

### 3.2. Epidermal Morphological Characteristics of Petals

The size and shape of epidermal cells varied among the six cultivars. Epidermal cells viewed from above in three cultivars with long FL exhibited an elongated cell shape, and the epidermal cells were arranged in parallel, with a larger cell area (Figures 2A–C and 3A–C). Rectangular cell shapes were observed on the epidermis of 'Sihelian', 'Zhihong' and 'Shanhutai', especially in the adaxial direction (Figure 2D–F). The pattern of anticlinal cells was straight to curved in all studied cultivars.

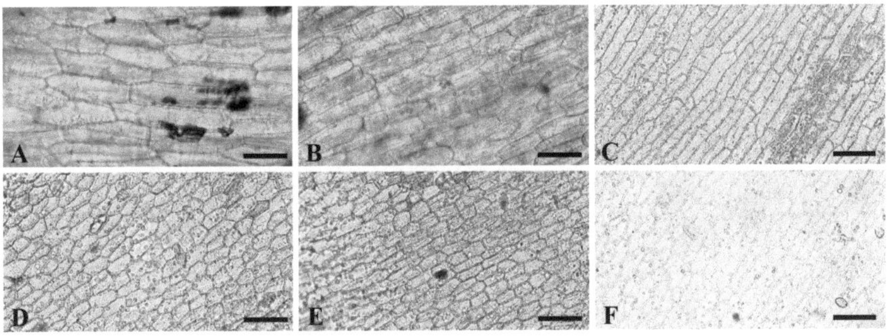

**Figure 2.** Top view of the abaxial epidermis of the petals of the six tested tree peony cultivar samples. (**A**) 'Souvenir de Maxime Cornu'; (**B**) 'Haihuang'; (**C**) 'Changshoule'; (**D**) 'Sihelian'; (**E**) 'Zhihong'; (**F**) 'Shanhutai'. Bar = 50 μm.

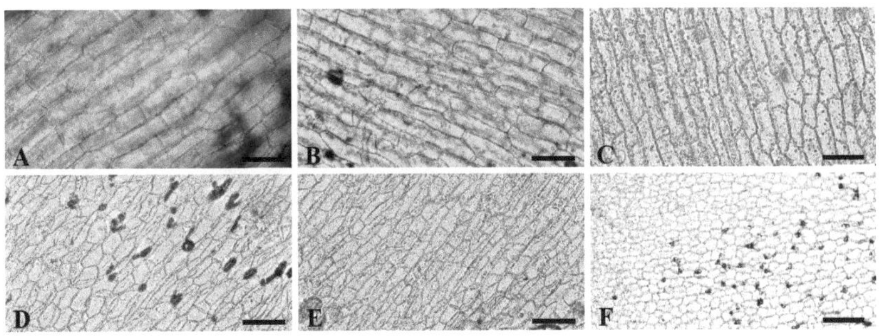

**Figure 3.** Top view of the adaxial epidermis of the petals of the six tested tree peony cultivar samples. (**A**) 'Souvenir de Maxime Cornu'; (**B**) 'Haihuang'; (**C**) 'Changshoule'; (**D**) 'Sihelian'; (**E**) 'Zhihong'; (**F**) 'Shanhutai'. Bar = 50 μm.

Consistent with our hypothesis, the surface of the epidermis was covered with smooth (Figure 4B) or sculpted cuticles (Figure 5D,E) to various degrees. Within the observed cultivars, we recognized two types of epidermal cuticle. The smooth parallel filamentous cuticle was present on the outer surfaces of the petal epidermal cells in 'Haihuang', 'Souvenir de Maxime Cornu' and 'Changshoule' (Figure 4A,C) [42]. The curly and rough cuticle was recorded in 'Shanhutai', 'Sihelian' and 'Zhihong' (Figure 4D). The texture of graininess was present on both surfaces of 'Sihelian', 'Zhihong', and 'Shanhutai', but absent in 'Souvenir de Maxime Cornu', 'Haihuang', and 'Changshoule'. This diversity is particularly evident on the abaxial surface (Figure 5).

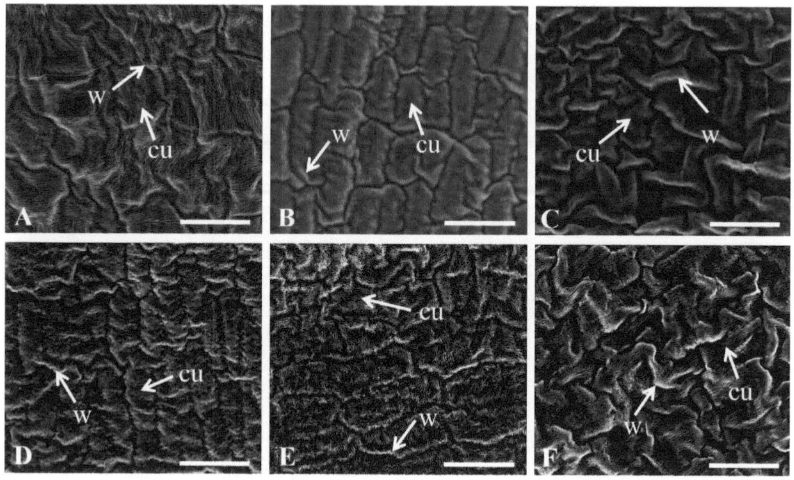


**Figure 4.** Scanning electron micrographs of the abaxial epidermis of the petals of the six tested tree peony cultivar samples. (**A**) 'Souvenir de Maxime Cornu'; (**B**) 'Haihuang'; (**C**) 'Changshoule'; (**D**) 'Sihelian'; (**E**) 'Zhihong'; (**F**) 'Shanhutai'. Abbreviations: cu: cuticle; W: wax. Bar = 50 μm.

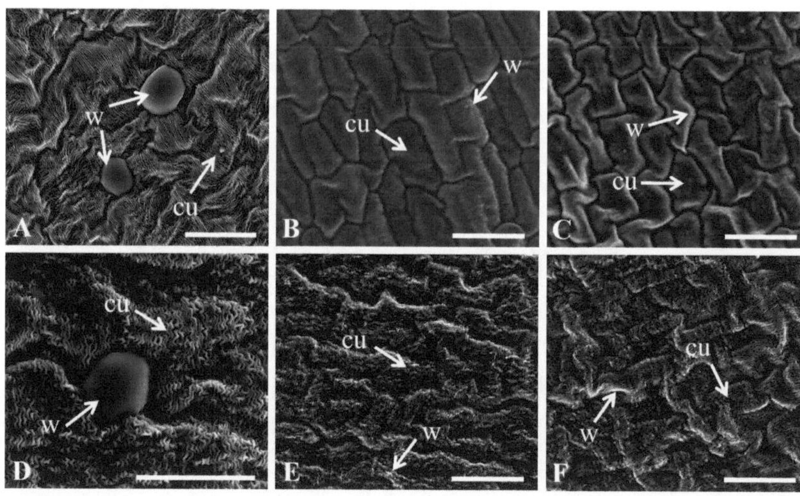


**Figure 5.** Scanning electron micrographs of the adaxial epidermis of the petals of the six tested tree peony cultivar samples. (**A**) 'Souvenir de Maxime Cornu'; (**B**) 'Haihuang'; (**C**) 'Changshoule'; (**D**) 'Sihelian'; (**E**) 'Zhihong'; (**F**) 'Shanhutai'. Abbreviations: cu: cuticle; W: wax. Bar = 50 μm.

The adaxial and abaxial epidermis are covered with cuticle (Figure 6). The thickest cuticle on the adaxial surface was observed in 'Jinge', followed by 'Haihuang' and 'Changshoule', and the thinnest cuticle was recorded in 'Shanhutai'. The thickest cuticle was observed on the abaxial surface in 'Haihuang', while the thinner cuticle was found on the epidermis of 'Sihelian', 'Zhihong' and 'Shanhutai'. Overall, the cuticle on the adaxial and abaxial epidermis of the petals with long FL was significantly thicker than that in petals with shorter FL (Table 2).

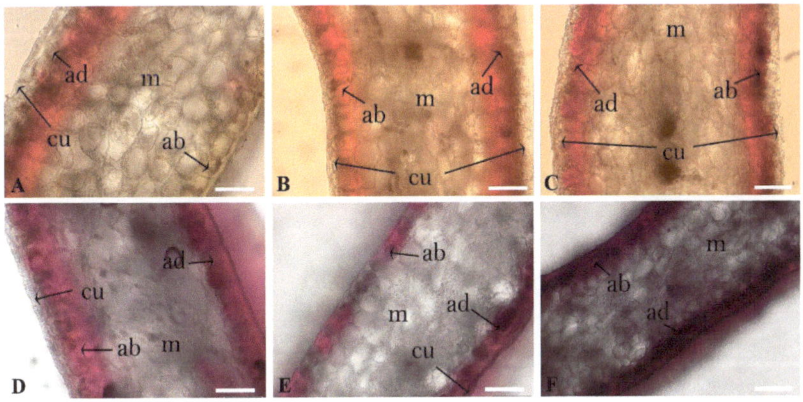


**Figure 6.** Cross-sectional micrographs of petals of the six tree peony cultivars. (**A**) 'Souvenir de Maxime Cornu'; (**B**) 'Haihuang'; (**C**) 'Changshoule'; (**D**) 'Sihelian'; (**E**) 'Zhihong'; (**F**) 'Shanhutai'. Abbreviations: ad: adaxial epidermis; ab: abaxial epidermis; cu: cuticle; m: mesophyll cells. (Bar = 50 μm).

**Table 2.** Petal anatomical characteristics of the six tree peony cultivars.

| Traits | 'Souvenir de Maxime Cornu' | 'Haihuang' | 'Changshoule' | 'Sihelian' | 'Zhihong' | 'Shanhutai' |
|---|---|---|---|---|---|---|
| Petal thickness (μm) | $183.3 \pm 10.9^{ab}$ | $198.1 \pm 9.2^{a}$ | $188.9 \pm 16.8^{ab}$ | $164.8 \pm 8.3^{b}$ | $75.3 \pm 5.1^{d}$ | $136.8 \pm 8.2^{c}$ |
| Cuticle thickness (μm) | $11.8 \pm 1.6^{a}$ | $4.2 \pm 0.6^{c}$ | $6.3 \pm 0.5^{b}$ | $3.7 \pm 1.2^{cd}$ | $1.6 \pm 0.6^{e}$ | $2.6 \pm 1.0^{d}$ |
| Number of cell layers | $10.5 \pm 1.2^{b}$ | $11.3 \pm 1.2^{b}$ | $18.2 \pm 2.0^{a}$ | $8.3 \pm 2.0^{bc}$ | $9.2 \pm 2.0^{bc}$ | $9.0 \pm 2.0^{bc}$ |
| Mesophyll thickness (μm) | $127.6 \pm 3.8^{ab}$ | $142.3 \pm 3.0^{a}$ | $137.0 \pm 3.2^{ab}$ | $120.3 \pm 1.8^{b}$ | $50.4 \pm 1.2^{d}$ | $98.5 \pm 3.5^{c}$ |
| Adaxial epidermis thickness (μm) | $23.5 \pm 0.6^{b}$ | $26.7 \pm 1.3^{a}$ | $22.7 \pm 0.8^{b}$ | $19.7 \pm 0.8^{bc}$ | $12.3 \pm 0.6^{d}$ | $18.9 \pm 3.5^{c}$ |
| Abaxial epidermis thickness (μm) | $20.3 \pm 1.2^{b}$ | $25.4 \pm 2.2^{a}$ | $22.1 \pm 1.3^{ab}$ | $18.2 \pm 2.1^{bc}$ | $11.0 \pm 1.2^{d}$ | $16.7 \pm 1.2^{c}$ |
| Vein density ($mm/mm^2$) | $2.3 \pm 0.2^{a}$ | $2.6 \pm 0.21^{a}$ | $2.0 \pm 0.3^{ab}$ | $1.6 \pm 0.1^{b}$ | $1.4 \pm 0.1^{c}$ | $1.5 \pm 0.1^{c}$ |
| Vessel number per vein | $8.0 \pm 1.00^{b}$ | $8.7 \pm 0.6^{b}$ | $12.2 \pm 0.10^{a}$ | $7.7 \pm 1.15^{bc}$ | $5.6 \pm 0.58^{c}$ | $7.9 \pm 0.5^{bc}$ |
| Vessel diameter (μm) | $7.8 \pm 0.2^{a}$ | $6.7 \pm 0.25^{b}$ | $6.2 \pm 0.4^{b}$ | $4.8 \pm 0.5^{c}$ | $5.3 \pm 0.3^{c}$ | $3.9 \pm 0.2^{d}$ |
| Petal fresh mass ($mg/cm^2$) | $24.6 \pm 0.3^{a}$ | $23.8 \pm 0.4^{a}$ | $19.8 \pm 0.9^{b}$ | $16.0 \pm 0.8^{c}$ | $12.9 \pm 0.1^{d}$ | $13.5 \pm 0.3^{d}$ |
| Petal dry mass ($mg/cm^2$) | $4.0 \pm 0.1^{a}$ | $4.2 \pm 0.1^{a}$ | $3.5 \pm 0.2^{b}$ | $3.0 \pm 0.1^{c}$ | $2.4 \pm 0.1^{d}$ | $2.2 \pm 0.2^{d}$ |

Note: Data shown are average ± standard deviation. Data with different lowercase letters in the same row indicate a significant difference at the 0.05 level.

No stomata were found on either surface of the petals in all examined cultivars (Figures 4 and 5).

### 3.3. Analysis of the Petal Cross-Sectional Anatomy

Petal thickness, cuticle thickness, and number of cell layers differed significantly ($p < 0.05$) among the studied cultivars (Table 2). The thickest petal (198.1 μm) was found in 'Haihuang', and the thinnest (75.3 μm) was in detected 'Zhihong', whereas the maximum petal cuticle thickness (11.8 μm) was reported in 'Souvenir de Maxime Cornu' and the minimum (1.6 μm) was reported in 'Zhihong' (Table 2). Consistent with our prediction, the number of petal cell layers of the three long-flowering tree peony cultivars ('Haihuang', 'Souvenir de Maxime Cornu' and 'Changshoule') was significantly higher ($p < 0.05$) than that of the three short-flowering tree peony cultivars ('Sihelian', 'Zhihong' and 'Shanhutai').

The petal mesophyll of the six tree peony cultivars was loosely arranged with many intercellular spaces with mesophyll thickness ranging between 50.4 and 142.3 μm (Figure 6, Table 2). The mesophyll thickness in the cultivars with long FL was significantly higher than in the cultivars with shorter FL. The values for mesophyll thickness of the petals did not differ significantly between cultivars with shorter FL (Table 2).

Veins (vascular bundles) were found in all of the observed tree peony cultivars. The mean density of the vein varied from 1.4 to 2.6. The maximum vein density was recorded for 'Haihuang' (2.6 per mm$^2$), followed by 'Souvenir de Maxime Cornu' (2.3 per mm$^2$) and the minimum in 'Zhihong' (1.4 per mm$^2$). The vessel number per vascular bundle and vessel diameter varied significantly ($p < 0.05$) with different tree peony cultivars (Table 2, Figure 7). The largest vessel number (12.2) was found in 'Changshoule', and the lowest number (5.6) was in 'Zhihong', while the maximum vessel diameter (7.8 μm) was reported in the 'Souvenir de Maxime Cornu' and the minimum (3.9 μm) in 'Shanhutai' (Table 2, Figure 7). The cultivars with long FL differed significantly from the cultivars with shorter FL in vein density, vessel number, and vessel diameter ($p < 0.05$).

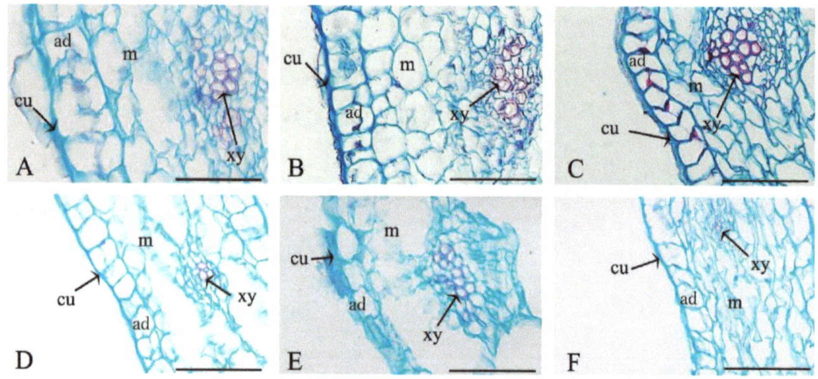


**Figure 7.** Cross-sections of petals from the six tree peony cultivars. (**A**) 'Souvenir de Maxime Cornu'; (**B**) 'Haihuang'; (**C**) 'Changshoule'; (**D**) 'Sihelian'; (**E**) 'Zhihong'; (**F**) 'Shanhutai'. Abbreviations: ad: adaxial epidermal cell; m: mesophyll; xy: xylem; cu: cuticle. Bar = 50 μm.

Among the six examined tree peony cultivars, we found broad range of diversity in the petal dry mass and the petal fresh mass. The maximum petal dry mass was reported in 'Haihuang' with 4.2 mg/cm$^2$, followed by the 'Souvenir de Maxime Cornu' with 4.0 mg/cm$^2$, and the minimum petal dry mass was recorded in 'Shanhutai', whereas the maximum petal fresh mass (24.6 mg/cm$^2$) was reported in 'Souvenir de Maxime Cornu' and the minimum (12.9 mg/cm$^2$) was reported in 'Zhihong' (Table 2).

### *3.4. The Correlations between Water-Related Traits and Floral Longevity*

Table 3 shows that significant relationships were found among traits associated with FL, petal dry mass, petal fresh mass, and flower maintenance. FL was significantly positively correlated with petal dry mass, petal fresh mass, mesophyll thickness, and vein density ($p < 0.05$), but not with xylem vessel number (Table 3). We also observed a significantly positive correlation between petal dry mass and petal fresh mass ($p < 0.05$).

**Table 3.** Pearson's correlations between water-related traits and floral longevity in tree peony cultivars.

| Traits | Floral Longevity | Petal Thickness | Cuticle Thickness | Mesophyll Thickness | Epidermis Thickness | Vein Density | Vessel Number | Vessel Diameter | Fresh Weight | Dry Weight |
|---|---|---|---|---|---|---|---|---|---|---|
| Floral longevity | 1 | | | | | | | | | |
| Petal thickness | 0.54 * | 1 | | | | | | | | |
| Cuticle thickness | 0.40 * | 0.32 | 1 | | | | | | | |
| Mesophyll thickness | 0.82 ** | 0.92 ** | 052 * | 1 | | | | | | |
| Epidermis thickness | 0.63 * | 0.94 ** | 0.22 | 0.68 * | 1 | | | | | |
| Vein density | 0.81 ** | 0.57 * | 0.54 * | 0.95 ** | 0.50 * | 1 | | | | |
| Vessel number per vein | 0.42 * | 0.51 * | 0.11 | 0.74 ** | 0.43 * | 0.70 | 1 | | | |
| Vessel diameter | 0.74 ** | 0.29 | 0.63 * | 0.76 ** | 0.32 | 0.54 * | 0.07 | 1 | | |
| Petal fresh mass | 0.80 ** | 0.75 ** | 0.43 * | 0.86 ** | 0.82 ** | 0.58 * | 0.23 | 0.73 ** | 1 | |
| Petal dry mass | 0.79 ** | 0.67 * | 0.42 * | 0.69 * | 0.73 ** | 0.56 * | 0.20 | 0.78 ** | 0.90 ** | 1 |

Note: '*' indicates significant differences ($p < 0.05$). '**' indicates very significant differences ($p < 0.01$).

### 3.5. Principal Component Analysis

The principal component analysis demonstrated that the first two components comprised 70.4% and 15.5% of the total variation, respectively (Figure 8). Variables that correlated mainly with component 1 were petal fresh mass, petal dry mass, abaxial epidermis thickness, adaxial epidermis thickness, and petal thickness. The parameters such as vessel number, number of cell layers and vein density were mainly positively loaded on component 2, while vessel diameter, petal fresh mass, petal dry mass, and cuticle thickness were negatively loaded on component 2. Species-loadings showed that the three tree peony cultivars, 'Haihuang', 'Souvenir de Maxime Cornu' and 'Changshoule' were grouped on the positive side, while Shanhutai', 'Sihelian' and 'Zhihong' clustered on the negative side, indicating that there was a significant difference in flower traits between cultivars with long FL and one with short FL. Scatter plots are often helpful in detecting patterns of variation, and as can be seen in the plot (Figure 8), the first and second components help to distinguish between long- and short-FL tree peony cultivars.

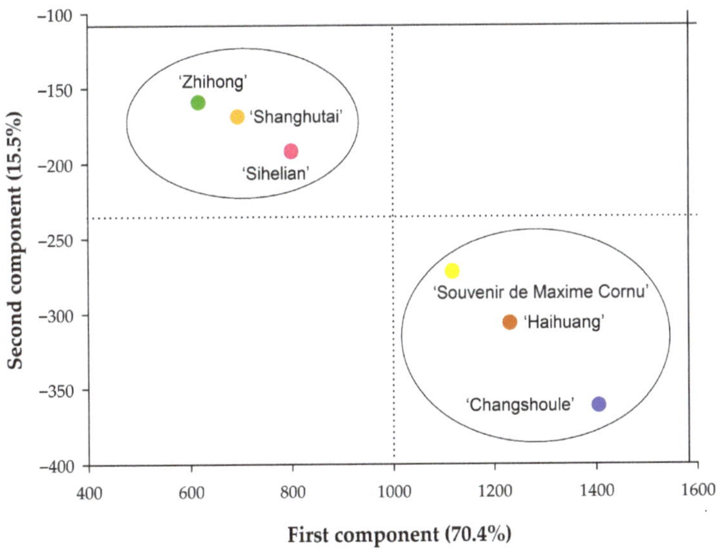


**Figure 8.** Scatter plot based on the first component and the second component.

## 4. Discussion

Using six tree peony cultivars with different FL, we examined the relationships between FL and petal anatomical traits. The principal novel findings of our study are the following: (1) long FL is closely coupled with the petal traits related to water conservation in tree peonies, such as mesophyll thickness, epidermis thickness, cuticle thickness, and no stomata; (2) the petals of the tree peony cultivars with long FL are basically accompanied by developed conducting tissues. These results suggest that most petal traits related to water conservation and supply, including the vein density, mesophyll thickness, and epidermis thickness, are closely related to floral longevity in the tree peonies, as demonstrated in orchid species [21].

Reducing water loss is important to maintain the water balance of whole plants and the flower turgor. Stomatal transpiration and cuticular transpiration are the two main pathways for water loss in flowers, leaves, and fruits [43–45]. Cuticular barriers play a key role in protecting plants against water loss in organs with closed stomata or no stomata [46]. The epidermis, including epidermal cells and epidermal cuticle, acts as a protective layer for water conservation, and its thickness is closely related to the lifespan of the flower or leaf; thicker epidermis was more tolerant to drought stress [37,47,48]. Zoran Ristic et al. [49]

reported that there is an inverse relationship between epidermal water loss and cell wall and cuticle thickness [50]. The epidermal cuticle of plant organs can greatly reduce excessive water evaporation by plant cells [51–53]. Studies have shown that plant cuticular wax has diverse crystal structures [54] and the content, distribution and size of wax crystals greatly influence water loss [55,56]. In addition, the smooth parallel filamentous cuticular wax layer can reflect sunlight to avoid the rise of leaf temperature and reduce radiation damage [57]. Our results showed that the petals of a flower with long FL were covered by thicker cuticle on the surface of the epidermis, while the cuticle of the petals of flowers with shorter FL is sparsely sculpted in curls. Furthermore, FL was significantly associated with epidermal and mesophyll thickness. This indicated that thicker epidermal cells and cuticular layer provide the ability of petal epidermis to function as a hydrophobic barrier, which in turn prolongs the FL of tree peony. The smooth parallel filamentous wax layer of long FL petals can reduce the radiation of sunlight to reduce water loss. There are no stomata on the petals of tree peonies. Tree peony cultivars with petals containing veins but without stomata represent an obvious case in which vein–stomatal coordination was absent. These findings are consistent with the results previously reported in leaves and flowers of different plant species [26,51,58,59], indicating that FL may be determined by water conservation and storage. Therefore, the improvement in FL of tree peony occurs mainly due to the reduction in the water loss of the flower and the improvement of the water storage capacity.

The high leaf dry mass is typical of stress-tolerant species [60,61]. Leaf dry mass correlates with water transport and utilization, and leaves with longer lifespan usually have higher dry mall per unit area [62]. High LMA is generally associated with greater drought resistance [60,61]. Our results showed that the FMA of the tree peony cultivars with longer FL was significantly higher than that of cultivars with shorter FL, indicating that FL is evolutionarily correlated with flower dry mass per unit area (Figure 3). This relationship seems to mirror a critical indicator of plant adaptive strategies [61].

In leaves and petals, water supply (vein density, number and diameter of vessels) and water storage are usually coordinated with water loss to keep the water balance [63–66]. Xylem vessels participate in the water transport and supply of water [59]. The size of the vascular bundle and the number of its vessels affect the conduction efficiency of plants [58]. We found the large diversity in petal vein density in the examined cultivars. Long FL was often accompanied by a greater number of vessels and a larger vessel diameter, probably suggesting that the flower with long FL has strong water supply capacity which can effectively meet the water requirements to maintain the petal turgor.

FL diversity usually reflects the adaptation of plants to their habitats [9,10,50]. Species with short FL usually bloom in weather conditions suitable for pollinator visiting, while species with long FL open in the environments that are not suitable for pollinator visiting [67,68]. Long FL can compensate for the lack of pollinators and improve reproductive success [14]. Tree peonies usually bloom in spring when pollinators could visit their flowers, which may be one of the explanations for the shorter blooming period of the six examined tree peony cultivars compared to *Paphiopedilum* [21].

## 5. Conclusions

Our results supported the hypothesis that the cost of maintaining floral function measured by the floral anatomical traits is correlated with FL. The floral anatomical details related to water conservation and water storage capacity including the number of layers of petal cells, the thickness of cuticle and epidermis, and the mesophyll thickness are beneficial in prolonging the flower longevity in tree peonies. In addition, the tree peony cultivars with longer FL have more developed vascular tissue, which can be more efficient in transporting the water and other nutrients needed to maintain the flower life. Our findings provided strong evidence for a functional association between FL and water conservation and supply capacity in tree peonies.

**Author Contributions:** Formal analysis, Y.G., W.J. and Y.Q.; investigation, Y.G., W.J., Y.W., Z.M., H.H. and S.H.; resources, W.J. and S.H.; data curation, Y.G., W.J. and Y.Q.; writing—original draft preparation, Y.G., W.J., S.Z. and Y.Q.; writing—review and editing, W.J. and S.H.; visualization, H.H.; project administration, W.J. and S.H.; funding acquisition, W.J. and S.H. All authors have read and agreed to the published version of the manuscript.

**Funding:** This work was funded by the National Key Research and Development Program of China (2018YFD1000401), Henan Provincial Science and Technology Plan Project (232102111107) and Zhongyuan Talent Plan Project (ZYYCYU202012129).

**Data Availability Statement:** Not applicable.

**Conflicts of Interest:** The authors declare no conflict of interest.

## References

1. Ashman, T.L.; Schoen, D.J. *Floral Longevity: Fitness Consequences. Floral Biology: Studies on Floral Evolution in Animal-Pollinated Plants*; Springer Science & Business Media: Berlin, Germany, 1996; p. 112.
2. Ashman, T.-L.; Schoen, D.J. How Long Should Flowers Live? *Nature* **1994**, *371*, 788–791. [CrossRef]
3. Ishii, H.S.; Sakai, S. Effects of Display Size and Position on Individual Floral Longevity in Racemes of *Narthecium asiaticum* (Liliaceae). *Funct. Ecol.* **2001**, *15*, 396–405. [CrossRef]
4. Sargent, R.; Roitberg, B. Seasonal Decline in Male-phase Duration in a Protandrous Plant: A Response to Increased Mating Opportunities? *Funct. Ecol.* **2000**, *14*, 484–489. [CrossRef]
5. Arathi, H.S.; Rasch, A.; Cox, C.; Kelly, J.K. Autogamy and Floral Longevity in *Mimulus guttatus*. *Int. J. Plant Sci.* **2002**, *163*, 567–573. [CrossRef]
6. Itagaki, T.; Sakai, S. Relationship between Floral Longevity and Sex Allocation among Flowers within Inflorescences in Aquilegia buergeriana var. *oxysepala* (*Ranunculaceae*). *Am. J. Bot* **2006**, *93*, 1320–1327. [CrossRef]
7. Johnson, S.D.; Neal, P.R.; Harder, L.D. Pollen Fates and the Limits on Male Reproductive Success in an Orchid Population. *Biol. J. Linn. Soc.* **2005**, *86*, 175–190. [CrossRef]
8. Jorgensen, R.; Arathi, H.S. Floral Longevity and Autonomous Selfing are Altered by Pollination and Water Availability in Collinsia Heterophylla. *Ann. Bot.* **2013**, *112*, 821–828. [CrossRef]
9. Rathcke, B.J. Floral Longevity and Reproductive Assurance: Seasonal Patterns and an Experimental Test with Kalmia latifolia (Ericaceae). *Am. J. Bot.* **2003**, *90*, 1328–1332. [CrossRef]
10. Vesprini, J.L.; Pacini, E. Temperature-Dependent Floral Longevity in Two Helleborus Species. *Plant Syst. Evol.* **2005**, *252*, 63–70. [CrossRef]
11. Weber, J.J.; Goodwillie, C. Variation in Floral Longevity in the Genus *Leptosiphon*: Mating System Consequences. *Plant Biol.* **2013**, *15*, 220–225. [CrossRef]
12. Yasaka, M.; Nishiwaki, Y.; Konno, Y. Plasticity of Flower Longevity in Corydalis ambigua. *Ecol. Res.* **1998**, *13*, 211–216. [CrossRef]
13. Giblin, D.E. Variation in Floral Longevity between Populations of Campanula Rotundifolia (*Campanulaceae*) in Response to Fitness Accrual Rate Manipulation. *Am. J. Bot.* **2005**, *92*, 1714–1722. [CrossRef]
14. Abdala-Roberts, L.; Parra-Tabla, V.; Navarro, J. Is Floral Longevity Influenced by Reproductive Costs and Pollination Success in Cohniella ascendens (Orchidaceae)? *Ann. Bot.* **2007**, *100*, 1367–1371. [CrossRef] [PubMed]
15. Stead, A.D. Pollination-Induced Flower Senescence: A Review. *Plant Growth Regul.* **1992**, *11*, 13–20. [CrossRef]
16. Nobel, P.S. Water Relations of Flowering of Agave deserti. *Bot. Gaz.* **1977**, *138*, 1–6. [CrossRef]
17. Ichimura, K.; Azuma, M. Characterization of Petal Senescent Types in Cut Dahlia and Extension of Their Vase Life by Treatment with Silver Thiosulfate Complex Followed by Glucose Plus Germicides. *Horticulturae* **2022**, *8*, 922. [CrossRef]
18. Jiang, Y.L.; Shi, J.L.; Du, X.H.; Dong, S.H.; Li, J.P. Preservation Effects of Different Preserv watives on Eustoma grandiflorum Cut Flower. *Mod. Agric. Sci. Technol.* **2022**, *13*, 80–84.
19. Sun, J.; Guo, H.; Tao, J. Effects of Harvest Stage, Storage, and Preservation Technology on Postharvest Ornamental Value of Cut Peony (*Paeonia lactiflora*) Flowers. *Agronomy* **2022**, *12*, 230. [CrossRef]
20. Van Doorn, W. The Postharvest Quality of Cut Lily Flowers and Potted Lily Plants. *II Int. Symp. Genus* **2011**, *900*, 255–264. [CrossRef]
21. Zhang, F.-P.; Feng, J.-Q.; Huang, J.-L.; Huang, W.; Fu, X.-W.; Hu, H.; Zhang, S.-B. Floral Longevity of Paphiopedilum and Cypripedium Is Associated with Floral Morphology. *Front. Plant Sci.* **2021**, *12*, 737. [CrossRef]
22. Yeats, T.H.; Rose, J.K.C. The formation and function of plant cuticles. *Plant Physiol.* **2013**, *163*, 5–20. [CrossRef] [PubMed]
23. Kirsch, T.; Kaffarnik, F.; Riederer, M.; Schreiber, L. Cuticular Permeability of the Three Tree Species *Prunus laurocerasus* L., *Ginkgo biloba* L. and *Juglans regia* L.: Comparative In-vestigation of the Transport Properties of Intact Leaves, Isolated Cuticles and Reconstituted Cuticular Waxes. *J. Exp. Bot.* **1997**, *48*, 1035–1045. [CrossRef]
24. Chang, W.; Zhang, S.-B.; Li, S.-Y.; Hu, H. Ecophysiological Significance of Leaf Traits in Cypripedium and Paphiopedilum. *Physiol. Plant.* **2011**, *141*, 30–39. [CrossRef] [PubMed]

25. Ke, Y.; Zhang, F.-P.; Zhang, Y.-B.; Li, W.; Wang, Q.; Yang, D.; Zhang, J.-L.; Cao, K.-F. Convergent relationships between flower economics and hydraulic traits across aquatic and terrestrial herbaceous plants. *Plant Divers.* **2023**, *1*, 6. [CrossRef]
26. Zhang, F.-P.; Yang, Y.-J.; Yang, Q.-Y.; Zhang, W.; Brodribb, T.J.; Hao, G.-Y.; Hu, H.; Zhang, S.-B. Floral Mass per Area and Water Maintenance Traits Are Correlated with Floral Longevity in *Paphiopedilum* (Orchidaceae). *Front. Plant Sci.* **2017**, *8*, 501. [CrossRef] [PubMed]
27. Ram, H.Y.M.; Rao, I.V.R. Physiology of flower bud growth and opening. *Proc. Indian Acad. Sci.* **1984**, *93*, 253–274. [CrossRef]
28. Patiño, S.; Grace, J. The cooling of convolvulaceous flowers in a tropical environment. *Plant Cell Environ.* **2002**, *25*, 41–51. [CrossRef]
29. Galen, C. It never rains but then it pours: The diverse effects of water on flower integrity and function. In *Reproductive Allocation in Plants*; Reekie, E., Bazzaz, F.A., Eds.; Elsevier Press: San Diego, CA, USA, 2005; pp. 77–95.
30. Roddy, A.B.; Dawson, T.E. Determining the water dynamics of flowering using miniature sap flow sensors. *Acta Hortic.* **2012**, *951*, 47–54. [CrossRef]
31. Roddy, A.B.; Guilliams, C.M.; Lilittham, T.; Farmer, J.; Wormser, V.; Pham, T.; Fine, P.V.; Feild, T.S.; Dawson, T.E. Uncorrelated evolution of leaf and petal venation patterns across the angiosperm phylogeny. *J. Exp. Bot.* **2013**, *64*, 4081–4088. [CrossRef]
32. Roddy, A.B.; Brodersen, C.R.; Dawson, T.E. Hydraulic conductance and the maintenance of water balance in flowers. *Plant Cell Environ.* **2016**, *39*, 2123–2132. [CrossRef]
33. Hong, D.Y.; Pan, K.Y. Paeonia Cathayana D.Y. Hong & K.Y. Pan, A new tree peony, with revision of *P. suffruticosa* ssp. *yinpingmudan*. *Acta Phytotax Sin.* **2007**, *45*, 285–288.
34. Chavarria, M.R.; Wherley, B.; Jessup, R.; Chandra, A. Leaf Anatomical Responses and Chemical Composition of Warm-Season Turfgrasses to Increasing Salinity. *Curr. Plant Biol.* **2020**, *22*, 100147. [CrossRef]
35. Olsen, J.T.; Caudle, K.L.; Johnson, L.C.; Baer, S.G.; Maricle, B.R. Environmental and Genetic Variation in Leaf Anatomy among Populations of *Andropogon gerardii* (Poaceae) Along a Precipitation Gradient. *Am. J. Bot.* **2013**, *100*, 1957–1968. [CrossRef]
36. Li, F.; Shi, L.; Feng, Y.; Wei, Q.; Chen, H.; Liu, D.; Qin, F.; Xu, H.; Qiu, Y.; Hu, B. The Observation of Ultra-Structure during the Petal Senescence of *Dracocephalum argunense* Fisch. *Ex Link. J. Food Agric. Environ.* **2012**, *10*, 732–735.
37. Jia, W.; Wang, Y.; Qi, Q.; He, S.; Mi, Z.; Zhu, X. Leaf Epidermal Morphology of Ten Wild Tree Peonies in China and Its Taxonomic Significance. *Horticulturae* **2022**, *8*, 502. [CrossRef]
38. Costa, V.B.S.; Pimentel, R.M.M.; Chagas, M.G.S.; Alves, G.D.; Castro, C.C. Petal micromorphology and its relationship to pollination. *Plant Biol.* **2017**, *19*, 115–122. [CrossRef]
39. Mamrutha, H.M.; Mogili, T.; Lakshmi, K.J.; Rama, N.; Kosma, D.; Kumar, M.U.; Jenks, M.A.; Nataraja, K.N. Leaf cuticular wax amount and crystal morphology regulate post-harvest water loss in mulberry (*Morus* species). *Plant Physiol. Biochem.* **2010**, *48*, 690–696. [CrossRef]
40. Zhou, J.; Liu, Z. Comparative Morphology of the Leaf Epidermis in *Ligusticum* (Apiaceae) from China. *Am. J. Plant Sci.* **2018**, *9*, 1105–1123. [CrossRef]
41. Ma, Y.; Sawhney, V.K.; Steeves, T.A. Staining of paraffin-embedded plant material in safranin and fast green without prior removal of the paraffin. *Can. J. Bot.* **1993**, *71*, 996–999. [CrossRef]
42. Richards, A.G.; Korda, F.H. Studies on arthropod cuticle. II. electron microscope studies of extracted cuticle. *Biol. Bull.* **1948**, *94*, 212–235. [CrossRef]
43. Harrap, M.J.M.; Rands, S.A. The Role of Petal Transpiration in Floral Humidity Generation. *Planta* **2022**, *255*, 78. [CrossRef] [PubMed]
44. Schuster, A.-C.; Burghardt, M.; Riederer, M. The Ecophysiology of Leaf Cuticular Transpiration: Are Cuticular Water Permeabilities Adapted to Ecological Conditions? *J. Exp. Bot.* **2017**, *68*, 5271–5279. [CrossRef] [PubMed]
45. Wu, Y.; Du, T.; Wang, L. Isotope Signature of Maize Stem and Leaf and Investigation of Transpiration and Water Transport. *Agric. Water Manag.* **2021**, *247*, 106727. [CrossRef]
46. Ding, F.; Wang, G.; Wang, M.; Zhang, S. Exogenous Melatonin Improves Tolerance to Water Deficit by Promoting Cuticle Formation in Tomato Plants. *Molecules* **2018**, *23*, 1605. [CrossRef] [PubMed]
47. Bhanot, V.; Fadanavis, S.V.; Panwar, J. Revisiting the Architecture, Biosynthesis and Functional Aspects of the Plant Cuticle: There is more Scope. *Environ. Exp. Bot.* **2021**, *183*, 104364. [CrossRef]
48. Gorb, E.V.; Gorb, S.N. Anti-Adhesive Effects of Plant Wax Coverage on Insect Attachment. *J. Exp. Bot.* **2017**, *68*, 5323–5337. [CrossRef]
49. Ristic, Z.; Jenks, M.A. Leaf Cuticle and Water Loss in Maize Lines Differing in Dehydration Avoidance. *J. Plant Physiol.* **2002**, *159*, 645–651. [CrossRef]
50. Trunschke, J.; Stöcklin, J. Plasticity of Flower Longevity in Alpine Plants is Increased in Populations from High Elevation Compared to low Elevation Populations. *Alp. Bot.* **2017**, *127*, 41–51. [CrossRef]
51. Griffiths, D.W.; Robertson, G.W.; Shepherd, T.; Birch, A.N.E.; Gordon, S.C.; Woodford, J.T. A comparison of the Composition of Epicuticular Wax from Red Raspberry (*Rubus idaeus* L.) and Hawthorn (*Crataegus monogyna* Jacq.) Flowers. *Phytochemistry* **2000**, *55*, 111–116. [CrossRef]
52. Kunst, L.; Samuels, A.L. Biosynthesis and Secretion of Plant Cuticular Wax. *Prog. Lipid Res.* **2003**, *42*, 51–80. [CrossRef]
53. Tomaszewski, D.; Zieliński, J. Epicuticular Wax Structures on Stems and Comparison between Stems and Leaves–A Survey. *Flora-Morphol. Distrib. Funct. Ecol. Plants* **2014**, *209*, 215–232. [CrossRef]

54. Koch, K.; Ensikat, H.-J. The Hydrophobic Coatings of Plant Surfaces: Epicuticular Wax Crystals and Their Morphologies, Crystallinity and Molecular Self-Assembly. *Micron* **2008**, *39*, 759–772. [CrossRef]
55. Riederer, M.; Schreiber, L. Protecting against Water Loss: Analysis of the Barrier Properties of Plant Cuticles. *J. Exp. Bot.* **2001**, *52*, 2023–2032. [CrossRef] [PubMed]
56. Schönherr, J. Characterization of Aqueous Pores in Plant Cuticles and Permeation of Ionic Solutes. *J. Exp. Bot.* **2006**, *57*, 2471–2491. [CrossRef] [PubMed]
57. Jagadish, S.K.; Way, D.A.; Sharkey, T.D. Plant Heat Stress: Concepts Directing Future Research. *Plant Cell Environ.* **2021**, *44*, 1992–2005. [CrossRef]
58. Van Doorn, W.G.; Van Meeteren, U. Flower Opening and Closure: A Review. *J. Exp. Bot.* **2003**, *54*, 1801–1812. [CrossRef]
59. Zhu, G.L.; Deng, R.H.; Wei, X.Z. Leaf epidermal micromorphology of Ziziphus jujuba var. spinosa in response to a gradient of drought stress. *Acta Ecol. Sin.* **2016**, *36*, 5193–5203.
60. Wright, I.J.; Reich, P.B.; Westoby, M.; Ackerly, D.D.; Baruch, Z.; Bongers, F.; Cavender-Bares, J.; Chapin, T.; Cornelissen, J.H.C.; Diemer, M.; et al. The Worldwide Leaf Economics Spectrum. *Nature* **2004**, *428*, 821–827. [CrossRef]
61. Poorter, H.; Niinemets, Ü.; Poorter, L.; Wright, I.J.; Villar, R. Causes and Consequences of Variation in Leaf Mass per Area (LMA): A Meta-Analysis. *New Phytol.* **2009**, *182*, 565–588. [CrossRef]
62. Fu, P.-L.; Jiang, Y.-J.; Wang, A.-Y.; Brodribb, T.; Zhang, J.-L.; Zhu, S.-D.; Cao, K.-F. Stem Hydraulic Traits and Leaf Water-Stress Tolerance are Coordinated with the Leaf Phenology of Angiosperm Trees in an Asian Tropical Dry karst Forest. *Ann. Bot.* **2012**, *110*, 189–199. [CrossRef]
63. Cal, A.J.; Sanciangco, M.; Rebolledo, M.C.; Luquet, D.; Torres, R.O.; McNally, K.L.; Henry, A. Leaf Morphology, rather than Plant Water Status, Underlies Genetic Variation of Rice Leaf Rolling under Drought. *Plant Cell Environ.* **2019**, *42*, 1532–1544. [CrossRef] [PubMed]
64. Yamane, K.; Hayakawa, K.; Kawasaki, M.; Taniguchi, M.; Miyake, H. Bundle Sheath Chloroplasts of Rice are more Sensitive to drought Stress than Mesophyll Chloroplasts. *J. Plant Physiol.* **2003**, *160*, 1319–1327. [CrossRef] [PubMed]
65. Martínez-Vilalta, J.; Poyatos, R.; Aguadé, D.; Retana, J.; Mencuccini, M. A New Look at Water Transport Regulation in Plants. *New Phytol.* **2014**, *204*, 105–115. [CrossRef] [PubMed]
66. Akram, N.A.; Shafiq, S.; Ashraf, M.; Aisha, R.; Sajid, M.A. Drought-Induced Anatomical Changes in Radish (*Raphanus sativus* L.) Leaves Supplied with Trehalose through Different Modes. *Arid. Land Res. Manag.* **2016**, *30*, 412–420. [CrossRef]
67. Primack, R.B. Longevity of Individual Flowers. *Annu. Rev. Ecol. Syst.* **1985**, *16*, 15–37. [CrossRef]
68. Steinacher, G.; Wagner, J. Flower Longevity and Duration of Pistil Receptivity in High Mountain Plants. *Flora-Morphol. Distrib. Funct. Ecol. Plants* **2010**, *205*, 376–387. [CrossRef]

*agronomy*

MDPI

*Article*

# MIKC-Type MADS-Box Gene Family Discovery and Evolutionary Investigation in *Rosaceae* Plants

**Yue Qin [1,2], Gaopu Zhu [1], Fangdong Li [1], Lin Wang [1], Chen Chen [1] and Han Zhao [1,*]**

[1] Research Institute of Non-Timber Forestry Chinese Academy of Forestry, Zhengzhou 450003, China; qinyue870711@163.com (Y.Q.); zhugaopu@163.com (G.Z.); lifangdong66@163.com (F.L.); 174816279@163.com (L.W.)

[2] College of Forestry, Nanjing Forestry University, Nanjing 210037, China

* Correspondence: zhaohan@caf.ac.cn; Tel.: +86-136-0765-0203

**Abstract:** MADS-box is an important transcriptional regulatory element in plant growth. The MIKC-type MADS-box genes play important roles. However, the identification and evolutionary investigation of MIKC-type MADS-box family members in *Rosaceae* have been inadequate. Therefore, based on whole-genome data from *Prunus dulcis, Prunus salicina, Prunus armeniaca, Prunus persica, Prunus mira,* and *Amygdalus nana*, we depicted the evolution and divergence patterns of MIKC-type MADS-box family genes. In this study, we found 222 MIKC-type MADS-box genes from six *Rosaceae* species. These genes were classified into five clades, and only motif 1 was identified across all MIKC-type MADS-box proteins, except PdMADS42 and PmiMADS16. The structural properties of these genes significantly varied in sequence lengths between species, despite the high levels of similarity in exon lengths and numbers. MIKC-type MADS-box genes were found to have mostly been limited through purifying selection processes. Remarkably divergent regions were found inside the MIKC-type MADS-box genes' domains, where clade III displayed more conserved activities and may have retained more original functions over the evolutionary process; clade I, on the other hand, may have undergone substantial functional limitations in a specific functional role. These findings provide the groundwork for future research into the molecular evolutionary processes of the plant MIKC-type MADS-box gene family.

**Keywords:** selective evolutionary pressure; MIKC-type MADS-box genes; *Rosaceae* plants; gene evolution; functional differences

**Citation:** Qin, Y.; Zhu, G.; Li, F.; Wang, L.; Chen, C.; Zhao, H. MIKC-Type MADS-Box Gene Family Discovery and Evolutionary Investigation in *Rosaceae* Plants. *Agronomy* **2023**, *13*, 1695. https://doi.org/10.3390/agronomy13071695

Academic Editor: Antonio Lupini

Received: 24 April 2023
Revised: 21 June 2023
Accepted: 22 June 2023
Published: 25 June 2023



## 1. Introduction

Flower organs are the basis of plant evolution and classification, and have always played important roles in plant development and biological research [1]. The MADS-box gene family is a class of transcription factors widely distributed among animals, plants, fungi, and other organisms [2,3]. MADS-box derives its name from the first initials of fore groups of genes: the mini chromosome maintenance 1 genes from yeast (MCM1), the agamous genes from *Arabidopsis Thaliana* (AG), deficiens genes from snapdragon (DEF), and serum response factor genes from humans (SRF). These genes all contain a MADS-box conserved region of about 180 bp at the N-terminal, while identifying the similar DNA sequences [4,5]. The plant MADS-box genes were discovered from the mutant study of the model plants *Arabidopsis* and Snapdragon. MADS-box genes are widely distributed throughout the genome and expressed at different development stages. Their functions are involved in all aspects of plant growth and development, especially the development and regulation of flower organs, including initiation, differentiation, morphogenesis, and gene transcription regulation, and have become some of the most relevant transcription factor family in plants [6]. The MADS-box gene family in plants can be divided into Type I and Type II. Due to the relatively simple structure and functional redundancy of Type I genes, studies on plant MADS-box genes mostly focus on Type II genes, also known as

MIKC-type genes [1]. The encoded protein consists of four domains: MADS domain, I domain (intervening domain), K domain (keratin-like domain), and C-terminal domain. [7]. The MADS-box domain and region I are associated with the formation of protein dimers, the K-box domain mediates protein interactions, and the C-terminal is associated with transcription factor activation and the formation of higher-order protein complexes [8,9].

With the development of sequencing technology and bioinformatics, more and more plants have been thoroughly studied through genome or transcriptome sequencing. Many MADS-box gene family members and their functions within the flower organs of various species have been identified and revealed, including genes controlling flowering time and flowering inhibition (FLC, SVP, SOC1, AGL24), genes regulating floral meristem differentiation (CAL and FUL), and genes governing floral organ characteristics [10–13]. The pathway controlling flowering initiation and the genes involved, response pathway, and expression regulation mechanism were also analyzed. For example, overexpression of BpAP1 gene in European silver birch can cause early flowering [14], while overexpression of PvMADS5, a SEP homologous gene in bamboo, causes delayed flowering in *Arabidopsis* plants [15]. These studies have improved understanding of the MADS-box gene family, but many members are yet to be discovered, and the regulation mechanism of flower organ development requires further research.

*Prunus dulcis, Prunus salicina, Prunus armeniaca, Prunus persica, Prunus mira,* and *Amygdalus nana* are important economic forestry species in *Rosaceae*. The study of genes related to flower regulation is of great value not only in phylogeny and taxonomy, but also in genetic improvement and cultivation techniques. In previous studies, researchers have conducted the identification and function analysis of the MADS-box gene family in *Prunus persica, Prunus armeniaca,* and *Prunus dulcis*, obtaining a number of genes, illustrating the genes' structure and sequence features, and analyzing the gene functions. However, studies investigating its evolution between different species are limited. Thus, in this study, we used genome-wide data of six *Rosaceae* plant species (*Prunus dulcis, Prunus salicina, Prunus armeniaca, Prunus persica, Prunus mira,* and *Amygdalus nana*) to screen for MIKC-type MADS-box gene family members. We also looked at the evolutionary relationship, areas of functional divergence, and amino acid structures of the discovered MIKC-type MADS-box genes. These findings are not only important for elucidating the adaptive history of the *Rosaceae* species, but also for predicting the functional area of the MIKC-type MADS-box genes.

## 2. Materials and Methods

### 2.1. Gene Family Identification

Genomic sequence and annotation data for *Prunus dulcis, Prunus salicina, Prunus armeniaca, Prunus persica,* and *Prunus mira* were obtained from the GDR database (rosaceae.org), and the *Amygdalus nana* genome assembly was obtained through our research facility (private information). We utilized the sequences of *Arabidopsis* MADS-box proteins as a search basis to retrieve MIKC-type MADS-box proteins, employing the local alignment searching modality (BLASTP) with an E value of $1 \times 10^{-10}$ [16,17]. Then, HMMER v3.3.2 with the hidden Markov model (HMM) profile (PF00319 and PF01486) was employed to explore the proteomes with parameters 'hmmsearch—notextw—acc—cut_ga_' [18]. Next, we compared the MADS-box domain using SMART (smart.embl.de) with the Pfam database in the normal smart mode.

### 2.2. Sequence Analysis

We used MAFFT 7.310 with parameters 'mafft—maxiterate 1000—globalpair input > output' to perform multiple protein sequence alignment [19]. The conserved motifs of the MIKC-type MADS-box proteins were investigated using the MEME Suite 5.5.0 with parameters 'meme julei_6_species. fasta—protein—oc.—nostatus—time 14,400—mod anr—nmotifs 10—minw 6—maxw 100—objfun classic—markov_order 0′ [20]. WebLogo 2.8.2

was used to display the sequences [21], and Tbtools software v1.106 was used to visualize the MEME and gene structure results.

### 2.3. Phylogenetic and Conservative Analyses

ProtTest 3.4 with model 'LG + I + G + F' was used to construct the optimum protein substitution model [22]. Using the neighbor-joining technique and MEGA Version 10.2.2 software, we built a phylogenetic tree using JTT + G with 1000 bootstrap repetitions [23]. Then, we used R-ggtree software 1.8.1 to enhance the phylogenetic tree [24–26].

### 2.4. Selective Evolutionary Pressure Analysis

The Codeml program EasyCodeml Version 1.4 was used to analyze the selective evolutionary pressure at the branch, site, and branch-site models (BM, SM, and BSM, respectively) [27]. Branch models are used to identify the adaptive development of several branches. Site models can discover the $\omega$ ratio between the sites and positive selected sites. Provided that the $\omega$ ratio varies between the sites and branches, BSMs may be used [28–31].

### 2.5. Functional Difference Analyses

The DIVERGE 3.0 tool was used to analyze the functional divergences and distance of the gene family members [32]. Type-I mutations result from functional deviations following gene duplication, subsequently causing distinct evolutionary rates to be shown by duplicate genes. The residues in duplicate genes that showed striking variations in amino acid properties across copies tested using type-II functional divergences [33,34]. Clusters i and j had a type-I functional distance equal to $dF(i,j) = -\ln(1 - \theta ij)$. Wang and Gu proved that the dF is additive under a scenario of independence, such that the functional branch dimension for a given gene cluster x is bF(x), and the functional branch dimension of clusters i and j is bF(i) plus bF(j). In the case of a known ij, the cluster branch of the functional branch length may be calculated using the normal least squares technique; if bF = 0, then the duplicate site of the gene has evolved at a pace very close to that of the ancestor gene [35].

## 3. Results

### 3.1. Phylogenetic and Sequence Analyses

We identified 222 MIKC-type MADS-box proteins from six *Rosaceae* species' whole genome data. Of those identified, 54, 27, 23, 69, 23, and 26 MIKC-type MADS-box genes could be detected in *Prunus dulcis* (PdMADS1-54), *Prunus salicina* (PsMADS1-27), *Prunus armeniaca* (PaMADS1-23), *Prunus persica* (PpMADS1-69), *Prunus mira* (PmiMADS1-23), and *Amygdalus nana* (AnMADS1-26), respectively. The amino acid lengths ranged from 162 aa to 670 aa.

Phylogenetic analysis revealed five distinct groups (clades I–V) within the MIKC-type MADS-box gene family. There was a significant difference in the number of genes across the five clades, with the largest and smallest clades (clade I and III) having 110 and 6, respectively (Table 1, Figure 1). The six species' genes were found in nearly every clade. The genes of *Prunus dulcis* and *Prunus persica* were the most abundant in clades I, IV and V. *Prunus persica* had most genes in clade II, but *Prunus armeniaca* was missing. The six species had the same amount of genes in clade III (Table 1, Figure 2). The results suggested that the MIKC-type MADS-box genes in *Rosaceae* are nearly identical, but serve distinct purposes.

**Table 1.** The MIKC-type MADS-box genes of six *Rosaceae* species.

| Species | Common Name | MIKC-Type MADS-Box Gene Numbers | Clade | | | | |
|---|---|---|---|---|---|---|---|
| | | | I | II | III | IV | V |
| *Amygdalus nana* | Short almond | 26 | 14 | 3 | 1 | 1 | 7 |
| *Prunus mira* | Light walnut | 23 | 11 | 3 | 1 | 2 | 6 |
| *Prunus persica* | Peach | 69 | 26 | 9 | 1 | 5 | 28 |
| *Prunus armeniaca* | Apricot | 23 | 15 | 0 | 1 | 2 | 5 |
| *Prunus salicina* | Plum | 27 | 14 | 3 | 1 | 2 | 7 |
| *Prunus dulcis* | Almond | 54 | 30 | 3 | 1 | 7 | 13 |
| Total | | 222 | 110 | 21 | 6 | 19 | 66 |

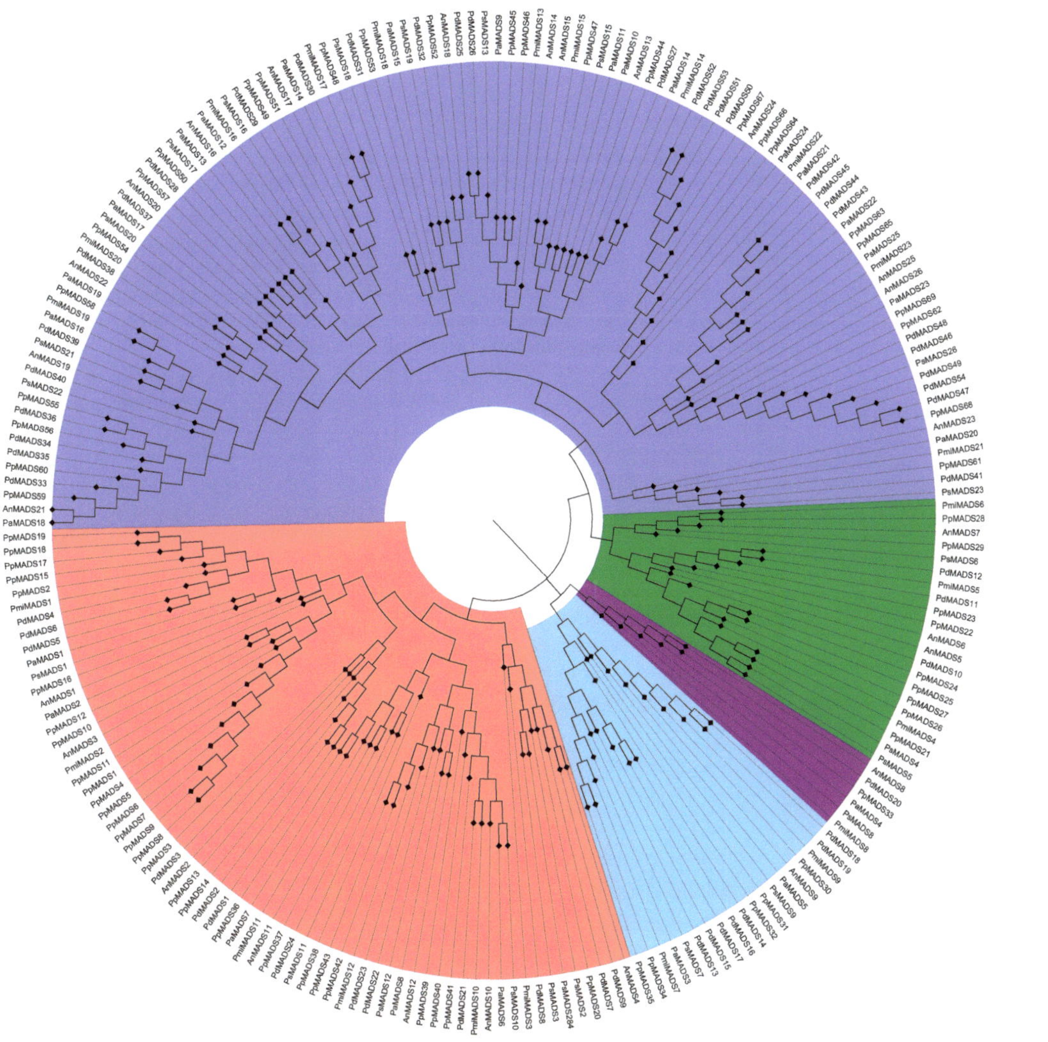

**Figure 1.** NJ phylogenetic tree of the MIKC-type MADS-box gene family from *Prunus mira* (PmiMADS), *Prunus armeniaca* (PaMADS), *Prunus persica* (PpMADS), *Prunus dulcis* (PdMADS), *Prunus salicina* (PsMADS), and *Amygdalus nana* (AnMADS) in *Rosaceae* plants. Blue is clade I, green is II, purple is III, light blue is IV, and pink is V.

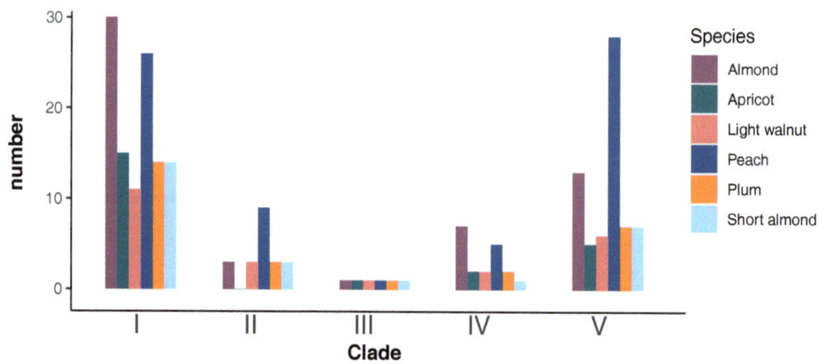


**Figure 2.** The distribution of MIKC-type MADS-box genes from six *Rosaceae* species in different clades. Species are represented by different colored boxes.

### 3.2. *Motif and Structure Analyses*

Ten conserved motifs, ranging from 21 to 80 amino acids, were found in this investigation. Only Motif 1 was found in all of the MIKC-type MADS-box proteins (Figure 3), except PdMADS42 and PmiMADS16. Furthermore, MIKC-type MADS-box proteins had substantially identical sequences for 10 amino acid residues (alanine, cysteine, glutamic acid, glycine, lysine, leucine, glutamine, arginine, serine, and threonine) (Figure 4). On the other hand, some MIKC-type MADS-box proteins lacked certain motifs; for example, Motifs 1 and 6 did not exist in PmiMADS16, and Motifs 4 and 6 did not exist in AnMADS23. However, there were similar or identical arrangement orders and motif types within the same clades. The results showed that the six *Rosaceae* species' MIKC-type MADS-box proteins were relatively conserved within a specific clade; the structural analysis produced similar results.

The results of the exon-intron structure analysis showed that the amino acid lengths ranged from 163 to 671. Among them, PsMADS3, PsMADS4, PsMADS7, PsMADS8, and PsMADS24 had no introns (Figure 5). The remaining MIKC-type MADS-box genes had 2–42 exon and 1–41 intron structures. Genes with 8 exons accounted for 47.75 percent of the total (106/222), most genes contained 7 or 8 exons, and only one gene contained 42 exons. This proves that the structures are specialized after gene duplication. These genes also have somewhat complicated architectures, including lengthy initial exons. In conclusion, the conserved motifs, gene architectures, and evolutionary connections of the MIKC-type MADS-box proteins were quite similar. Additionally, the diversification of gene roles was suggested due to the differences in the retained motifs and gene architectures among clades.

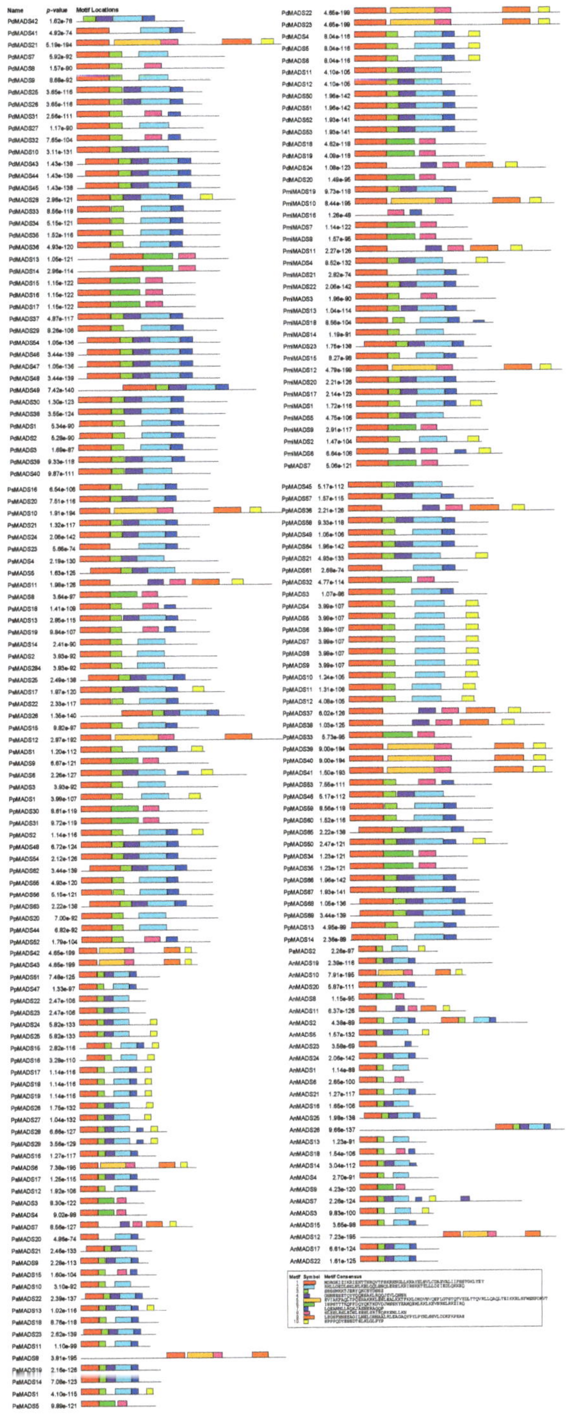

**Figure 3.** The conserved motifs of 222 MIKC-type MADS-box family members. Motifs 1 to 10 are represented by different colored boxes.

**Figure 4.** Sequence logo of Motif 1. Amino acid residues are represented by different colored letters.

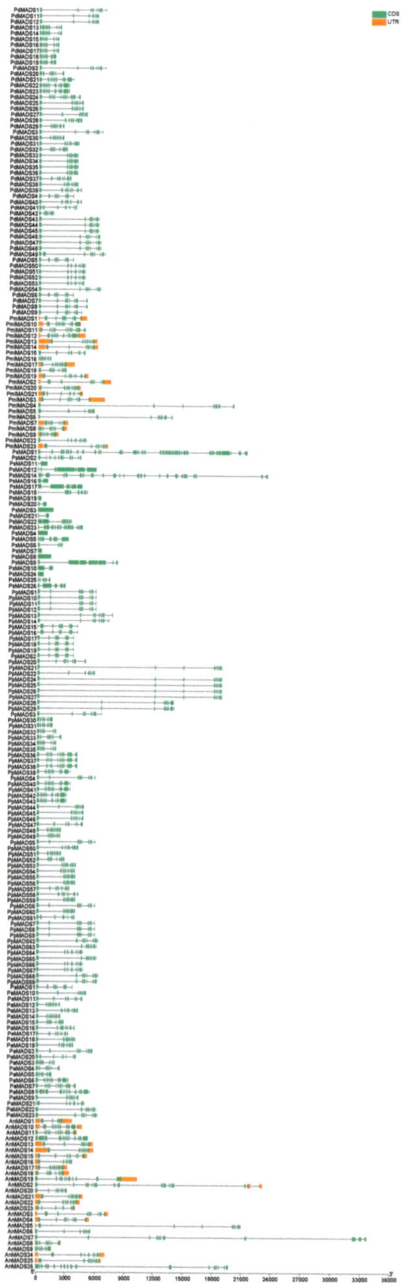

**Figure 5.** Gene structure of 222 MIKC-type MADS-box gene family members. The green boxes represent exons, and the gray lines represent introns.

### 3.3. Selective Pressure Analysis

We used EasyCodeML to analyze the selective pressure, used the branch model to test the evolutionary rate difference, and used the branch-site model to increase the detection of positive selection sites in the foreground branches. The results showed that the genetic differentiation had no significant differences ($p > 0.05$), and no positive sites were obtained according to the branch model and branch-site model results (Tables 2 and 3). Thus, we assumed the same selection pressure for each point and used the site model for testing the selection site positivity in these sequences. We found that most sites were subjected to purifying and neutral selections, and only the M8 model identified one site as having a significant positive selection site (Table 4). Although we obtained the positive selection sites, the MIKC-type MADS-box genes mainly underwent purifying selection events.

**Table 2.** Branch model parameters of selective pressure analysis.

| Model | np/aa | ln L | Parameters Estimates | | | Compared Model | *p*-Value |
|---|---|---|---|---|---|---|---|
| Two ratio Model 2 | 448 | −3172.993 | $\omega$: | $\omega_0 = 1.533$ | $\omega_1 = 0.768$; $\omega_2 = 999.000$; $\omega_3 = 0.862$; $\omega_4 = 2.208$; $\omega_5 = 2.452$ | Model 0 vs. Two ratio Model 2 | >0.05 |
| Model 0 | 443 | −3175.884 | $\omega$= | 1.558 | | | |

**Table 3.** Branch-site model parameters of selective pressure analysis.

| Model | np/aa | ln L | Parameters Estimates | | | | | Compared Model | *p*-Value | Positive Sites |
|---|---|---|---|---|---|---|---|---|---|---|
| | | | Site class | 0 | 1 | 2a | 2b | | | |
| Model A | 446 | −3166.858 | f<br>$\omega_0$<br>$\omega_1$ | 0.000<br>0.229<br>0.229 | 0.000<br>1.000<br>1.000 | 0.156<br>0.229<br>999.000 | 0.844<br>1.000<br>999.000 | Model A vs. Model A null | >0.05 | Not allowed |
| Model A null | 445 | −3165.206 | 1 | | | | | | | Not allowed |

**Table 4.** Site model parameters of selective pressure analysis.

| Model | np | ln L | Parameters Estimates | Compared Model | *p*-Value | Positive Selected Sites |
|---|---|---|---|---|---|---|
| $M_0$ | 443 | −3144.848 | $\omega = 0.318$ | $M_0$ vs. $M_3$ | <0.001 | Not allowed |
| $M_3$ | 447 | −3054.877 | *p*: 0.266, 0.519, 0.216<br>$\omega$: 0.095, 0.432, 0.955 | | | Not allowed |
| $M_{1a}$ | 444 | −3071.511 | *p*: 0.330, 0.670<br>$\omega$: 0.193, 1.000 | $M_{1a}$ vs. $M_{2a}$ | <0.001 | Not allowed |
| $M_{2a}$ | 446 | −3062.176 | *p*: 0.289, 0.528, 0.183<br>$\omega$: 0.204, 1.000, 2.045 | | | Not allowed |
| $M_7$ | 444 | −3055.573 | $p = 0.760$, $q = 1.042$ | $M_7$ vs. $M_8$ | >0.05 | Not allowed |
| $M_8$ | 446 | −3053.688 | $p_0 = 0.778$<br>$p = 0.852$ | | | 11 E 0.616 |

### 3.4. Functional Divergence Analysis

Gene families comprise multiple homologous genes with similar structures but different functions. In order to see whether changes in amino acid sequences within the MIKC-type MADS-box genes led to adaptive functional variation, we investigated the functional divergence of five distinct groups. The results suggest that, except for clades I/V, most type I coefficient $\theta$ values across the clades were statistically > 0, which indicates the presence of type I functional divergences across the clades. Six clades (I/II, I/III, II/III, II/IV, III/IV, and III/V) were remarkable ($p < 0.001$) for type I functional divergence. Clades III/V had the highest functional divergence coefficient θ (= 1.150), whereas clades I/V had the lowest (=−0.002). Other clades' type I θ values also varied, indicating that their rates of evolution were unique. In addition, type I functionally divergent sites were not detected in clades I/II, I/V, II/V, III/V, and IV/V (Table 5).

**Table 5.** Functional divergence analysis of MIKC-type MADS gene family in *Rosaceae*.

| Clade | θ ± SE | | MFE z-Scores | *p*-Value | | Amino Acid Position with Q(k) > 0.9 |
|---|---|---|---|---|---|---|
| | Type I | Type II | | Type I | Type II | |
| I/II | 0.579 ± 0.156 | −0.112 ±0.189 | −4.115715 | <0.001 | >0.05 | — |
| I/III | 0.871 ± 0.158 | 0.005 ± 0.182 | −6.191108 | <0.001 | >0.05 | 193,299,307,316.332,341,343, 346,359,370,371,374,386,397,400,403,407 |
| I/IV | 0.327 ± 0.225 | −0.081 ± 0.207 | −1.505518 | >0.001 | >0.05 | 190,301,302,310,311,316.346,349.361, 388,403 |
| I/V | −0.002 ± 0.263 | −0.660 ± 0.379 | 0.009046 | >0.001 | >0.05 | — |
| II/III | 1.058 ± 0.159 | 0.299 ± 0.050 | −7.305373 | <0.001 | >0.05 | 190,192,193,195,272,273,290,292,295, 297,298,299,300,301,303,305,307,314, 316,318,320,321,332,342,343,345,347, 349,356,358,359,366,370,371,373,374, 385,386,387,395,396,397,403,404,405, 406,409,411 |
| II/IV | 0.874 ± 0.239 | 0.193 ± 0.070 | −3.826513 | <0.001 | >0.05 | 193,271,272,290,296,298,299,301,302, 305,310,311,313,316,319,320,331,341, 347,350,352,361,370,371,373,382,388, 394,397,403,409 |
| II/V | 0.831 ± 0.288 | −0.221 ± 0.203 | −2.974702 | >0.001 | >0.05 | — |
| III/IV | 1.085 ± 0.232 | 0.232 ± 0.063 | −4.882459 | <0.001 | >0.05 | 190,193,195,271,292,296,301,302,307, 310,311,313,314,319,321,331,332,341, 343,345,347,349,350,352,358,359,361, 374,382,386,387,388,394,395,404,406 |
| III/V | 1.150 ± 0.279 | −0.233 ± 0.205 | −4.220686 | <0.001 | >0.05 | — |
| IV/V | 0.205 ± 0.410 | −0.425 ± 0.248 | −0.50314 | >0.001 | >0.05 | — |

### 3.5. Functional Distance Analysis

Type I functional divergence-related cluster analyses were unable to determine which groups of duplicated genes were hampered by functional bottlenecks. Therefore, we compared the bF values across the five clades and found that they changed, indicating that unique roles evolved in the MIKC-type MADS-box gene family throughout time. Furthermore, the MIKC-type MADS-box genes tended to be conserved, as shown by bF values above 0. The lowest bF value (−2.453) was found in clade III, suggesting that the functions of the genes in this clade were more conserved and may have kept more of their original forms during evolution. With the longest functional branch length (0.405), clade I likely underwent considerable alterations to its functional restrictions in order to fulfill its specialized functional purpose (Table 6).

**Table 6.** Type I functional distance analysis of MIKC-type MADS gene family in *Rosaceae*.

| Clade | bF |
|---|---|
| I | 0.405 |
| II | −1.270 |
| III | −2.453 |
| IV | −0.801 |
| V | −0.403 |

## 4. Discussion

The MADS-box gene family plays an essential role in plant reproductive growth, and it also provides a research hotspot. However, most studies on the MADS-box gene family in *Rosaceae* mainly focus on identification and functional analysis in only one species [36–39]. In order to understand the variation pattern of MADS-box gene structure and function among different species, our study performed MIKC-type MADS-box gene identification between six key species in *Rosaceae*, and investigated their evolutionary relationship. These results are of great significance in understanding the MADS-box family function.

In the present study, we identified 222 MIKC-type MADS-box genes within six *Rosaceae* species. Among them, there were 54, 27, 23, 69, 23, and 26 genes in *Prunus dulcis*, *Prunus salicina*, *Prunus armeniaca*, *Prunus persica*, *Prunus mira*, *and Amygdalus nana*, respectively, which was almost consistent with other findings in *Rosaceae* [36,37]. However, the number of MIKC-type MADS-box family members was quite different from other families. For instance, comparing with 232 members in legumes (92 genes in *Glycine max*, 45 genes in *Medicago truncatula*, 50 genes in *Vitis vinifera*, and 45 genes in *Arabidopsis thaliana*), the number was generally smaller in *Rosaceae* [40]. In the studies of conserved motifs and structure analysis, we found that no motif was shared by all MIKC-type MADS-box proteins, however Motif 1 was shared by almost all proteins, with the exception of PdMADS42 and PmiMADS16. These results suggested that Motif 1 was conserved across members of this gene family, and that PdMADS42 and PmiMADS16 had distinct purposes. Additionally, different species had varied sequence lengths, although the exon number and length were highly similar. Identified from *Prunus salicina*, the PsMADS3, PsMADS4, PsMADS7, PsMADS8, and PsMADS24 had no introns, and the gene annotation did not determine the intron structure. The result showed that intron-involving alleles or inversion might have deleted these introns due to homologous recombination, after the reverse transcription of mature mRNA [41].

Phylogenetic analysis revealed five distinct clades within the MIKC-type MADS-box gene family of the six *Rosaceae* species. The six species genes could be found in almost every clade, but *Prunus armeniaca* did not exist in clade II. The results showed that the six species could not be distinguished by them, and the genes in clade II might have some special functions which were missing in *Prunus armeniaca*. According to previous research, we found that *Prunus armeniaca* had some differences in the phenotypic characteristics of its flowers and fruit compared with the other five species. For instance, the petal was round-to-obovate, and the calyx tube was cylindrical with short fluff at base; moreover, the florescence and fruit stage were relatively early in *Prunus armeniaca* [42]. With further analysis and verification, we may be able to obtain several specific genes associated with these traits.

Using three models (branch, branch-site, and site models) to perform selective pressure analysis of the six species, we obtained only one positive site in the site model, and there were no positive sites in the branch and branch-site models. It suggested that various places undergo various selection forces, but the evolution and function of MIKC-type MADS-box genes were conserved, and purifying selection limited the MIKC-type MADS-box gene family in *Rosaceae* [43,44]. Inadequate site-related positive selection data may have been eliminated because they were reduced by the vast majority of the remainder of sites being under the evolutionary condition of neutral or purifying selection [45]. In addition, we found the results of the branch and branch-site models indicated no significant differences, which might be due to the six species being sibling species, or the conformity of the genomes and annotation level among the six species; nevertheless, further analysis is required to clearly understand the reasons.

Functional divergence analysis may be utilized to outline the putative residues of amino acids that cause the functional diversification of gene families, which occurs as a result of gene duplication. In this investigation, we found Type I functionality was shown to vary significantly among clades, and obtained locations that had functional divergence ($Q(k) > 0.90$). It suggested that the functional divergence of the current *Rosaceae* gene family was mostly attributable to the differing proportions of evolution following genetic duplications. In addition, members of clade III kept more of their original roles throughout evolution, whereas those of clade I specialized theirs to serve a specific purpose. In future, we can research the gene functions of clade I to analyze their different functions throughout their evolution.

In order to better study the gene mutations in *Rosaceae*, our research facility has assembled the genome-wide sequence of *Amygdalus nana* for the first time. *Amygdalus nana* is a treasured wild species of the Tertiary period deciduous forests in the ancient Mediterranean [46]. At present, there are only several natural populations in China, Kazakhstan, and Russia, which are listed as endangered plants in China [47]. The kernel of *Amygdalus nana*, a vital species of non-timber forestry, has an oil content of 51.1%, similar to that of almond and higher than that of oil crops, such as rapeseed, cottonseed, and soybean [48]. In addition, having a strong ability to adapt to an extremely harsh environment, especially with cold and drought tolerance, it can grow in the hilly and mountainous areas at −35 °C and annual precipitation of 50mm. Therefore, further research on *Amygdalus nana* may shed light on the evolution of and relationships between *Rosaceae* species.

## 5. Conclusions

In this study, the evolution of MIKC-type MADS-box gene family members in six *Rosaceae* plants was analyzed for the first time. 222 MIKC-type MADS-box family genes were identified from *Prunus dulcis*, *Prunus salicina*, *Prunus armeniaca*, *Prunus persica*, *Prunus mira*, and *Amygdalus nana*. The results of sequence and phylogenetic analyses showed that these genes were divided into five groups, the six species genes could be found in almost every clade, but *Prunus armeniaca* did not exist in clade II. Using selective evolutionary pressure and functional difference analyses, a number of positive selection sites and functional divergence locations were obtained. In addition, the conservation between different clades was evaluated, and it was concluded that the genes of clade I were specialized to serve a specific purpose. These results provide great help in the further mining of functional genes for key traits.

**Author Contributions:** Formal analysis, Y.Q., L.W. and C.C.; data curation, Y.Q., F.L. and H.Z.; writing—original draft preparation, Y.Q. and G.Z.; writing—review and editing, H.Z. and F.L.; visualization, L.W. and C.C.; project administration, H.Z. and G.Z. All authors have read and agreed to the published version of the manuscript.

**Funding:** This work was funded by the Key Research and Development Program of Xinjiang Uygur Autonomous Region (2022292937) and the National Key Research and Development Program of China (2022YFD2200400).

**Data Availability Statement:** Required data might be provided following a proper request from the senior author.

**Conflicts of Interest:** The authors declare no conflict of interest.

## References

1. Wang, Y.; Mu, Y.X.; Wang, J. Research progress of floral development regulation by MADS-box gene family. *Acta Agric. Zhejiangensis* **2021**, *33*, 1149–1158.
2. Kim, S.H.; Mizuno, K.; Fujimura, T. Isolation of MADS-box genes from sweet potato [*Ipomoea batatas* (L.) Lam.] expressed specifically in vegetative tissues. *Plant Cell Physiol.* **2002**, *43*, 314–322. [CrossRef] [PubMed]
3. Zhao, H.B.; Jia, H.M.; Wang, Y.; Wang, G.Y.; Zhou, C.C.; Jia, H.J.; Gao, Z.S. Genome-wide identification and analysis of the MADS-box gene family and its potential role in fruit development and ripening in red bayberry (*Morella rubra*). *Gene* **2019**, *717*, 144045. [CrossRef]
4. Yanofsky, M.F.; Ma, H.; Bowman, J.L.; Drews, G.N.; Feldmann, K.A.; Meyerowitz, E.M. The protein encoded by the Arabidopsis homeotic gene agamous resembles transcription factors. *Nature* **1990**, *346*, 35–39. [CrossRef] [PubMed]
5. Schwarzsommer, Z.; Hue, I.; Huijser, P.; Flor, P.J.; Hansen, R.; Tetens, F.; Lonnig, W.E.; Saedler, H.; Sommer, H. Characterization of the antirrhinum floral homeotic MADS-box gene deficiens-evidence for DNA-bingding and autoregulation of its persistent expression throughout flower development. *EMBO J.* **1992**, *11*, 251–263. [CrossRef]

6. Dreni, L.; Zhang, D. Flower development: The evolutionary history and functions of the AGL6 subfamily MADS-box genes. *J. Exp. Bot.* **2016**, *67*, 1625–1638. [CrossRef]
7. Alvarez-Buylla, E.R.; Pelaz, S.; Liljegren, S.J.; Gold, S.E.; Burgeff, C.; Ditta, G.S.; Pouplana, L.R.D.; Martinez-Castilla, L.; Yanofsky, M.F. An ancestral MADS-box gene duplication occurred before the divergence of plants and animals. *Proc. Natl. Acad. Sci. USA* **2000**, *97*, 5328–5333. [CrossRef]
8. Riechmann, J.L.; Meyerowttz, E.M. MADS domain proteins in plant development. *Biol. Chem.* **1997**, *378*, 1079–1101.
9. Zhao, X.Y.; Xian, D.Y.; Song, M.; Tang, Q.L. Research progress of MIKC-type MADS-box protein regulation on flowering. *Biotechnol. Bull.* **2014**, *7*, 8–15.
10. Susanne, S.; Alice, K.; Sirui, P.; Lars, S.J.; Rainer, M. Genome-wide analysis of MIKC-type MADS-box genes in wheat: Pervasive duplications, functional conservation and putative neofunctionalization. *New Phytol.* **2020**, *225*, 511–529.
11. Chunmei, H.; Can, S.; Jaime, A.T.S.; Li, M.Z.; Duan, J. Genome-wide identification and classification of MIKC-type MADS-box genes in *Streptophyte lineages* and expression analyses to reveal their role in seed germination of orchid. *BMC Plant Biol.* **2019**, *19*, 223.
12. Jiang, S.C.; Pang, C.Y.; Song, M.Z.; Wei, H.L.; Fan, S.-l.; Yu, S.X. Analysis of MIKCc-type MADS-box gene family in *Gossypium hirsutum*. *J. Integr. Agric.* **2014**, *13*, 1239–1249. [CrossRef]
13. Gao, W.; Zhang, L.; Xue, C.L.; Zhang, Y.; Liu, M.J.; Zhao, J. Expression of E-type MADS-box genes in flower and fruits and protein interaction analysis in Chineses Jujube. *Acta Hortic. Sin.* **2022**, *49*, 739–748.
14. Huang, H.J.; Wang, S.; Jiang, J.; Liu, G.; Li, H.; Chen, S.; Xu, H. Overexpression of BpAP1 induces early flowering and produces dwarfism in *Betula plantyphylla* × *Betula pendula*. *Physiol. Plant* **2014**, *151*, 495–506. [CrossRef] [PubMed]
15. Liu, S.; Qi, T.; Ma, J.; Ma, L.; Liu, X. Cloning and functional analysis of SEP-like gene from *Phyllostachys violascens*. *J. Nucl. Agric. Sci.* **2016**, *30*, 1453–1459.
16. Parenicová, L.; Folter, S.D.; Kieffer, M.; Horner, D.S.; Colombo, L. Molecular and phylogenetic Analyses of the complete MADS-box transcription factor family in Arabidopsis new openings to the MADS World. *Plant Cell* **2003**, *15*, 1538–1551. [CrossRef]
17. Zhang, T.; Hu, Y.; Jiang, W.; Fang, L. Sequencing of allotetraploid cotton(*Gossypium hirsutum* L. acc. TM-1) provides a resource for fiber improvement. *Nat. Biotechnol.* **2015**, *33*, 531–537. [CrossRef]
18. Prakash, A.; Jeffryes, M.; Bateman, A.; Finn, R.D. The HMMER Web server for protein sequence similarity search. *Curr. Protoc. Bioinform.* **2017**, *60*, 3–15. [CrossRef]
19. Katoh, K.; Standley, D.M. A simple method to control over-alignment in the MAFFT multiple sequence alignment program. *Bioinformatics* **2016**, *32*, 1933–1942. [CrossRef]
20. Timothy, L.B.; James, J.; Charles, E.G.; William, S.N. The MEME Suite. *Nucleic Acids Res.* **2015**, *43*, W39–W49.
21. Crooks, G.E.; Hon, G.; Chandonia, J.M.; Brenner, S.E. WebLogo: A sequence logo generator. *Genome Res.* **2004**, *14*, 1188–1190. [CrossRef] [PubMed]
22. Darriba, D.; Taboada, G.L.; Doallo, R.; Posada, D. Prottest 3: Fast selection of best-fit models of protein evolution. *Bioinformatics* **2011**, *27*, 1164–1165. [CrossRef] [PubMed]
23. Sudhir, K.; Glen, S.; Michael, L.; Christina, K.; Koichiro, T. MEGA X: Molecular Evolutionary Genetics Analysis across Computing Platforms. *Mol. Biol. Evol.* **2018**, *35*, 1547–1549.
24. Yu, G. Using ggtree to visualize data on tree-like structures. *Curr. Protoc. Bioinform.* **2020**, *69*, e96. [CrossRef]
25. Yu, G.; Lam, T.T.Y.; Zhu, H.; Guan, Y. Two methods for mapping and visualizing associated data on phylogeny using ggtree. *Mol. Biol. Evol.* **2018**, *35*, 3041–3043. [CrossRef]
26. Yu, G.; Smith, D.K.; Zhu, H.; Guan, Y.; Lam, T.T.Y. ggtree: An R package for visualization and annotation of phylogenetic trees with their covariates and other associated data. *Methods Ecol. Evol.* **2017**, *8*, 28–36. [CrossRef]
27. Gao, F.; Chen, C.; Arab, D.A.; Du, Z.; He, Y.; Ho, S.Y.W. EasyCodeML: A visual tool for analysis of selection using CodeML. *Ecol. Evol.* **2019**, *9*, 3891–3898. [CrossRef]
28. Yang, Z.H. PAML4: Phylogenetic analysis by maximum likelihood. *Mol. Biol. Evol.* **2007**, *24*, 1586–1591. [CrossRef]
29. Yang, Z.H.; Nielsen, R. Synonymous and nonsynonymous rate variation in nuclear genes of mammals. *J. Mol. Evol.* **1998**, *46*, 409–418. [CrossRef]
30. Yang, Z.H.; Nielsen, R. Codon-substitution models for detecting molecular adaptation at individual siyes along specific lineages. *Mol. Biol. Evol.* **2002**, *19*, 908–917. [CrossRef]
31. Yang, Z.H.; Wong, W.S.; Nielsen, R. Bayes empirical bayes inference of amino acid sites under positive selection. *Mol. Biol. Evol.* **2005**, *22*, 1107–1118. [CrossRef] [PubMed]
32. Gu, X.; Zou, Y.Y.; Su, Z.X.; Huang, W.; Zhou, Z.; Arendsee, Z.; Zeng, Y.W. An update of DIVERGE software for functional divergence analysis of protein family. *Mol. Biol. Evol.* **2013**, *30*, 1713–1719. [CrossRef] [PubMed]
33. Gu, X.; Velden, K.V. DIVERGE: Phylogeny-based analysis for functional-structural divergence of a protein family. *Bioinformatics* **2002**, *18*, 500–501. [CrossRef]
34. Gu, X. A simple statistical method for estimating type-II(cluster-specific) functional divergence of protein sequences. *Mol. Biol. Evol.* **2006**, *23*, 1937–1945. [CrossRef] [PubMed]
35. Wang, Y.F.; Gu, X. Functional divergence in the caspase gene family and altered funcional constrains: Statistical analysis and prediction. *Genetics* **2001**, *158*, 1311–1320. [CrossRef]

36. Christina, E.W.; Elisa, V.; Sergio, J.T.; Lgnazio, V.; Douglas, G.B. A genome-wide analysis of MADS-box genes in peach [*Prunus persica* (L.) Batsch]. *BMC Plant Biol.* **2015**, *15*, 41.
37. Hisayo, Y.; Tomomi, O.; Hiroaki, J.; Yukari, H.; Ryuta, S.; Ryutaro, T. Expressional regulation of PpDAM5 and PpDAM6, peach (*Prunus persica*) dormancy-associated MADS-box genes, by low temperature and dormancy-breaking reagent treatmen. *J. Exp. Bot.* **2011**, *62*, 3481–3488.
38. Liu, X.Y.; He, Z.J.; Qiu, Y.M. Screening and Bioinformatics analysis of almond MADS-box Gene family. *Mol. Plant Breed.* **2022**, *20*, 1477–1486.
39. Chen, C.; Zhu, G.P.; Zhao, H.; Liu, H.M.; Luo, Y.; Xu, W.Y.; Huang, M.Z.; Wu, Y.T.N.; Wang, L. Genome-wide identification of MADS-box Gene family and expression analysis in *Prunus sibirica*. *Mol. Plant Breed.* **2020**, *18*, 6575–6585.
40. Zhang, Y.; Wang, J.; Yu, Z.; Xu, Q.; Zhang, L.; Pan, Y. Bioinformatics analysis of MIKC-type MADS-box gene family in legumes. *Chin. J. Oil Crop Sci.* **2022**, *44*, 798–809.
41. Wu, K.L.; Guo, Z.J.; Wang, H.H.; Li, J. The WRKY family of transcription factors in rice and Arabidopsis and their origins. *DNA Res.* **2005**, *12*, 9–26. [CrossRef] [PubMed]
42. Yv, T.; Lu, L.; Ku, T.; Li, C.; Chen, S. *Flora Reipublicae Popularis Sinicae*; Science Press: Beijing, China, 1986; Volume 38, pp. 11–40.
43. Vallender, E.J.; Lahn, B.T. Positive selection on the human genome. *Hum. Mol. Genet.* **2004**, *13*, R245–R254. [CrossRef] [PubMed]
44. Zhang, J.Q.; Zhu, K.Y.; Shi, X.H. Adaptive evolution and identification analysis of the MADS-box gene family in *Paeonia lactiflora*. *Mol. Plant Breed.* **2019**, *17*, 6959–6966.
45. Tian, Y.X.; Wang, Q.G.; Zhang, H.; Zhou, N.N.; Yan, H.J.; Jian, H.Y.; Li, G.S.; Tang, K.X.; Qiu, X.Q. Genome-wide identification and evolutionary analysis of MLO gene family in Rosaceae plants. *Hortic. Plant J.* **2022**, *8*, 110–122. [CrossRef]
46. Ma, S.; Wang, C.; Sun, F.; Wei, B.; Nie, Y. Genetic diversity of an endangered plant *Amygdalus ledebouriana* in Xinjiang. *Sci. Silvae Sin.* **2019**, *9*, 71–80.
47. Yan, G.; Xu, Z. Study on the wild fruit tree diseases of Tianshan mountains and their distribution in Xinjiang. *Arid. Zone Res.* **2001**, *18*, 47–49.
48. Wu, Y.M.; Wu, Y.A.; Wu, Y.X.; Cao, Z. The advances of the studies on almond (*Amygdalus communis* L.) A literature review. *J. Gansu Agric. Univ.* **1996**, *1*, 86–92.

agronomy

MDPI

Article

# *PsFT*, *PsTFL1*, and *PsFD* Are Involved in Regulating the Continuous Flowering of Tree Peony (*Paeonia* × *lemoinei* 'High Noon')

**Limei Zhang [1], Fangyun Cheng [1,*], He Huang [1], Ziwen Geng [1] and Chaoying He [2]**

[1] State Key Laboratory of Efficient Production of Forest Resources, Beijing Key Laboratory of Ornamental Plants Germplasm Innovation & Molecular Breeding, National Engineering Research Center for Floriculture, Key Laboratory of Genetics and Breeding in Forest Trees and Ornamental Plants of Ministry of Education, Peony International Institute, School of Landscape Architecture, Beijing Forestry University, Beijing 100083, China; limeizhang92@163.com (L.Z.)

[2] State Key Laboratory of Systematic and Evolutionary Botany, Institute of Botany, Chinese Academy of Sciences, Beijing 100093, China; chaoying@ibcas.ac.cn

* Correspondence: chengfy@bjfu.edu.cn

**Abstract:** Tree peonies are an economically important crop with flowers of high ornamental value. Most tree peony cultivars in gardens are once-flowering, and the continuous flowering (CF) trait has been revealed only in a few tree peony cultivars, such as 'High Noon' ('HN'). However, the molecular mechanism underlying its CF remains unclear. Here, we demonstrated that PsTFL1 functions as a floral inhibitor via the ectopic expression of *PsTFL1* in transgenic *Arabidopsis thaliana* plants. Our findings suggest that PsFT and PsTFL1 interact with PsFD, and the detected interactions may occur in the nucleus. Compared with the non-CF variety, the gene expression patterns of *PsFT*, *PsTFL1*, and *PsFD* during the flower development indicate that these three genes may be related to the CF habit in tree peony 'HN'. These findings will aid future investigations of CF behavior and promote the breeding of tree peonies and other perennial woody plants.

**Keywords:** tree peony; continuous flowering; *FT*; *TFL*; *FD*

**Citation:** Zhang, L.; Cheng, F.; Huang, H.; Geng, Z.; He, C. *PsFT*, *PsTFL1*, and *PsFD* Are Involved in Regulating the Continuous Flowering of Tree Peony (*Paeonia* × *lemoinei* 'High Noon'). *Agronomy* **2023**, *13*, 2071. https://doi.org/10.3390/agronomy13082071

Academic Editors: Jinzhi Zhang and Avi Sadka

Received: 2 July 2023
Revised: 26 July 2023
Accepted: 2 August 2023
Published: 6 August 2023



## 1. Introduction

The transition from vegetative to reproductive development (flowering) requires environmental signals, such as the photoperiod and vernalization, which trigger the expression and regulation of genes and their integration in the shoot apical meristem (SAM). Continuous flowering (CF), a unique reproductive process, has been reported to be regulated by flowering integrator genes [1–3]. Some flowering integrator genes include *FLOWERING LOCUS T* (*FT*), *TERMINAL FLOWER1* (*TFL1*), *SUPPRESSOR OF OVEREXPRESSION OF CONSTANS1* (*SOC1*), and *LEAFY* (*LFY*) [4–7], and the roles of *FT* and *TFL1* have been widely studied in *Arabidopsis* [8], *Antirrhinum* [9], *Rosa* [1,10,11], *Jatropha* [12], and many other species. The FT and TFL1 belong to the phosphatidyl ethanolamine-binding protein (PEBP) family; however, FT promotes flowering, and TFL1 represses it [1,4,8–13]. Structural studies of FT and TFL1 have revealed that they both exist in an external loop along with an adjacent peptide, serving as a potential ligand-binding site. The structures formed by the rest of the proteins play a key role in determining the antagonistic functions of FT and TFL1 [14]. The molecular antagonistic mechanism between these two genes has also been investigated in some plants. In contrast to the role of *FT* as a florigen that promotes plant flowering, the overexpression or RNAi silencing of *TFL1* genes in rape [15], ragweed [16], apple [17], and pear [18] results in delayed or early flowering.

FD, a member of the *basic region leucine zipper* (*bZIP*) family, interacts with FT and can have a major effect on early flowering [14]. FT has been reported to be a florigen in previous studies [10,14,19,20], and an FT/FD heterodimer formed from the combination of FT with

FD binds to the promoter of *APETALA1* (*AP1*) to activate flowering initiation [19–22]. In rice, 14-3-3 (the FT-like protein in rice) and FD form a hexametric structure, which has been referred to as the 'florigen activation complex' [13]. However, in *Arabidopsis*, *TFL1* represses the floral transition and maintains the indeterminacy of the shoot apex [17,23,24]. In contrast to loss-of-function phenotypes, overexpression of *TFL1* delays flowering and inhibits the transition of the inflorescence meristem (IM) to the floral meristem (FM), resulting in an extended IM stage [25,26]. Therefore, *TFL1* as a negative regulator regulates the phase transition from vegetative growth to reproductive growth and from IM to FM in the shoot apical meristem (SAM). *TFL1* is involved in transcriptional repression mechanisms that control FM identity genes, which are also FT-targeted [27]. Studies of the fusion of *TFL1* with activator or repressor domains have revealed that *TFL1* does not possess the functional domains commonly found in transcription factors or regulators. Therefore, TFL1 and FT act as adhesive proteins rather than transcriptional regulators and interact with FD-like proteins [14].

In Chinese rose and woodland strawberry, the recessive mutation of *KSN* (homolog of *TFL1*) results in continuous flowering and determinate/indeterminate growth, making the *TFL1* homologs provide promising candidate genes for examining the remontancy of flowering [10]. The TFL1 protein sequence is highly similar to FT, and TFL1 shares 71% of amino acid residues, including 55% identical residues, with FT [27]. Because of a retrotransposon insertion, the transcription of *RoKSN* is blocked in CF roses, and the absence of the floral repressor induces continuous blooming [11]. In *Fragaria vesca*, a 2-bp deletion in the coding region of the *TFL1* homolog introduces a frame shift that leads to CF [11]. In *Rosa*, fluorescence resonance energy transfer (FRET) assays have been used to clarify competition between RoFT and RoKSN for interaction with RoFD. The FRET results suggest that RoFT and RoKSN are able to interact strongly with RoFD, with a FRET value of >10% [10]. Advances in our understanding of CF in model plants can provide insights into the remontancy and CF of other species such as tree peonies.

Tree peony species, or Mudan, is a woody species in the genus *Paeonia*. Known as the 'king of flowers', it has been cultivated in China since the Tang Dynasty (618 A.D.). More than 1000 cultivars of tree peonies have been released worldwide [28], and they can be divided into three flowering classes: once-flowering (OF), autumn-flowering (AF), and continuous-flowering (CF) classes [5,28–30]. Particularly, the CF trait has drawn more attention, but the molecular regulatory mechanisms underlying CF in tree peonies remain elusive. We previously cloned *PsFT* and showed that its overexpression can promote flowering in transgenic *Arabidopsis thaliana* plants, suggesting that *PsFT* may be associated with flowering [29,30]. In CF roses, some studies have revealed that the interaction between *FT*, *TFL1*, and *FD* in flower bud differentiation is related to the regulatory mechanism of CF [1,10]. Does such a regulatory mechanism exist in CF tree peonies? In this study, we isolated *PsTFL1* and *PsFD* in tree peonies and analyzed the sequences of *FT*/*TFL1*/*FD* and the structures of the proteins that they encode. We also studied the expression patterns of *PsFT*, *PsTFL1*, and *PsFD* during bud differentiation in CF and OF tree peonies and explored their protein–protein interactions. These findings provide insights into the understanding of the molecular mechanism of CF in tree peonies and will aid future studies concerning the CF habits of plants, especially in ornamental species such as roses and tree peonies.

## 2. Materials and Methods

### 2.1. Plant Materials

Two classes of tree peonies with different flowering traits were used in this study: CF and OF. The CF cultivar *P.* × *lemoinei* 'HN' was obtained from the Jiufeng Experiment Station of Beijing Forestry University (Haidian District, Beijing, China, 40°03′ N, 116°05′ E). *P. delavayi* var. *lutea*, a potential CF resource, was obtained from GuoSe Peony Nursery (Yanqing district, Beijing, China, 40°45′ N, 115°97′ E). The OF cultivars *P. suffruticosa* 'LYH' were also obtained from GuoSe Peony Nursery. Changes in gene expression during the floral differentiation process in the common buds of CF 'HN', the potential CF potential

*P. delavayi* var. *lutea*, and the OF 'LYH' were analyzed. The flower buds were collected from 25 May to 5 September every 10 days in CF 'HN', OF 'LYH', and the potential CF potential *P. delavayi* var. *lutea* was analyzed also. Common flower bud initiation coincides with the enlargement of the meristem, and bract primordia become visible in early July, according to Zhou et al. [29]. The apical buds used for expression analysis were immediately frozen in liquid nitrogen. Changes in gene expression in the CF bud of 'HN' were also characterized during the differentiation process. CF buds in 'HN' were divided into four periods (bud sprouting, S1; bud development and elongation, S2; flower establishment, S3; and flower expansion, S4) according to their morphological characteristics. To compare the expression patterns in common buds in the dormant stage, we collected common buds from 'HN'.

### *2.2. Sequence Isolation*

Total RNA was extracted using an Aidlab RNA Isolation Kit (Aidlab Biotechnologies Co., Ltd., Beijing, China), and cDNA was synthesized using the FastQuant RT Kit with DNase (Tiangen Biotech, Beijing, China). The raw gene fragments screened from the transcriptome sequencing database were sequenced in a previous study [29]. Polymerase chain reaction (PCR) and the rapid amplification of cDNA ends (SMARTer® RACE 5′/3′Kit) were used to obtain *TFL1* and *FD* from buds. The primers used are presented in Table S1. The *FT* gene was sequenced in a previous study [30].

### *2.3. Protein Structure and Phylogenetic Analysis*

The Online SWISS-MODEL (https://swissmodel.expasy.org/interactive, (accessed on 18 August 2022)) software was used to predict the secondary structures of PsFT, PsTFL1, and PsFD. The sequences from other species were analyzed along with sequences from 'HN'. The complete nucleotide sequences, protein sequences, and corresponding annotation information for all these species were downloaded from the JGI and NCBI databases. The BLAST method was used to identify homologous genes of *FT* in 10 species (or cultivars), *TFL1* in 11 species (or cultivars), and *FD* in 14 species (or cultivars) from the protein database (Table S2). We used RA × ML software v1.0 to construct the maximum-likelihood (ML) phylogenetic tree. The tree topology was visualized using online software (https://itol.embl.de/, (accessed on 9 May 2023)). Support (BS) values for each node were calculated based on 10,000 bootstrap replicates.

### *2.4. Expression Analysis*

To investigate the expression levels of *PsFT*, *PsTFL1*, and *PsFD* at different times during flower bud differentiation, we conducted real-time fluorescent quantitative PCR using a CFX96 Real-Time PCR System (Bio-rad, Hercules, CA, USA) in a 20-μL reaction volume comprising 10 μL of FastStart Universal SYBR Green II Master Mix (TaKaRa, Tokyo, Japan), 6 μL of $ddH_2O$, 1 μL of 10 mM forward and reverse primers, and 2 μL of diluted cDNA as the template for target genes. Real-time PCR was performed with the following thermal cycling conditions 95 °C for 10 min, followed by 40 cycles of 2 min at 95 °C, 5 s at 95 °C, and 30 s at annealing temperature (*PsFT*: 55 °C; *PsTFL1*: 50 °C; *PsFD*: 60 °C). Melting curve analysis was conducted using the default settings of the CFX96 Real-Time PCR System (Bio-rad, Hercules, CA, USA). The annealing temperature ranged from 50–60 °C (*PsFT*: 55 °C, *PsTFL1*: 50 °C, *PsFD*: 60 °C, and reference gene: 60 °C). There were three replicates for each sample. The qRT-PCR results were analyzed using the $2^{-\Delta\Delta Ct}$ method [31], and all primers used in this study are listed in Table S1.

### *2.5. Subcellular Localization*

The full-length open reading frame (ORF) of three genes (*PsFT*, *PsTFL1*, and *PsFD*) were inserted into the 35S GFP vector. Three vectors (*35S::FT-GFP*, *35S::TFL1-GFP*, and *35S::FD-GFP*) were generated and transformed into tobacco epidermal cells for two days under low-light density and room-temperature conditions (Invitrogen, Carlsbad, CA, USA). Autofluorescence showed the chlorophyll fluorescence signal was excited at a wavelength of 640 nm and emitted at a wavelength of 675 nm. The protoplasts were observed via a laser-scanning confocal microscope (Leica, Wetzlar, Germany).

### 2.6. Plant Transformation

Full-length *PsTFL1* ORF was cloned onto the vector pEarleyGate 100. The resulting construct, *PsTFL1-PEG100*, was transformed into *A. thaliana* (Col-0) using the floral dip method [30]. After vernalization for 2 days at 2 °C, the transformants ($T_0$ lines) from the dipped plants that survived exposure to the Basta solution (1/500) were transplanted into a substrate consisting of vermiculite and nutrient soil (*v:v* = 1:1) in a growth chamber at 21 °C under long-day (LD) conditions (16 h photoperiod, light intensity: 50 $mmol{\cdot}m^{-2}{\cdot}s^{-1}$). Morphological observations of 15 plants from the second generation ($T_2$) were recorded. The number of rosette leaves and the day when the first flower bloomed were determined.

### 2.7. Yeast 2-Hybrid (Y2H) Analysis

Y2H analysis was used to clarify the interactions among PsFT, PsTFL1, and PsFD using the Y2H gold system (Clontech, Mountain View, CA, USA). The ORFs of *PsFT* and *PsTFL1* (without signal peptide) were introduced into the pGBKT7 vectors as bait. The ORFs of *PsFD* and *PsTFL1* (without signal peptide) were introduced into pGADT7 as prey. Transformed yeast cells were assayed for growth on synthetic dropout SD/-Leu with agar, SD/-Leu/-Trp with agar, SD/-Leu/-Trp/-His with agar, and SD/-Ade/-His/-Leu/-Trp with agar containing X-α-galactosidase (X-α-gal) and aureobasidin A (AbA).

### 2.8. Bimolecular Fluorescence Complementation (BiFC)

The coding regions of *PsTFL1* and *PsFT* were introduced into the pCAMBIA1300-35S-YC155 vector and *PsFD* into the pCAMBIA1300-35S-NY173 vector using a Gateway cloning system (Invitrogen, Carlsbad, CA, USA). The loaded vectors were transferred into *Agrobacterium tumefaciens* (GV3101) and then infiltrated into the abaxial air spaces of 4-week-old *Nicotiana benthamiana* plants. Fluorescent signals and bright-field pictures were taken using a confocal laser scanning microscope (LeicaSP8, Wetzlar, Germany).

### 2.9. Co-IP Assay

To confirm the heterodimization of PsFT-PsFD and PsTFL1-PsFD, we performed protein co-immunoprecipitation assays. The full-length coding sequences of *PsFT* and *PsTFL1* were recombined into the pBinGFP2 vector to be expressed as *PsFT-GFP* and *PsTFL1-GFP*, respectively, and *PsFD* was recombined into the pCAMBIA vector to be expressed as *PsFD-Myc*. Three-week-old *N. benthamiana* plants were transiently co-expressed with anti-flag antibodies (Abmart) using a Pierce Crosslink IP kit (Thermo, Shanghai, China) following the manufacturer's instructions. Finally, a 10% SDS-polyacrylamide gel immunoblot analysis was performed with *anti-GFP* (Abmart) and *anti-Myc* (Abmart) antibodies.

### 2.10. Sequencing Analysis and Primer Synthesis

All resultant constructs were commercially sequenced, and all primers (Table S1) were synthesized by RuiBiotech (Beijing, China).

## 3. Results

### 3.1. Comparison of Tree Peony Flowering Traits

Most tree peony cultivars in gardens are OF; they generally complete floral differentiation before winter and then flower in the following spring with dormancy release through winter. Common buds continue to flower in the following spring after winter dormancy. The CF trait has been revealed only in a few tree peony cultivars, such as 'High Noon' ('HN'), which can continuously bloom throughout the annual growth cycle from spring through summer to autumn as the buds differentiate continuously [29,30] (Figure 1). We described the buds in the two classes of cultivars blooming in spring (late April to early May) as common buds and those blooming in other specific times as CF buds. Our previous studies indicated that the bract primordia of common buds become visible in early July both in CF and in OF cultivars, which corresponds to the initiation of flower bud differentiation, and the CF buds of CF cultivars; additionally, the common buds in all cultivars differentiate simultaneously [29]. These varieties differ significantly in their flowering times, which makes them an excellent system for exploring the molecular regulatory mechanisms underlying CF in tree peonies.

**Figure 1.** Different blooming modes of tree peonies. (**a**) OF genotype 'Luo Yang Hong' ('LYH') that only flowers in spring. After the first flower appears in early May, scale buds enter the bud primordium differentiation stage in late May. The bud primordium differentiation stage is from late May to November, and floral induction occurs in early July. The dormancy stage is in mid-November to the following March. Budding growth is initiated in March of the following year, and flowering is initiated in early May. (**b**) CF genotype ('HN'). After the first flower appears in early May, a portion of the buds continue germinating, and floral induction occurs in early July. CF bud differentiation is completed shortly after, and flowering occurs between early August and early October. The overwinter buds enter the floral induction stage in early July, the dormancy stage is from mid-November to the following March, and the germination stage begins in early March of the following year. Anthesis is initiated in early May. Common flowering is indicated by a red box; CF is indicated by yellow box.

### 3.2. Sequencing Analyses of FD and TFL1 in Tree Peonies

The complete full-length sequences of *PsFD* and *PsTFL1* were amplified using the RACE method. Multiple sequence alignment showed detailed information on the coding sequences and corresponding amino acid (aa) sequences of these two genes (Figure S1). PsTFL1 and PsFD contained 171 aa and 249 aa, respectively. Phylogenetic analysis revealed homologous genes in each clade—*FT-like*, *TFL1-like*, and *FD-like* (Figure S2)—and these homologous genes belonged to their own clade. The PsTFL1 protein showed high sequence similarity to the PsFT protein, and both proteins, belonging to the PEBP family, had greater than 70% sequence conservation (Figure S1a,b). They were not only very similar in the overall protein fold but also in the putative ligand binding and neighboring effector sites (Figure 2). However, there were substantial differences in the *FD-like* clade among species (Figure S1c). PsFD, a member of the basic region leucine zipper (bZIP) protein family, contained a chained mode structure (Figure 2). Moreover, the heterodimers could be formed as PsFD-PsFT and PsFD-PsTFL1 with high confidence (Figure 2).

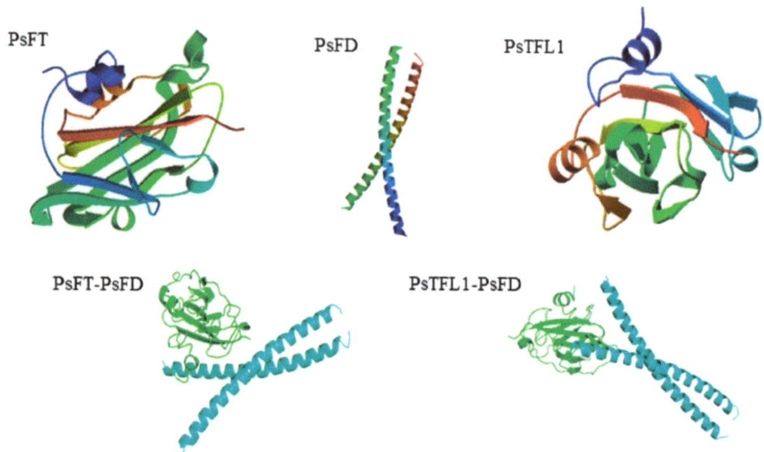


**Figure 2.** Secondary protein structures of PsFT, PsTFL1, and PsFD; interaction model construction of PsFD-PsFT and PsFD-PsTFL1. (Interaction confidence analysis: PsFD-PsFT: 0.8671; PsFD-PsTFL1: 0.8551).

### 3.3. *Expression of PsTFL1, PsFD, and PsFT in Different Flowering Cultivars*

The qRT-PCR results of common buds in 'HN', 'LYH', and *P. delavayi* var. *lutea* revealed that *PsFT* expression increased continuously from late June to mid-July and then decreased from mid-July to early September (Figure 3a). The expression of *PsFD* from early June to mid-August was consistent with that of *PsFT*, although its variation was observed in the three peony cultivars (or species) (Figure 3b). These findings indicate that the expression of *PsFT* and *PsFD* increased as floral bud differentiation was initiated with bract primordia formation, as in a previous study [29]. The expression level of *PsFT* and *PsFD* was increased in the process of flower bud differentiation, despite differences in degrees among the three cultivars (species). However, the expression of *PsTFL1* was only detected in the OF cultivar 'LYH', and the expression started to be elevated around August 5, reached the peak around August 15, and decreased until it became silent around September 5, which was exactly the CF stage (Figure 3c). However, this phenomenon was not observed in CF 'HN' and potential CF *P. delavayi* var. *lutea*, thus implying the essential role of *PsTFL1* in CF.

We further analyzed the expression patterns of *PsTFL1*, *PsFT*, and *PsFD* in different floral buds in HN and found that during the flower bud differentiation of CF 'HN', the expression of *PsTFL1* gradually decreased as the expression of *PsFT* increased (Figure 3d–g). In 'HN' common buds in dormancy, the expression of *PsFT* was hardly detected, but the expression of *PsFD* and *PsTFL1* was high (Figure 3e–g). During CF bud differentiation, the increased expression of *PsFT* and *PsFD* was accompanied by a decrease in *PsTFL1* expression. The results indicate that *PsFT* and *PsFD* were at peak expression during the S3 stage of the CF buds in CF 'HN', while the expression of *PsTFL1* gradually decreased from the S1 to S4 stages for the CF buds.

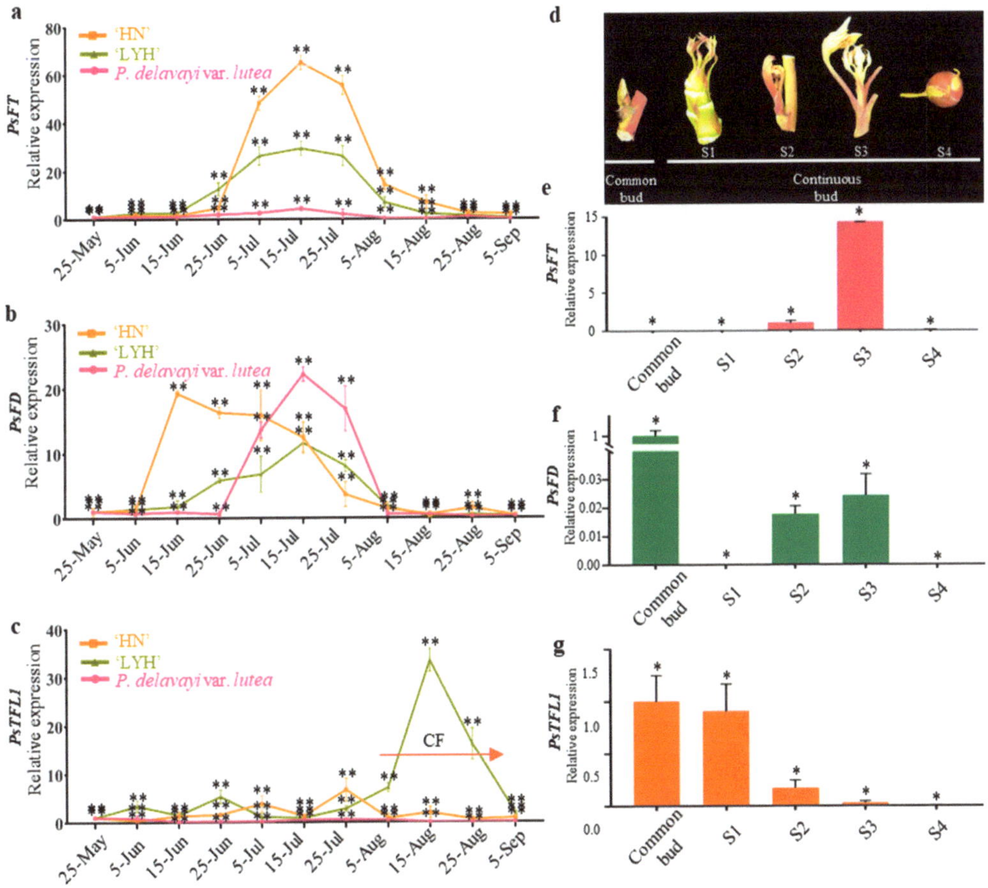


**Figure 3.** Expression patterns of *PsFT*, *PsFD*, and *PsTFL1* in buds during flower differentiation. (**a**) *PsFT*. (**b**) *PsFD*. (**c**) *PsTFL1*. Red arrow indicates the period of CF flowering. (**d**) The morphological structure of common and CF bud in 'HN'. (**e–g**) Expression pattern of *PsFT* (**e**), *PsFD* (**f**), and *PsTFL1* (**g**) in common and CF buds of 'HN'. The common buds of OF 'HN' and the S1, S2, S3, and S4 stages of CF 'HN' were collected in mid-September during continuous flowering. The curves show the mean values of three biological replicates for each sample. Data are the means of three biological replicates, and three technical replicates were performed for each biological replicate. Error bars show the standard deviation. Horizontal axis: plant data; vertical axis: gene expression levels relative to *PsUBIQUITIN*. Significance levels are calculated via analysis of variance (ANOVA) and indicated by asterisks: * ($p < 0.05$); ** ($p < 0.01$).

### 3.4. Subcellular Localization of PsTFL1, PsFD, and PsFT

To reveal the subcellular localization of PsTFL1, PsFD, and PsFT, we made the indicated constructs to express fusion proteins with GFP driven by a 35S promoter (Figure 4). In this experiment, the fused GFP fluorescence signal could indicate the subcellular localization of the proteins of interest. In this study, laser confocal microscopy revealed that the PsFT-GFP and PsTFTL1-GFP proteins were primarily expressed in the nucleus and cytoplasm, while the PsFD-GFP protein was expressed in the nucleus (Figure 4). The GFP control (*35S::GFP*) demonstrated the subcellular localization of the GFP proteins alone, which was observed in the nucleus and cytoplasm. Considering the autofluorescence signal and the merged image (Figure 4), we concluded that PsFT and PsTFL1 are localized in the cytoplasm and nucleus, whereas PsFD is localized in the nucleus.

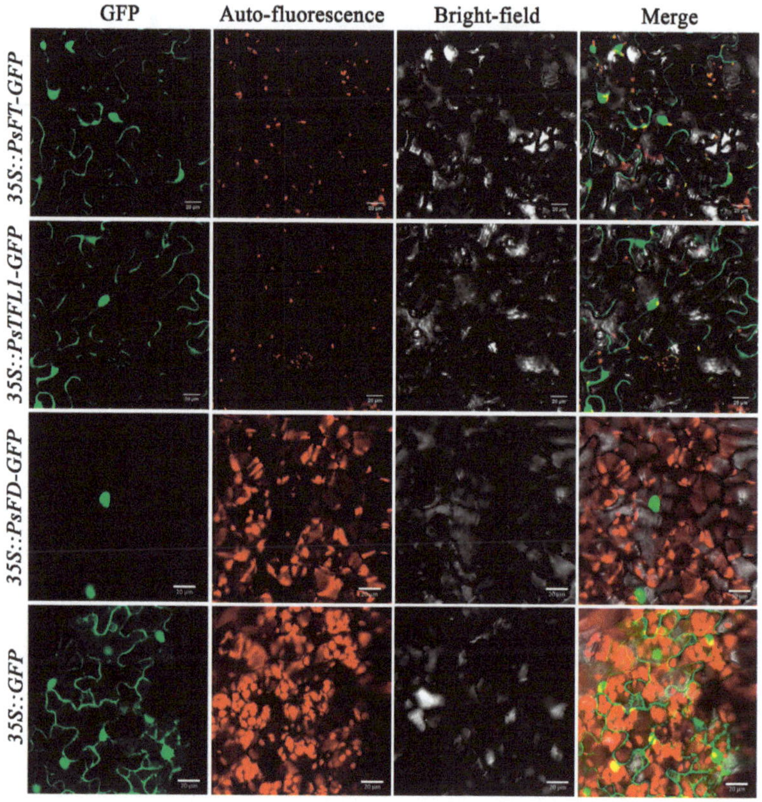


**Figure 4.** Subcellular localization of PsFT, PsTFL1, and PsFD in tobacco epidermal cells. GFP, Green fluorescence protein; Auto-fluorescence, chlorophyll fluorescence signal; Merge, merged image of GFP, auto-fluorescence, and bright-field images. Bars = 20 μm.

### 3.5. Ectopic Expression of PsTFL1 in A. thaliana

To examine the role of *PsTFL1* in the bud differentiation regulations of CF 'HN', we generated *PsTFL1*-overexpressing transgenic *A. thaliana* plants, and 15 independent kan-resistant $T_1$ plants were obtained. The $T_2$ lines were again filtrated with a kan solution, and morphological observations were taken from 18 plants from the second generation ($T_2$). Under LD conditions, transgenic $T_2$ lines exhibited later flowering characters and more rosette leaves than WT plants (Figure 5a,b). As quantified, the average number of rosette leaves in transgenic *Arabidopsis* lines was around 30, and the time from germination to flowering was around 64 days. In contrast, the number of rosette leaves in wild-type (WT) *Arabidopsis* plants was around 17, and the time from germination to flowering was around 46 days (Figure 5c). Therefore, the ectopic expression of *PsTFL1* in *A. thaliana* revealed the function of *PsTFL1* in delaying blooming.

c

| Genotype | No. of plants | Days to flowering | Rosette leaf |
|---|---|---|---|
| **Col-0** | 18 | 46.5±1.79 | 17.33±1.68 |
| **TFL1-OX** | 18 | 64.22±3.3 | 30.56±3.62 |

**Figure 5.** *PsTFL1* from tree peonies induced late flowering in transgenic *A. thaliana*. (**a**) The transgenic line that constitutively expressed *PsTFL1* exhibited the late-flowering phenotype in contrast to the wild-type (WT) and vector controls under long-day conditions. (**b**) Vertical view. Bars = 0.5 cm. (**c**) The statistics of days to flowering and rosette leaves in WT and over expression of *PsTFL1* in *A. thaliana*.

### *3.6. PsFT and PsTFL1 Physically Interact with PsFD*

To investigate the direct interaction between PsFT and PsTFL1 with PsFD, as hypothesized (Figure 2), we performed yeast two-hybrid (Y2H) assays. The results showed that the yeast cells co-transformed with PsFT-PsFD or PsTFL1-PsFD exhibited survival and growth, indicating that PsFT and PsTFL1 can interact with PsFD in vitro, while PsFT did not interact with PsTFL1 (Figure 6a). We further conducted BiFC and Co-IP assays to further clarify the observed interactions between the above pairs of proteins. In BiFC assays, a yellow fluorescent protein fluorescence signal was observed in the epidermal cells of tobacco leaves when *PsFD-YFPn* and *PsFT-YFPc*, as well as *PsFD-YFPn* and *PsTFL1-YFPc*, were co-expressed compared with the set negative controls (Figure 6b), indicating that PsFD physically interacts with PsFT and PsTFL1 in plant cells. In Co-IP assays, we found that PsFT and PsTFL1 were recombined into pBinGFP2 vectors to yield GFP-PsFT and GFP-PsTFL1, respectively, and PsFD was recombined into pCAMBIA to yield the CAM-PsFD vector (Figure 6c). GFP-PsFT and GFP-PsTFL1 co-immunoprecipitated with CAM-PsFD at a protein size from 55 to 72 kD in both *anti-GFP* and *anti-myc* media (Figure 6c). These results confirmed the existence of protein–protein interactions between PsFD-PsFT and PsFD-PsTFL1.

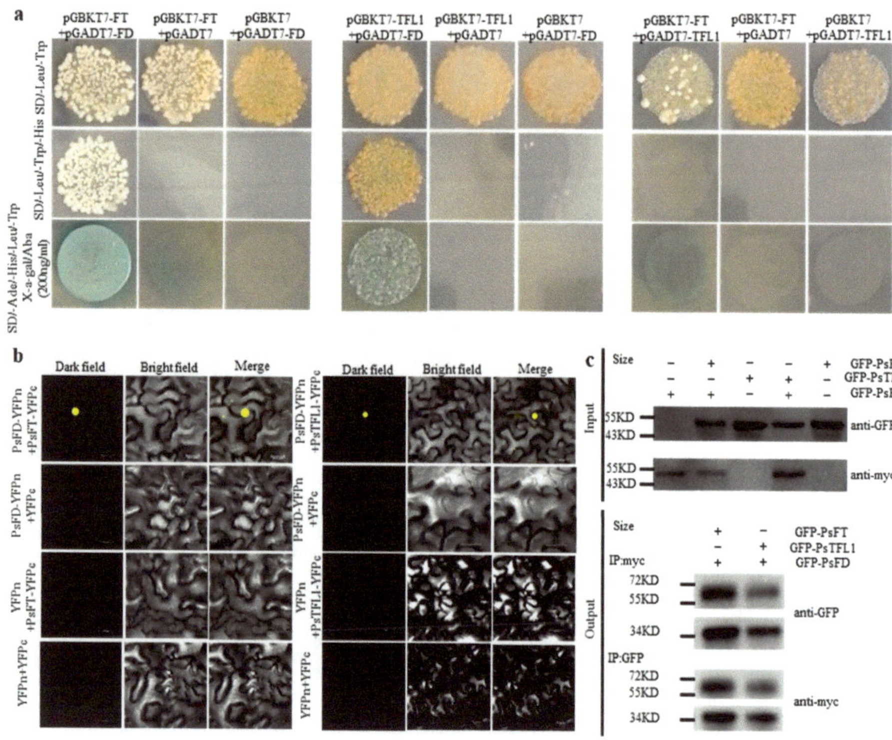

**Figure 6.** Protein–protein interactions among PsFT, PsTFL1, and PsFD. (**a**) Yeast two-hybrid assay. Yeast transformed with combinations of PsFT-PsFD and PsTFL1-PsFD with PsFT and PsTFL1, as pGBKT7-bait and PsFD as pGADT7-bait were cultured on both nonselective media (SD/leu/-trp) and selective media (X-a-gal/Aba). (**b**) BiFC assays were conducted to confirm the interaction between PsFT-PsFD and PsTFL1-PsFD in tobacco leaves. Dark-field, bright-field, and merged images are shown, Bars = 25 μm. (**c**) Co-immunoprecipitation assays. Input: PsFD-pCAMBIA and PsFT-pBinGFP2, as well as PsFD-pCAMBIA and PsTFL1-pBinGFP2, were co-expressed in *N. benthamiana*. Immunoprecipitated proteins (IPs) were analyzed via immunoblotting and probing with either *anti-GFP* (*α-GFP*) or *anti-myc* (*α-myc*). Proteins were extracted and subjected to Western blotting. Output: The bands observed between 55 kDa and 72 kDa in the output are the result of the binding between PsFT-PsFD and PsTFL1-PsFD in the Co-IP experiment.

## 4. Discussion

Clarifying the regulatory mechanisms of plant flowering has implications for crop production and breeding [32]. Many studies have examined the molecular mechanism underlying the CF phenotype to promote improvement in the ornamental and economic value of plants. *FT*, *TFL1*, and *FD* have been shown to be involved in the differentiation of CF buds in some perennial plants such as *R. rugosa* [1], *R. hybrid* RI [10,11], and *Fragaria vesca* [11], but whether these genes play an important role in CF tree peonies as well remains unclear. The results of this study proposed the regulatory roles of *PsFT*, *PsTFL1*, and *PsFD* in CF 'HN' tree peonies.

### *4.1. The Different Roles of PsFT-PsFD and PsTFL1-PsFD in Flowering Control*

The *FT* and *TFL1* genes are members of the *PEBP* family, and PEBP proteins are generally well conserved [8]. FT is a mobile protein that triggers flowering, and its sequence is highly conserved across species [12,20,33]. In this study, we found that the key amino acid residues of PsFT and PsTFL1 are conserved, but they are different in PsFD. Moreover, they are not only very similar in the overall protein fold, but they are also similar in the putative ligand binding site. This structure can act as a switch between two downstream possibilities in *Arabidopsis*, FT converts FD into a strong activator, while TFL1 converts FD into a strong repressor [14]. This similar site suggested that the biological functions of these molecules are similar [8,14,34]. PsFT and PsTFL1 may possess a buckle structure that binds with PsFD. The similarity in their binding site and interaction confidence analysis, with PsFD–PsFT = 0.8671 and PsFD–PsTFL1 = 0.8551, indicates that PsFT and PsTFL1 can both interact with PsFD.

In *A. thaliana*, both *FT* and *TFL1* were localized in the nucleus and cytoplasm and interacted with *FD* in the nucleus [27,35]. In this study, subcellular localization assays showed that *PsFD* was localized in the nucleus, and *PsFT* and *PsTFL1* were localized both in the nucleus and cytoplasm. These results indicated that there is overlap in the localization of *PsFT*, *PsTFL1*, and *PsFD* in the cell, which is necessary for their interactions, and that PsFT and PsTFL1 can interact with PsFD in the nucleus in tree peonies, similar to their function in *A. thaliana* [8,35]. In this study, our Y2H, BiFC, and Co-IP assays revealed interactions between PsFT, PsTFL1, and PsFD and confirmed protein–protein interactions between PsFT and PsFD and between PsTFL1 and PsFD. Furthermore, the differential structure of the functional groups of PsFT and PsTFL1 might have different functions in tree peonies. We reported previously that the overexpression of the *PsFT* gene promoted flowering in transgenic *A. thaliana* [30]. In this study, we confirmed that *PsTFL1* delayed flowering and resulted in more rosette leaves in transgenic *Arabidopsis* plants. In line with the findings in *Arabidopsis* [14], we suggested that PsFT-PsFD or PsTFL1-PsFD may promote or delay flowering in tree peonies.

### 4.2. The Role of PsFT, PsTFL1, and PsFD in Regulating CF

As a florigen, *FT* promotes flower induction or regulates blooming [12], and its homologs have often been reported in apple [36,37], *Arabidopsis* [8,20], poplar [38], citrus [39], loquat [40], and other species. In loquat, there are two copies of *FT*; the expression of *EjFT1* is involved in bud sprouting and leaf development in mid-May, and *EjFT2* is involved in floral bud induction in mid-June [40]. However, in 'HN', we only detected a single copy of *PsFT*, an *FT-like* gene. The expression of *PsFT* in common bud differentiation increased continuously from late June to mid-July and then decreased from mid-July to early September. *PsFT* is also expressed during CF in CF cultivars. Our previous study showed that transferring *PsFT* from 'HN' into *A. thaliana* significantly advanced flowering compared with WT plants [30]. The *PsFT* sequenced from tree peony 'HN' was detected to be expressed during common bud differentiation, CF flowering. *PsFT* may act as a florigen in tree peonies.

In this work, a single-copy *TFL1* homolog sequence was identified in 'HN'. We found that the expression of *PsTFL1* remained low in the CF 'HN' and CF potential *P. delavayi* during common bud differentiation from early June to late August but increased as bud differentiation finished in the OF 'LYH' when a peak appeared from 5 August to 15 August. *PsTFL1* was always expressed both in the common buds in dormant stages of all cultivar classes, but its expression gradually decreased from S1 to S4 in CF buds of 'HN'. This might suggest that the low expression of *PsTFL1* in CF tree peonies could allow them to flower continuously under suitable conditions. The overall expression of *KSN* (a homologous gene of *TFL1* in rose) has been shown to inhibit the floral transition in a CF rose, and the expression of *KSN* was significantly repressed in the CF cultivar compared with OF varieties [1,10,11]. Therefore, we deduced that *PsTFL1* might inhibit bud differentiation, and the decrease in its expression during floral differentiation might be related to the flowering of CF 'HN' in mid-August.

We found that the expression of *PsFD* was first upregulated in CF 'HN' on 5 June, followed by 'LYH' and *P. delavayi* on 15 June and 25 June. After entering the floral differentiation stage, the expression of *PsFD* first increased, and this was followed by an increase in the expression of *PsFT*. Therefore, *PsFD* might mediate the expression of *PsFT*, thereby regulating the process of floral differentiation. In the development and flowering of 'HN', the expression of both *PsFD* and *PsFT* increased, and the expression of *PsTFL1* decreased. *PsFT*, *PsTFL1*, and *PsFD* may play a role in inducing CF in 'HN'. In dormant buds of 'HN', we found that *PsFT* was weak expressed; however, *PsTFL1* and *PsFD* were highly expressed. These findings suggest that *PsFD* may be involved in the expression of *PsFT* and *PsTFL1* in CF cultivars. The function of *FD* is not limited to the regulation of flowering time, but it also involves several aspects of plant development [36]. FD may function as a molecular connector that can form an FD–FT or FD–TFL1 florigen activation or inhibition complex by binding with FT and TFL1 [14].

According to the expression curves during the bud differentiation of CF 'HN', the expression of *PsTFL1* was gradually reduced, while the expression of *PsFT* was increased, which coincides with the increased expression of *PsFD*. In common buds from 'HN', a high expression of *PsTFL1* coincides with the absence of *PsFT* expression, indicating that *PsTFL1* and *PsFD* could be related to the inhibition of flowering in the common buds of 'HN'. Therefore, *PsFT* likely plays a key role in flower bud differentiation and flowering in tree peony, and it is inhibited by *PsTFL1*. The results of our study showed that *PsFT*, *PsTFL1*, and *PsFD* may be related to the CF trait in the tree peony 'HN'.

In our study, the results of the qRT-PCR assays suggested that *PsFD* and *PsFT* could be expressed during common bud differentiation, common flowering, and CF flowering in 'HN'. The expression level of *PsTFL1* remained low during the common floral differentiation of CF 'HN' and CF potential *P. delavayi* compared with OF 'LYH'. Furthermore, *PsTFL1* expression gradually decreased from S1 to S4 in the CF buds of 'HN'. We also found that the expression of *PsFT* and *PsFD* increased during CF in CF 'HN'. The regulation of *PsFD*, *PsFT*, and *PsTFL1* could be correlated with CF flowering in 'HN'. Ectopic expression in *A. thaliana* revealed that PsFT acts as a florigen and PsTFL1 acts as a repressor protein. Protein–protein interaction assays revealed that PsFT and PsTFL1 could interact with PsFD. Our findings suggest that the regulation of *PsFT* and *PsTFL1* may be involved in the flowering or lack thereof in tree peonies during common and CF flowering.

## 5. Conclusions

In this study, we demonstrated that PsTFL1 functions as a floral inhibitor in tree peonies. Our findings suggest that PsFT and PsTFL1 can interact with PsFD to form a complex, and the gene regulation patterns of *PsFD*, *PsFT*, and *PsTFL1* may be involved in the CF flowering of 'HN'. However, the mechanisms underlying the differences in expression require further clarification.

**Supplementary Materials:** The following supporting information can be downloaded at: https://www.mdpi.com/article/10.3390/agronomy13082071/s1, Figure S1. Alignments of the deduced protein sequences. a. PsFT and its homologous proteins from other plants. b. PsTFL1 and its homologous proteins from other plants. c. PsFD and its homologous proteins from other plants. Figure S2. Maximum likelihood phylogenetic tree of the indicated putative proteins. a. PsFT and PsTFL1. b. PsFD with 10,000 bootstrap replicates (G-box binding factor 3 of *Zea mays* as an outgroup of PsFD). Table S1. Sequences of oligonucleotide primers used in this study. Table S2. Information of genes used in multiple sequence alignment.

**Author Contributions:** Conceived and designed the experiments: F.C. and L.Z. Performed the experiments: L.Z. Analyzed data: L.Z., F.C. and C.H. Drafted the paper: L.Z. and Z.G. Modified the paper: F.C., H.H. and C.H. Approved the paper: F.C. All authors have read and agreed to the published version of the manuscript.

**Funding:** The National Natural Science Foundation of China (Grant No.31972446). The funders had no role in study design, data collection, analysis, decision to publish, or manuscript preparation.

**Acknowledgments:** Thank you to Ruiyuan Biotech Company (Nanjing, China) for the assistance in protein model construction and interaction confidence analysis.

**Conflicts of Interest:** The authors declare that they have no known competing financial interest or personal relationship that could have appeared to influence the work reported in this paper.

## References


1. Bai, M.; Liu, J.Y.; Fan, C.G.; Chen, Y.Q.; Chen, H.; Lu, J.; Sun, J.J.; Ning, G.G. *KSN* heterozygosity is associated with continuous flowering of *Rosa rugosa* Purple branch. *Hortic. Res.* **2021**, *8*, 26. [CrossRef] [PubMed]
2. Simpson, G.G.; Gendall, A.R.; Dean, C. When to switch to flowering. *Annu. Rev. Cell Dev. Biol.* **1999**, *15*, 19–25. [CrossRef]
3. Srikanth, A.; Schmid, M. Regulation of flowering time: All roads lead to Rome. *Cell Mol. Life Sci.* **2011**, *68*, 2013–2037. [CrossRef] [PubMed]
4. Yi, X.; Gao, H.; Yang, Y.; Yang, S.; Luo, L.; Yu, C.; Wang, J.; Cheng, T.; Zhang, Q.; Pan, H. Differentially expressed genes related to flowering transition between once- and continuous-flowering Roses. *Biomolecules* **2022**, *12*, 58. [CrossRef] [PubMed]
5. Wang, S.L.; Beruto, M.; Xue, J.Q.; Zhu, F.Y.; Liu, C.J.; Yan, Y.M.; Zhang, X.X. Molecular cloning and potential function prediction of homologous *SOC1* genes in tree peony. *Plant Cell Rep.* **2015**, *34*, 1459–1471. [CrossRef]
6. Blumel, M.; Dally, N.; Jung, C. Flowering time regulation in crops-what did we learn from *Arabidopsis*? *Curr. Opin. Biotechnol.* **2015**, *32*, 121–129. [CrossRef]
7. Andrés, F.; Coupland, G. The genetic basis of flowering responses to seasonal cues. *Nat. Rev. Genet.* **2012**, *13*, 627–639. [CrossRef]
8. Coelho, C.P.; Minow, M.A.; Chalfun-Júnior, A.; Colasanti, J. Putative sugarcane *FT/TFL1* genes delay flowering time and alter reproductive architecture in *Arabidopsis*. *Front. Plant Sci.* **2014**, *5*, 221. [CrossRef]
9. Banfield, M.J.; Brady, R.L. The structure of *Antirrhinum* centroradialis protein (CEN) suggests a role as a kinase regulator. *J. Mol. Biol.* **2000**, *297*, 1159–1170. [CrossRef]
10. Randoux, M.; Davière, J.M.; Jeauffre, J.; Thouroude, T.; Pierre, S.; Toualbia, Y.; Perrotte, J.; Reynoird, J.P.; Jammes, M.J.; Oyant, L.H.; et al. *RoKSN*, a floral repressor, forms protein complexes with *RoFD* and *RoFT* to regulate vegetative and reproductive development in rose. *New Phytol.* **2014**, *202*, 161–173. [CrossRef]
11. Iwata, H.; Gaston, A.; Remay, A.; Thouroude, T.; Jeauffre, J.; Kawamura, K.; Oyant, L.H.; Araki, T.; Denoyes, B.; Foucher, F. The *TFL1* homologue *KSN* is a regulator of continuous flowering in rose and strawberry. *Plant J.* **2012**, *69*, 116–125. [CrossRef] [PubMed]
12. Li, C.Q.; Luo, L.; Fu, Q.T.; Niu, L.; Xu, Z.F. Isolation and functional characterization of *JcFT*, a *FLOWERING LOCUS T* (*FT*) homologous gene from the biofuel plant *Jatropha curcas*. *BMC Plant Biol.* **2014**, *14*, 125. [CrossRef]
13. Taoka, K.; Ohki, I.; Tsuji, H.; Furuita, K.; Hayashi, K.; Yanase, T.; Yamaguchi, M.; Nakashima, C.; Purwestri, Y.A.; Tamaki, S.; et al. 14-3-3 proteins act as intracellular receptors for rice Hd3a florigen. *Nature* **2011**, *476*, 332–335. [CrossRef]
14. Ahn, J.H.; Miller, D.; Winter, V.J.; Banfield, M.J.; Lee, J.H.; Yoo, S.Y.; Henz, S.R.; Brady, R.L.; Weigel, D. Divergent external loop confers antagonistic activity on floral regulators *FT* and *TFL1*. *EMBO J.* **2006**, *25*, 605–614. [CrossRef] [PubMed]
15. Guo, Y.; Hans, H.; Christian, J.; Molina, C. Mutations in single *FT*- and *TFL1*-paralogs of rapeseed (*Brassica napus* L.) and their impact on flowering time and yield components. *Front. Plant Sci.* **2014**, *5*, 282. [CrossRef] [PubMed]
16. Kardailsky, I.; Shukla, V.K.; Ahn, J.H.; Dagenais, N.; Christensen, S.K.; Nguyen, J.T.; Chory, J.; Harrison, M.; Weigel, D. North European invasion by common ragweed is associated with early flowering and dominant changes in *FT/TFL1* expression. *J. Exp. Bot.* **2018**, *69*, 2647–2658. [CrossRef]
17. Flachowsky, H.; Szankowski, I.; Waidmann, S.; Peil, A.; Tränkner, C.; Hanke, M.V. The *MdTFL1* gene of apple (*Malus* × *domestica* Borkh.) reduces vegetative growth and generation time. *Tree Physiol.* **2012**, *32*, 1288–1301. [CrossRef]
18. Freiman, A.; Shlizerman, L.; Golobovitch, S.; Yablovitz, Z.; Korchinsky, R.; Cohen, Y.; Samach, A.; Chevreau, E.; Roux, P.L.; Patocchi, A.; et al. Development of a transgenic early flowering pear (*Pyrus communis* L.) genotype by RNAi silencing of *PcTFL1-1* and *PcTFL1-2*. *Planta* **2012**, *235*, 1239–1251. [CrossRef]
19. Yamaguchi, A.; Kobayashi, Y.; Goto, K.; Abe, M.; Araki, T. *TWIN SISTER OF FT* (*TSF*) acts as a floral pathway integrator redundantly with *FT*. *Plant Cell Physiol.* **2005**, *46*, 1175–1189. [CrossRef]
20. Wigge, P.A.; Kim, M.C.; Jaeger, K.E.; Busch, W.; Schmid, M.; Lohmann, J.U.; Weigel, D. Integration of spatial and temporal information during floral induction in *Arabidopsis*. *Science* **2005**, *309*, 1056–1059. [CrossRef]
21. Khan, M.R.; Ai, X.Y.; Zhang, J.Z. Genetic regulation of flowering time in annual and perennial plants. *Wiley Interdiscip. Rev. RNA* **2014**, *5*, 347–359. [CrossRef]
22. Harig, L.; Beinecke, F.A.; Oltmanns, J.; Muth, J.; Müller, O.; Rüping, B.; Twyman, R.M.; Fischer, R.; Prüfer, D.; Noll, G.A. Proteins from the FLOWERING LOCUS T-like subclade of the PEBP family act antagonistically to regulate floral initiation in tobacco. *Plant J.* **2012**, *72*, 908–921. [CrossRef]
23. Alvarez, J.; Guli, C.L.; Yu, X.H.; Smyth, D.R. Terminal flower: A gene affecting inflorescence development in *Arabidopsis thaliana*. *Plant J.* **1992**, *2*, 103–116. [CrossRef]
24. Shannon, S.; Meeks-Wagner, D.R.A. mutation in the *Arabidopsis TFL1* gene affects inflorescence meristem development. *Plant Cell* **1991**, *3*, 877–892. [CrossRef] [PubMed]
25. Kardailsky, I.; Shukla, V.K.; Ahn, J.H.; Dagenais, N.; Christensen, S.K.; Nguyen, J.T.; Chory, J.; Harrison, M.; Weigel, D. Activation tagging of the floral inducer *FT*. *Science* **1999**, *286*, 1962–1965. [CrossRef] [PubMed]
26. Hanzawa, Y.; Money, T.; Bradley, D. A single amino acid converts a repressor to an activator of flowering. *Proc. Natl. Acad. Sci. USA* **2005**, *21*, 7748–7753. [CrossRef]

27. Hanano, S.; Goto, K. *Arabidopsis TERMINAL FLOWER1* is involved in the regulation of flowering time and inflorescence development through transcriptional repression. *Plant Cell* **2011**, *23*, 3172–3184. [CrossRef]
28. Cheng, F.Y. Advances in the breeding of tree peonies and a cultivar system for the cultivar group. *Int. J. Plant Breed.* **2007**, *1*, 89–104.
29. Zhou, H.; Cheng, F.Y.; Wang, R.; Zhong, Y.; He, C.Y. Transcriptome comparison reveals key candidate genes responsible for the unusual reblooming trait in tree peonies. *PLoS ONE* **2013**, *8*, e79996. [CrossRef]
30. Zhou, H.; Cheng, F.Y.; Wu, J.; He, C.Y. Isolation and functional analysis of *Flowering Locus T* in tree peonies (*PsFT*). *J. Am. Soc. Hortic. Sci.* **2015**, *140*, 265–271. [CrossRef]
31. Livak, K.J.; Schmittgen, T.D. Analysis of relative gene expression data using real-time quantitative PCR and the 2(-Delta Delta C(T)) method. *Methods* **2001**, *25*, 402–408. [CrossRef]
32. Eshed, Y.; Lippman, Z.B. Revolutions in agriculture chart a course for targeted breeding of old and new crops. *Science* **2019**, *366*, eaax0025. [CrossRef]
33. Laurie, R.E.; Diwadkar, P.; Jaudal, M.; Zhang, L.; Hecht, V.; Wen, J.; Tadege, M.; Mysore, K.S.; Putterill, J.; Weller, J.L.; et al. The *Medicago FLOWERING LOCUS* homolog, *MtFTa11*, is a key regulator of flowering time. *Plant Physiol.* **2011**, *156*, 2207–2224. [CrossRef]
34. Yang, L.; Wang, H.N.; Hou, X.H.; Zou, Y.P.; Han, T.S.; Niu, X.M.; Zhang, J.; Zhao, Z.; Todesco, M.; Balasubramanian, S.; et al. Parallel evolution of common allelic variants confers flowering diversity in *Capsella rubella*. *Plant Cell* **2018**, *30*, 1322–1336. [CrossRef] [PubMed]
35. Abe, M.; Kobayashi, Y.; Yamamoto, S.; Daimon, Y.; Yamaguchi, A.; Ikeda, Y.; Ichinoki, H.; Notaguchi, M.; Goto, K.; Araki, T. FD, a bZIP Protein mediating signals from the floral pathway integrator *FT* at the shoot apex. *Science* **2005**, *309*, 1052–1056. [CrossRef] [PubMed]
36. Cheng, X.F.; Li, G.F.; Krom, N.; Tang, Y.H.; Wen, J.Q. Genetic regulation of flowering time and inflorescence architecture by *MtFDa* and *MtFTa1* in *Medicago truncatula*. *Plant Physiol.* **2021**, *185*, 161–178. [CrossRef] [PubMed]
37. Kotoda, N.; Hayashi, H.; Suzuki, M.; Igarashi, M.; Hatsuyama, Y.; Kidou, S.I.; Igasaki, T.; Nishiguchi, M.; Yano, K.; Shimizu, T.; et al. Molecular characterization of *FLOWERING LOCUT T-like* genes of apple (*Malus* × *domestica* Borkh.). *Plant Cell Physiol.* **2010**, *5*, 561–575. [CrossRef]
38. Hsu, C.Y.; Adams, J.P.; Kim, H.; No, K.; Ma, C.; Strauss, S.H.; Drnevichc, J.; Vanderveldea, L.; Ellisa, J.D.; Ricea, B.M.; et al. *FLOWERING LOCUS T* duplication coordinates reproductive and vegetative growth in perennial poplar. *Proc. Natl. Acad. Sci. USA* **2011**, *108*, 10756–10761. [CrossRef]
39. Endo, T.; Shimada, T.; Fujii, H.; Kobayashi, Y.; Araki, T.; Omura, M. Ectopic expression of an *FT* homolog from Citrus confers an early flowering phenotype on trifoliate orange (*Poncirus trifoliata* L. Raf.). *Transgenic Res.* **2005**, *14*, 703–712. [CrossRef]
40. Reig, C.; Gil-Muñoz, F.; Vera-Sirera, F.; García-Lorca, A.; Martínez-Fuentes, A.; Mesejo, C.; Pérez-Amador, M.A.; Agustí, M. Bud sprouting and floral induction and expression of *FT* in loquat [*Eriobotrya japonica* (Thunb.) Lindl.]. *Planta* **2017**, *246*, 915–925. [CrossRef]

MDPI AG
Grosspeteranlage 5
4052 Basel
Switzerland
Tel.: +41 61 683 77 34

*Agronomy* Editorial Office
E-mail: agronomy@mdpi.com
www.mdpi.com/journal/agronomy

www.ingramcontent.com/pod-product-compliance
Lightning Source LLC
Chambersburg PA
CBHW041349210926
51850CB00013B/3028